MINING DISTRICTS OF NEVADA

ISBN 0-913814-48-2.

Printed in the U.S.A.

MINING DISTRICTS
and
MINERAL RESOURCES
of
NEVADA

by

FRANCIS CHURCH LINCOLN

This is a photographic reproduction of the 1923 first edition which was published by the Nevada Newsletter Publishing Company in Reno. The advertisements in the first edition are omitted in this reissue.

Published by Stanley Paher

Nevada Publications
Box 15444
Las Vegas, Nevada

NEVADA - OREGON
Denio
STATE LINE
IDAHO - NEVADA
STATE LINE
CALIFORNIA - NEVADA STATE LINE
Three Creek
Owyhee
Rowland
Jarbridge
Mountain City
O'Neil
Contact
Jarbridge Pk.
Alt. 10800
McDermitt
HUMBOLDT
NAT'L
FOR.
Whiterock
Gold Creek
Charleston
Alkali Lake
Vya
Quinn River Crossing
Platora
Rebel Creek
Paradise Mt.
Alt. 9600
Paradise Valley
North Fork
Amos
Tuscarora
Seetoya Pk.
Alt. 9546
Metropolis
Willow Point
Midas
Death
Wells
P.P.O.
Tobar
V.Hy.
BLACK ROCK DESERT
Winnemucca
Golconda
Halleck
Elko
Valmy
Dunphy
Carlin
Snow Water Lake
Spruce Mt.
Alt. 11041
Sulphur
Mill City
Imlay
Battle Mtn.
Beowawe
Palisade
Bullion
Lamoille
Arthur
Lee
Jiggs
Ruby Valley
Franklin Lk.
Harrison Pass
Alt. 9500
Ruby Lake
Currie
Gerlach
SMOKE CR. DESERT
SAN MATEO DESERT
Oreana
Rochester
Blackburg
Packard
Dry Salt Marsh
Flanigan
DESERT
Lovelocks
ALKALI DESERT
Hobson
PYRAMID LAKE
Pyramid
WINNEMUCCA LAKE
Humboldt Lake
Birch
Cherry Cr.
Schellbourne
Carson Sink
Mt. Callahan
Alt. 10208
Diamond Pk.
Alt. 10634
Salt Lake
Steptoe
TOIYABE NATIONAL FOREST
Wadsworth
Derby
Reno
Fernley
Hazen
Leeteville
Stillwater
Sparks
Verdi
Fallon
Lahontan
Steamboat Sps.
Virginia City
Salt Wells
Wonder
Alpine
New Pass
Canyon
Austin
L.Hy.
Eureka
McGill
Illipah
Kimberly
Ruth
Ely
Hamilton
Carson Lake
Churchill
Dayton
Carson City
Sand Sps.
Frenchmans Sta.
Bermond
Westgate
Eastgate
Bunker Hill Pk.
NEVADA NAT'L FOR.
Barnes
Conners
Preston
Lund
Genoa
Minden
Gardnerville
Ludwig
Wabuska
Yerington
Schurz
Rawhide
Campbells
Burnt Cabin
Gold Park
Ione
Millett
ALKALI FLAT
TOQUIMA RANGE
HOT CR. RANGE
Duckwater
White Pine Pk.
Alt. 11277
Mason
Rand
WALKER LAKE
Toiyabe Dome
Alt. 11778
Round Mountain
Currant
Mt. Grafton
Alt. 10964
Wellington
Mt. Grant
Alt. 11303
Hawthorne
Luning
Belmont
Manhattan
Bald Mt.
Alt. 9274
MONO N. F.
Coleville
Sweetwater
Mina
Hot Creek
Timber Mt.
Alt. 10280
N.M.E.
Nyala
NEVADA NAT'L FOR.
Bridgeport
Bodie
MONO NAT'L FOR.
Millers
Columbus Salt Marsh
Tonopah
Arrowhead
Sharp
Basalt
Coaldale
YOSEMITE River
Mono Lake
MONO LAKE
Tioga Pass
Alt. 9941
Boundary Pk.
Alt. 13145
Arlemont
Silverpeak
Goldfield
Dry Lake
Caliente
Hiko
Yosemite Nat'l Park
El Portal
Mammoth
Cuprite
Lida
Alt. 7406
Devil's Post Pile N. M.
Wawona
SIERRA NAT'L FOR.
INYO
Hornsilver
Bonnie Claire
Alamo
Pahrahagat
Bass Lake
Oakhurst
Coarsegold
Bishop
Westgard Pass
Alt. 7276
Huntington Lake
Big Cr
Raymond
Bigpine
Alt. 14254
INYO NAT'L FOR.
Hot Springs
Beatty
Rhyolite
PANAMINT RANGE
DEATH VALLEY
Friant
Tollhouse
Trimmer
Carrara
Clovis
Sanger
Fresno
Gen. Grant Nat'l Park
Mt. Williamson
Alt. 14384
Independence
Manzanar
Moapa
Fowler
Parlier
Reedley
Orosi
SEQUOIA
Giant Forest
Mt. Whitney
Alt. 14501
Lone Pine
Indian Springs
Corn Creek
Selma
Kingsburg
Dinuba
Keeler
Emigrant Springs
Johnnie
DIXIE NAT'L FOR.
Red Mt.
Laton
Hanford
Visalia
Lemoncove
OWENS LAKE
Lemoore
Armona
Exeter
Lindsay
Balch Park
Olancha
Darwin
Alt. 5320
Wildrose Sp.
Death Valley
Ryan
Pahrump
Charleston Pk.
Alt. 11910
Las Vegas
Stratford
Tulare
SEQUOIA NAT'L FOR.
Camp Nelson
Strathmore
Porterville
Corcoran
Telescope Pk.
Alt. 11045
Ballarat
Tipton
Terra Bella
Little Lake
Tecopa
Goodsprings
Sloan
Hot Springs
Alpaugh
Ducor
Glenville
Woody
Trona
Saratoga Springs
Lone Willow Spring
Jean
Delano
Kernville
Inyokern
Freeman
McFarland
Isabella
Weldon
Wasco
Famoso
Valley Wells
Searchlight

PREFACE

Earlier Works. The object of this book is to present authentic information concerning the mining districts and mineral resources of Nevada. No comprehensive work of this character has appeared for a number of years. In the early days of Nevada mining, from 1866 to 1878, the field was covered by the federal reports of J. Ross Browne, Rossiter W. Raymond, and Clarence King; and the state reports of R. H. Stretch, A. F. White, and H. R. Whitehill. In 1909, E. E. Stuart, Inspector of Mines for Nevada, compiled "Nevada's Mineral Resources" which was printed by the State Printer at Carson City. J. M. Hill's "Mining Districts of Western United States", which included a section on Nevada, was published by the United States Geological Survey in 1912.

Plan. "Mining Districts and Mineral Resources of Nevada", as its name implies, is divided into two main parts:—one on mining districts and the other on mineral resources. The section on mining districts is modeled in a general way after Hill's bulletin, but gives more extended descriptions and includes mining as well as geological information. The map which accompanies this book shows the approximate location of each of the mining districts described. The section on mineral resources provides a cross reference by useful mineral substances to the section on mining districts, besides noting many localities not included under districts and containing a number of production tables and bibliographies of individual resources.

Acknowledgements. This work is essentially a compilation, and credit for its contents is due, except where especially noted, to the authorities listed in the bibliographies. The writings of the staff of the United States Geological Survey have furnished the bases for the majority of the geological descriptions; while the Division of Mineral Resources of the United States Geological Survey has contributed most of the data concerning productions. Authors writing for the technical press have been drawn upon very largely as may be seen by a glance at the bibliographies. On the historical side, the early reports mentioned in the first paragraph have proved most useful, as have also the histories of Thompson and West, Bancroft, and Davis. Weed's "Mines Handbook" has supplied much valuable data concerning operating companies. The accuracy and scope of the publication have been greatly increased by the corrections and additions which have been made to the original manuscript by Nevada mining men. The writer is glad of this opportunity to acknowledge his indebtedness to all who have assisted him either directly or indirectly in the preparation of this book, and to thank them for their help.

Reno, Nevada,
August, 1923.

ABBREVIATIONS EMPLOYED IN THE BIBLIOGRAPHIES

AAAS American Association for the Advancement of Science.

AIME American Institute of Mining and Metallurgical Engineers.

AJS American Journal of Science.

AMC American Mining Congress.

AR Annual Report.

B Bulletin.

B1866 Browne, J. R.,"A Report upon the Mineral Resources of the States and Territories West of the Rocky Mountains" (for 1866), Washington, Government Printing Office, (1867).

B1867 ———Same for 1867,

Ball 285 Ball, S. H., "Notes on the Ore Deposits of Southwestern Nevada and Eastern California", USGS B 285 (1906) 53-73.

Ball 308 ———"A Geologic Reconnaissance in Southwestern Nevada and Eastern California", USGS B 308 (1907).

Bancroft Bancroft, H. H., "History of Nevada, Colorado, and Wyoming, 1540-1888", San FFrancisco, (1890).

C&ME Chemical and Metallurgical Engineering.

Davis Davis, S. P., Editor, "The History of Nevada", Reno and Los Angeles, (1913).

E&MJ Engineering and Mining Journal, and Engineering and Mining Journal-Press.

EG Economic Geology.

Emmons408 Emmons, W. H., "A Reconnaissance of Some Mining Camps in Elko, Lander, and Eureka Counties, Nevada", USGS B 408 (1910).

GSA Geological Society of America.

Hill 507 Hill, J. M., "The Mining Districts of the Western United States", USGS B 507 (1912).

Hill 594 ———"Some Mining Districts in Northeastern California and Northwestern Nevada", USGS B 594 (1915).

Hill 648 ———"Notes on Some Mining Districts in Eastern Nevada", USGS B 648 (1916).

J Journal.

JG Journal of Geology.

M Monograph.

M&EW Mining and Engineering World.

M&M Mines and Minerals.

M&Met Mining and Metallurgy.

M&SP Mining and Scientific Press.

MI Mineral Industry.

MR1882 to MR1921 "Mineral Resources of the United States", for the years designated, by the U .S. Geological Survey.

MS Mining Science.

P Proceedings.

PP Profesional Paper.

QJGS Quarterly Journal of the Geological Society.

R Report.

R1868 Raymond, R. W., "Report on the Mineral Resources of the States Territories West of the Rocky Mountains" (for 1868), Washington, Government Printing Office, (1869).

R1869 ———Same for 1869.

R1870 ———Same for 1870.

R1871 ———Same for 1871.

R1872 ———Same for 1872.

R1873 ———Same for 1873.

R1874 ———Same for 1874.

R1875 ———Same for 1874.

Ransome414 Ransome, F. L., "Notes on Some Mining Districts in Humboldt County, Nevada", USGS B 414 (1909).

S Science.

SLMR Salt Lake Mining Review.

SMN1866 Stretch, R. H., "Annual Report of the State Mineralogist of the State of Nevada for 1866", Carson City, (1867).

SMN1867-8 White, A. F., "Report of the State Mineralogist of Nevada for the Years 1867 and 1868", Carson City, (1869).

SMN1869-70 ———Same for 1869 and 1870.

SMN1871-2 Whitehill, H. R., "Biennial Report of the State Mineralogist of the State of Nevada for the Years 1871 and 1872".

SMN1873-4 ———Same for 1873 and 1874.

SMN1875-6 ———Same for 1875 and 1876.

SMN1877-8 ———Same for 1877 and1878.

SMQ School of Mines Quarterly.

Schrader624 Schrader, F. C.;Stone, R. W.; and Sanford, S., "Usefu lMinerals of the United States", USGS B 624 (1917).

Abbrevations Employed in the Bibliographies — Concluded

Spurr208 — Spurr, J. E., "Descriptve Geology of Nevada South of the Fortieth Parallel and Adjacent Portions of California", USGS B 208 (1903).

StuartNMR — Stuart, E.E., "Nevada's Mineral Resources", Carson City, (1909).

T — Transactions.

TP — Technical Paper.

Thompson & West — Thompson & West, Publishers, "History of Nevada", Oakland, (1881).

USBM — United States Bureau of Mines.

USGE40th — King, C., Geologist-in-Charge, "United States Geological Exploration of the Fortieth Parallel", Washington, Government Printing Office.

USGS — United States Geological Survey.

USGSW100th — Wheeler, G. M., In charge, "United States Geographical Surveys West of the 100th Meridian", Washington, Government Printing Office.

Weed MH — Weed, W. H., Editor, "The Mines Handbook", Volume XV, Tuckahoe, N. Y., (1922).

WSP — Water Supply Paper.

Mining Districts *and* Mineral Resources *of* Nevada

SECTION ONE — MINING DISTRICTS

CHURCHILL COUNTY

ALPINE (Clan Alpine)
Gold, Silver, (Molybdenum)

Location. The Alpine District is at Alpine in W. Churchill Co. Fallon on the S. P. R. R. is 79 m. W.

History. The district was organized as the Clan Alpine District in 1864, and a mill built there in 1866, but the camp was abandoned shortly afterwards and has been inactive since. Molybdenite is said to occur at Scott's camp, near Alpine.

Bibliography. B1866 128 SMN1866 29 Hill 507 199
B1867 334 SMN1867-8 86 Thompson & West 366
Horton, F. W., "Molybdenum, Its Ores and Their Concentration", USBM B 111 (1916) 87.

BERNICE
Antimony, Silver

Location. The Bernice District is at Bernice on the W. slope of the Clan Alpine Range in N. E. Churchill Co. It lies to the E. of Dixie Marsh and is 90 m. by road E. N. E. of Fallon which is on the S. P. R. R.

History. W. W. Van Reed was the first to ship antimony ore from the district, his product going to the old Star & Mathews smelter in San Francisco, according to a letter from J. T. Reid, while the last important shipments were made by Sanders & Young from 1893 to 1896. W. W. Williams worked a group of silver mines in the eighties and nineties, erecting a 10-stamp mill and roaster to treat the ore. A little antimony ore containing silver was shipped from the district in 1906.

Geology. The crest of the Clan Alpine Range is composed of granitic rock, according to Mallery, while sedimentary rocks form the W. flank. Quartz-stibnite veins occur in the lower member of the sedimentary series which is a brown, indurated shale with occasional intercalated limestone strata. The veins are richest where they intersect these limestone strata, and stibnite also occurs disseminated with pyrite in the limestone, but not in sufficient quantity to constitute ore. Sphalerite is associated with the stibnite in the Antimony King vein, and pyrite occurs in the slate wall-rock.

Bibliography. MR1906 512 Hill 507 199.
Mallery, W., "Antimony Veins at Bernice, Nevada", M&SP 112 (1916) 556.

BOLIVIA see TABLE MOUNTAIN

BOYER see TABLE MOUNTAIN

BROKEN HILLS

Silver, Lead, Gold

Location. The Broken Hills District occupies a group of low hills between the Fairview and Ellsworth Ranges in S. E. Churchill Co., on the Nye Co. border. Fallon on the S. P. R. R. lies 63 m. to the N. W., and Luning on the same railroad, 48 m. to the S. W.

History. The district was discovered by James Stratford and Joseph Arthur in 1913, who worked there till 1920 when they sold out to the Broken Hills Silver Corp. George Graham Rice's Fidelity Finance & Funding Co. started to finance this corporation but got it into difficulties and a reorganization was effected in 1921. W. H. Kinnon, Engr. of the Kansas City - Nevada Cons. Ms. Co., whose property is located in the Bruner District, Nye Co., was appointed manager, and arrangements were made to treat the Broken Hill ore in the Bruner mill.

Production. The Broken Hills Mine produced about $70,000 from 1913 to 1920.

Geology. Volcanic tuff forms the main mass of the Broken Hills Range, according to a report by A. P. Thompson. This tuff is capped by basalt and underlain by andesite, which is only visible where faulted upward at the base of the range. The Broken Hills ore deposit is a vein in the andesite about 100 ft. from the fault. The vein filling consists of crushed and altered andesite with minor amounts of quartz. Rich silver sulphides occur in bunches in the vein associated with lead and gold; and silver-lead ore also occurs in stockwork lenses in the andesite outside the vein. Disseminated pyrite and sphalerite extend into the wall-rock.

Mines. The Broken Hill Silver Corp., was incorporated in Nevada in 1920 with a capital stock of 3,000,000 shares of 10 cts. par value of which 1,400,000 shares have been issued. The home office is at 259 N. Virginia St., Reno. E. Malley is Pres., and G. Ross is Sec.

Bibliography. MR1920 I 320. MR1921 I 379.
WeedMH 1153 Broken Hills Comb. M. Co., Broken Hills Silver Corp.
1198 Eureka Silver Lick Ex. Ms. Co.

BROWNS see TOY

CLAN ALPINE see ALPINE

COPPER KETTLE

Copper

Location. The Copper Kettle District is situated in Copper Kettle or Grimes Canyon on the W. slope of the Stillwater Range in N. Churchill Co. It adjoins the White Cloud District on the N.

History. The district was discovered 15 years ago. Welsh and Green own a large group of claims from which several carloads of copper ore were shipped 6 years ago.

Geology. The ore contains chalcocite and copper oxides and carbonates and occurs near a contact of diorite with overlying altered porphyry.*

Bibliography. USGS Carson Sink topographic map.

COPPER VALLEY (Ragged Top) see PERSHING COUNTY

COTTONWOOD CANYON see TABLE MOUNTAIN

DESERT (White Plains)

Gold

Location. The Desert District lies at the N. end of the Hot Springs Mts., on

*The data for this description was kindly furnished by J. T. Reid.

the N. E. flank of Desert Peak which has an altitude of 5,401 ft. in N. W. Churchill Co. It is 8 m. S. W. of White Plains (Huxley Station) which is on the S. P. R. R. The White Plains Flat saline district adjoins it on the N. E.

History. The district was known as early as 1863 when a 5-stamp mill was built there to treat its ores. This mill proving unsuccessful, another was constructed at the outlet of the Humboldt, 14 m. distant. The principal mine was the Desert Queen which was worked continuously up to 1885 and has been worked intermittently since.

Geology. The ore ocurs in small, blind lodes, according to Browne, and is a rich, free gold ore in a red hematite gangue.

Bibliography. B1867 334 MR1907 I 346 Hill 507 200
SMN 1866:29 MR1908 I 471
MR1909 I 394
USGS Carson Sing topographic map.

DIXIE MARSH
Salt, Borax, (Potash)

Location. Dixie Marsh lies to the E. of the Stillwater Range in N. E. Churchill Co. It covers about 9 sq. m. in the lowest part of the present valley, which was called the Osobb Valley by the geologists of the Fortieth Parallel Survey.

History. From 1861 to 1868, this marsh produced a large amount of salt which was shipped to the silver mills of Virginia City, Austin, Belmont, Unionville, and even as far as Silver City, Idaho. Some 10 cars of borax were produced by R. Nieschwander from the N. end of the marsh according to a letter from J. T. Reid. No production has been made for many years. Tests made by the U. S. Geological Survey and the Railroad Valley Co., in recent years showed the potash content of this marsh to be commercially negligible.

Geology. The marsh was formed by the desiccation of a lake not over 150 feet in depth, according to Young. The deposit consists of alternating layers of salt and saline mud. The salt contains a little sodium carbonate and sulphate, and the brine a trace of potash.

Bibliography. USGE40th 2 (1877) 707-8 MR1912 II 891 Potash
MR1917 II 419-421 "
USGS Carson Sink topographic map.
Free, E. E. "The Topographic Features of the Desert Basins of the U. S.", U. S. Dept. Agr. B 54 (1914) 31.
Gale, H. S., "Notes on the Quaternary Lakes of the Great Basin", USGS B 540 (1914) 402.
Hance, J. H., "Potash in Western Saline Deposits", USGS B 540 (1914) 463-4.
Phalen, W. C., "Salt Resources of the U. S.", USGS B 669 (1919) 140.
Young, G. J., "Potash Salts and Other Salines in the Great Basin Region" U. S. Dept. Agr. B 61 (1914) 54-6.

EAGLE MARSH see LEETE

EAGLEVILLE (Hot Springs)
Gold, Silver, (Barite)

Location. The Eagleville District is at Eagleville in S. Churchill Co. Fallon on the S. P. R. R. is 64 m. to the N. W.

History. Small productions were reported from the district in 1905 and 1908, and from 1915 to 1919 the Golden Extension Mine was shipping. Barite of good grade is said to occur in the district.

Bibliogaphy. MR1905 266 MR1917 I 264 Hill 507 200
MR1908 I 470 MR1918 I 229
MR1915 I 625 MR1919 I 384
MR1916 I 468
USGS Carson Sink topographic map.

EASTGATE

Silver, Gold, Lead

Location. The Eastgate District is situated at Eastgate on the W. slope of the Desatoya Range in S. E. Churchill Co. Fallon on the S. P. R. R. lies 60 m. to the W. N. W. The district adjoins the Westgate District on the E. and the Gold Basin District in Lander Co., on the S. W.

History. The district made small productions in 1908, 1917 and 1920. The Nevada Lincoln M. Co. incorporated in 1915 got into a legal controversy with the Nevada Wilson M. Co. in 1918 which is now said to be settled.

Geology. Mines Handbook states that the Nevada Lincoln property contains 3 gold-silver veins with ore occurring as bands and seams in a broken formation of quartz and talc.

Bibliography. MR1905 265 MR1908 I 470 Hill 507 200
MR1906 293 MR1917 I 264
MR1920 I 319
Weed M H 1281 Nevada Lincoln M. Co., 1289 Nevada Wilson M. Co.

FAIRVIEW

Gold, Silver, Lead, Copper

Location. The Fairview District is located at Fairview on the W. slope of Fairview Peak in S. Churchill Co. The town of Fairview is 42 m. E.S.E. of Fallon which is on the S. P. R. R. Fairview has an altitude of 4,600 ft., the mines are about 1,000 ft. higher, and Fairview Peak rises to a height of 8,250 ft.

History. The district was discovered by F. O. Norton in 1905. A boom ensued the following spring which gave the district a temporary population of 2,000. The principal mine is the Nevada Hills which was located by P. Langsden in January, 1906. The Nevada Hills M. Co. erected a 20-stamp mill at this mine which was operated from September, 1911, to June, 1917, when it was shut down for lack of ore. Since the Nevada Hills M. Co. ceased operations, a small amount of work has been done in the district by leasers and by other companies.

Geology. Fairview Peak is composed of Tertiary eruptives resting upon a basement of Paleozoic schists and limestones. The volcanic rocks in the order of their age are :—dacite tuff, earlier andesite, later andesite, tuff, and rhyolite, with some later tuffs and flows of minor importance, according to Greenan.

Ore Deposits. The ore deposits are fissure veins in and at the borders of the earlier andesite. The veins are intersected by normal faults and post mineral fracturing has occured in them. The Nevada Hills vein is from 1 ft. to 15 ft. in width. Its gangue consists of quartz, altered andesite, calcite, and smaller amounts of pyrolusite and rhodochrosite; and its ore minerals are argentite, ruby silver, horn silver, pyrite, chalcopyrite, galena, tetrahedrite, sphalerite, silver, and gold.

Mines. Three companies have recently been active in the Fairview District.

The Dromedary Hump Cons. Ms. Co., with a capital stock of 2,000,000 shares is in charge of E.W.Stratton., the Chalk Mt. Silver-Lead Ms. Co., is under the management of E. M. Dawes of Reno. L. C. Smith is Pres. of the Nevada Hills M. Co., Reorg., with office at 225 N. Center St., Reno. This company owns the Nevada Hills Mine and has a capital stock of 1,250,000 shares of 10 cts. par value of which 184,000 remain in the treasury. Its stock is listed on the San Francisco Exchange.

Bibliography. MR1905 116,266 MR1911 I 667 MR1917 I 265
MR1906 122,293 MR1912 I 785 MR1918 I 230
MR1907 I 344-5 MR1923 I 817 MR1919 I 384
MR1908 I 471 MR1914 I 670 MR1920 I 319
MR1909 I 394 MR1915 I 625 MR1921 I 379
MR1910 I 50/ MR1916 I 468-9
USGS Carson Sink topographic map.

Eames, L. B., "Agitation at the Nevada Hills", Discussion, M&SP 108 (1914) 386.

Fleming, J. B., "Nevada Hills Surface Equipment", E&MJ 91 (1911) 1001-2.

Greenan, J. O., "Geology of Fairview, Nevada", E&MJ 97 (1914) 791-3.

Hill 507 200

James, A., "Agitation at the Nevada Hills", Discussion, M&SP 108 (1914) 624.

Lawson, A. C., "Geology of the Nevada Hills", Abstract, GSA B 23 (1912) 74.

Megraw, H. A., "Cyaniding Silver at Nevada Hills Mill", E&MJ 95 (1913) 645-8.

Rice, C. T., "Gold ad Silver at Fairview, Nevada", E&MJ 82 (1906) 729-730.

Scott, W. A., "The Nevada Hills Mill", M&SP 104 (1912) 143.

Stuart NMR 7,108-9.

Todd, R. B., "Rapid Development of Fairview, Nevada", E&MJ 84 (1907) 1157.

Weed MH 1187 Dromedary Hump Cons. Ms. Co.
1200 Fairview Red Rock Cons. Ms. Co.
1258 Minnesota Nevada Investment Co.
1280 Nevada Hills M. Co., Reorg.

Zalinski, E. R., "The Mines of the Fairview District, Nevada", E&MJ 83 (1907) 699-703.

HOLY CROSS

Silver, Gold, Copper, Lead, Manganese

Location. The Holy Cross District is situated in S. W. Churchill Co., on the border of Lyon and Mineral Cos. Schurz on the S. P. R. R. is 12 m. to the S. W.

Geology. The manganese deposit on the Bullion Group extends for 3,000 ft. copper and lead, were made from the district from 1923 to 1921. A few tons of manganese ore were shipped by R. Z. Hodges from the Bullion Group in 1918.

Geology. The manganese deposit on the Bullion Group extends for 3,000 feet along a shear sone in rhyolite and rhyolite tuff 50 ft. in width, and is associated with manganiferous silver veins. Stringers and small masses of psilomelane and pyrolusite replacing sheared rhyolite constitute the ore, which contains from 5% to 15% of manganese and is amenable to concentration, according to Pardee and Jones.

Bibliography. MR1913 I 817 MR1916 I 469 MR1920 I 328
MR1914 I 670 MR1918 I 230. MR1921 I 379,388
MR1915 I 625 MR1919 I 385
USGS Carson Sink topographic map.

Pardee, J. T. & Jones, E. L., Jr.,"Deposits of Manganese Ore in Nevada", USGS B 710 (1920) 233.

HOT SPRINGS see EAGLEVILLE

HOT SPRINGS MARSH see LEETE

I. X. L. (Silver Hill)
Silver, Gold, (Lead, Copper)

Location. The I. X. L. District is located in the Stillwater Range in central Churchill Co. I. X. L. Canyon is on the E. slope of the range, 70 m. S. E. of the town of Lovelock which is on the S. P. R. R.; while Silver Hill in Cox Canyon across the range on the W. flank is but 45 m. from Lovelock and is sometimes considered as a separate district. The Shady Run Run District adjoins the I. X. L. on the N. E.

Geology. The veins occur between a granite foot-wall and a slate hanging-wall, according to Thompson & West, and the silver and gold are found both free and with galena. A letter from J. T. Reid states that the copper ores are in highly altered andesite.

Bibliography. B1866 128 SMN1865 29 MR1906 293
B1867 333-4 MR1907 I 345-6
USGS Carson Sink topographic map.
Hill 507 200. Thompson & West 364.
WeedMH 1334 Silver Range Ms. Co.

JESSUP
Gold, Silver, Diatomaceous Earth

Location. The Jessup District is at Jessup in N. W. Churchill Co. White Plains, (Huxley Station), on the S. P. R. R. lies 10 m. to the S. E. and the town of Lovelock on the same railroad, 35 m. to the N. E. White Plains Flat adjoins the Jessup District on the S. E., the Juniper Range District on the N. W., the Copper Valley District on the N., and the Toy District on the N. E.

History. The district was discovered in 1908 and has made small shipments of gold ores and of silver ores intermittently since that date. Diatoomaceous earth deposits occur both to the N. and to the S. of Jessup, and small shipments have been made from the latter locality according to a letter from J. T. Reid.

Bibliography. MR1908 I 471 MR1911 I 667 MR1914 I 670
MR1909 I 394 MR1912 I 785 MR1917 I 265
MR1910 I 508 MR1913 I 817 MR1921 I 379
USGS Carson Sink topographic map.
Hill 507 200. Stuart NMR 122.

JUNIPER RANGE see PERSHING COUNTY

LA PLATA see MOUNTAIN WELLS

LAKE

Lead, Silver, Antimony, Limestone, (Niter)

Location. The Lake District lies to the E. of Humboldt Lake on the W. flank of the Humboldt Lake Range, and extends from N. Churchill Co. into S. Pershing Co.

History. The James Say lead-silver-antimony deposits produced a few cars of ore in the sixties, seventies, and eighties, according to data supplied by J. T. Ried They are located 3 m. E. of Ocala which is on the S. P. R. R. Say called atention to peculiar saline deposits in this neighborhood in 1868, and William Silverwood located niter deposits to the N. E. in the Pershing Co. end of the district that same year. These deposits attracted attention again recently, but were found too low grade to pay to work. Shell limestone which occurs at the S. W. extremity of the Humboldt Lake Range was shipped to California as a fertilizer some 20 years ago and has been burned locally for lime.

Geology. The lead-silver-antimony lodes outcrop in Jurassic shales The niter deposits consist of small amounts of sodium nitrate in fissures and crevices in rhyolitic rock, according to Gale.

Bibliography. MR1882 599.

Gale, H. S., "Nitrate Deposits", USGS B 523 (1912) 19-25; Abstract, "Nitrate Deposits in Nevada", M&EW 37 (1912) 1048-1050.

Russell, I. C., "Geological History of Lake Lahontan", USGS M 11 (1885) 109-110.

Van Wagenan, T. H., "Nitrate Deposits, Humboldt County, Nevada", M&SP 84 (1902) 63.

LEETE (Eagle Marsh, Hot Springs Marsh)

Salt, Borax, Gold, Silver, Lead

Location. The Leete District is at Leete on the S. P. R. R. in N. W. Churchill Co. on the Lyon Co. border. The salt producing section of the district is sometimes called the Eagle Salt Marsh; while that which produces borax is known as the Hat Springs Borax Marsh.

History. Leete discovered the Eagle Salt Marsh in 1870. The Eagle Salt Works began operations and the following year was supplying salt to all the mills of Virginia City. It continued to produce up to 1913. A company worked the Hot Springs Borax Marsh in 1871 but failed to make a financial success. The Nezelda Mine situated 6 m. N. W. of Leete was worked for gold, silver, and lead in the eighties, according to a letter from J. T. Reid.

Production. The Eagle Salt Works began by producing 3,000 tons of salt annually. From 1879 to 1884, it produced 334,400 tons, including table salt of which 200 tons per year were made in 1883 and 1884. The subsequent production was on a decreasing scale.

Geology. The Leete saline deposit is a desert mud plain formed in Quaternary time by the desiccation of a portion of ancient Lake Lahontan. The salt is obtained by solar evaporation of a brine with specific gravity of 1.2115 which comes nearly to the surface. The saline content is mainly sodium chloride with small amounts of magnesium chloride, calcium chloride, and calcium sulphate, and traces of potassium chloride, silica, alumina, and ferric oxide. The Nezelda veins are in rhyolite, dacite, and andesite.

Bibliography. R1872 186 Borax; 187-8 Salt
SMN1871-2 15 Borax; 18-19 Salt
SMN1875-6 6
USGS Wadsworth topographic map
MR1882 546
MR1883-4 847
MR1911 II 858
MR1912 II 919
MR1913 II 300
MR1914 II 301

Hanks, H. G., "Repart on the Borax Deposits of California and Nevada", Cal. State M. Bur. 3rd AR (1883) II 45.

Phalen, W. C., "Salt Resources of the U. S." USGS M 11 (1885) 233.

Russell, I. C., "Geological History of Lake Lahontan", USGS M 11, (1885) 233.

Thompson & West 364.

MOUNTAIN WELLS (La Plata)
Silver

Location. The Mountain Wells District is at Mountain Wells on the E. slope of the Stillwater Range in central Churchill Co. Fallon on the S. P. R. R. is 30 m. to the W.

History. The district was discovered and the town of La Plata founded in 1862. La Plata was made the county seat and by 1866 there were 3 mills in the district. The ore supply became exhausted at about the time of the White Pine rush, the district was deserted for the new discovery, and the county seat was moved to Stillwater in 1868. A little silver ore was shipped from the district by prospectors in 1919 and 1920.

Bibliography. B1866 128
B1867 334
SMN1866 29
SMN1867-8 86
MR1919 I 385
MR1920 I 319
USGS Carson Sink topographic map.

Thompson & West 366.

NEW PASS see LANDER COUNTY

PARRAN see WHITE PLAINS FLAT

RAGGED TOP see COPPER VALLEY, PERSHING COUNTY

SALT WELLS MARSH see SAND SPRINGS MARSH

SAND SPRINGS MARSH (Salt Springs Marsh)
Salt, Borax, (Potash)

Location. Sand Springs Flat begins at Salt Wells 15 m. S. E. of the town of Fallon, which is on the S. P. R. R.; and extends to Sand Springs, 30 m. S. E. of Fallon. It has an area of 37 sq. m. and an altitude of 3,960 ft. The flat was formerly known as Alkali Valley, while the Carson Sink topographic map of the U. S. Geol. Survey calls it Fourmile Flat and Eightmile Flat. The section of the flat from which borax was obtained was known as Salt Wells Marsh.

History. The Sand Springs salt deposit was discovered in 1863, and the Sand Springs Co. began shipping to the silver mills of the Comstock Lode that same year. Production practically ceased upon the discovery of the more accessible Eagle Salt Marsh in the Leete District in 1870. William Throop discovered borax at Salt Wells, and the American Borax Co. erected borax works there in 1870. In 1871, this plant and a smaller one built by another company, were in operation, but they were abandoned a few years later. In recent years, the U. S. Geol. Survey tested the brines of Sand Springs Marsh for potash, and found the content to be commercially unimportant.

Production. The salt production, which began in 1863, reached 150 tons per month in 1866, and 250 tons in 1867, practically ceasing in 1870. Borax was produced at the rate of 20 tons per month in 1871, but production stopped in a few years.

Geology. Sand Springs Flat was occupied by an arm of Lake Lahontan 440 ft. in depth in Quaternary time. The desiccation of this lake and the faulting and tilting of its former bed to the E. resulted in the formation of a saline deposit 12½ sq. m. in extent on the lower portion of the flat near Sand Springs. The richer portion of the borax deposit covered 400 acres and had a borax content of about 10%, although occasionally reaching 30%.

Bibliography. B1866 94-5
B1867 310-1
R1872 186
SMN1869-70 16
SMN1871-2 15-17 Borax; 19 Salt.
SMN1873-4 17 Borax
MR1882 544-5 Salt; 570 Borax
MR1885 639
MR1907 II 664
MR1911 II 858
MR1912 II 919
MR1913 II 300
MR1914 II 300-1
MR1915 II 273
USGS Carson Sink topographic map.

Free, E. E.,"The Topographic Features of the Desert Basin of the U. S.", U. S. Dept. Agr. B 54 (1914) 16.

Hance, J. H., "Potash in Western Saline Deposits", USGS B 540 (1914) 462-3.

Hanks, H. G., "Report of the Borax Deposits of California and Nevada", Cal. State M. Bur. 3rd AR (1883) II 45.

Phalen, W. C., "Salt Resources of the U. S.", USGS B 669 (1919) 137-8.

Russell, I. C., "Geological History of Lake Lahontan", USGS M 11 (1885) 234-5.

Young, G. J., "Potash Salts and Other Salines in the Great Basin Region", U. S. Dept. Agr. B 61 (1914) 39,59.

SHADY RUN

Gold, Silver, Lead

Location. The Shady Run District is situated at Shady Run on the W. slope of the Stillwater Range in Central Churchill Co., adjoining the White Cloud District on the S. W. and the I. X. L. District on the N. E. It is about 40 m. by road S. E. of Lovelock which is on the S. P. R. R. and an equal distance from Fallon which is on the same railroad. The district embraces Fondaway, Shady Run, Shanghai and Mill Canyons on the W. slope of the range and Pike Hallow at the summit*.

History. The Shady Run District is one of the early discoveries which never got beyond the prospecting stage. The Zinn Bros. prospected Fondaway Canyon; T. Sullivan and F. Smith, Shady Run Canyon;K. B. Jenkins, a property at the mouth of Shanghai Canyon; and Humphrey and Wisner, Pike Hollow. A number of Virginia City miners built a small custom mill at the mouth of Mill Canyon in the eighties, but it operated for a brief period only. Some silver-lead ore was shipped from Pike Hollow.

Geology. Gold prospects in quartzite near instrusive quartz porphyry occur at Fondaway Canyon; and seams containing free gold ore found in

*The data for this description was kindly furnished by J. T. Reid.

the quartzite of Shady Run Canyon. A flat gold-bearing vein is situated at the top of the butte at the mouth of Shanghai Canyon. The ores of the district are generally base and carry considerable lead.

Bibliography. USGS Carson Sink topographic map. Stuart NMR 110.

SODA LAKE

Soda, (Borax)

Location. The Soda Lakes are in W. Churchill Co., 6 m. N. W. of the town of Fallon which is on the S. P. R. R. Big Soda Lake has an area of 268½ acres, while Little Soda Lake is smaller and of variable size. The rim of Big Soda Lake rises 80 feet above the surrounding desert, and according to Russell was 165 feet higher than the surface of the lake when the lake had a depth of 147 feet.

History. Little Soda Lake was discovered by Asa L. Kenyon in 1855 and sold by him in 1868. An unsuccessfpl attempt to extract borax from the brine of Big Soda Lake was made several years prior to 1870. The production of soda from Litle Soda Lake began in 1868 and from Big Soda Lake in 1875; and the production from these lakes was at its height in 1887, but ceased shortly thereafter. The brines of Soda Lakes have recently been diluted by infiltrating irrigation waters from the neighboring Truckee-Carson Project.

Production. Little Soda Lake produced from 400 to 500 tons of soda annually from 1868 to 1885 and was producing at the rate of 300 tons per year in 1887. Big Soda Lake only produced 200 or 300 tons of soda from 1875 to 1885, but in 1887 was producing at the rate of 450 tons per year. No production has been made from either lake for many years.

Geology. The Soda Lakes occupy the craters of low basaltic cones. They are believed to be fed by seepage from the neighboring Carson River and their high saline content to be due to the gradual concentration of this supply by evaporation. According to Chatard, the specific gravity of brine from Big Soda Lake is 1.0995 and a sample taken from a depth of 100 ft. contained 125.644 grams of solids per liter having the following hypothetical composition:—

Sodium Chloride	54.88%
Sodium Sulphate	14.96%
Sodlum Bicarbonate	12.44%
Sodium Carbonate	12.32%
Potassium Chloride	4.06%
Magnesium Carbonate	.75%
Borax	.34%
Silica	.25%
	100.00%

Bibliography. R1872 186-7 Soda & Borax
SMN1869-70 15-16
SMN1871-2 18
SMN1873-4 15-17
SMN1875-6 6-7
MR1882 570 Borax, 601 Soda
MR1893 728-9
MR1917 II 320
MR1918 II 176, 177, 178
USGS Carson Sink topographic map
USGE 40th 1 (1878) 510-4; 2 (1877) 746-750
Chatard, T. M., USGS B 9 (1884) 25 Water Analyses.
————"Natural Soda, Its Occurance and Utilization", USGS B 60 (1890) 46-53.

Hanks, H. G., "Report on the Borax Deposits of California and Nevada", Cal. State M. Bur. 4rd AR (1883) II 45.

Knapp, S. A., "Occurrence and Recovery of Sodium Carbonate in the Great Basin Region", MI 7 (1898) 627-631.

——————"Occurence and Treatment of Carbonate of Soda Deposits of the Great Basin Region", M&SP 77 (1898) 448.

Russell, I. C., "Geological History of Lake Lahontan" USGS M 11 (1885) 73-80.

Thompson & West, 363-4.

Young, G. J., "Potash Salts and Other Salines in the Great Basin Region", U. S. Dept. Agr. B 61 (1914) 64.

TABLE MOUNTAIN (Boyer, Cottonwood Canyon, Bolivia)

Nickel, Cobalt, Copper, Gold, Lead, Silver, Antimony (Kaolin, Oil Shale, Gypsum)

Location. The Table Mountain District is at Table Mountain in N. E. Churchill Co. and S. E. Pershing Co. From the town of Lovelock on the S.P.R.R. it is 25 m. E. S.E. to New York Canyon on the W. side of the range at the N. end of the district and 44 m. to Boyer's Ranch on the E. flank. The nickel and cobalt properties are in Cottonwood Canyon a few miles N. W. of the ranch; the copper deposits lie to the S.; the new gold discovery is to the N.; the lead-silver properties are further N. in Pershing Co.; while the kaolin is at the extreme N. of the district in New York Canyon. The Table Mountain District was known as the Bolivia District in the early days.

History. Alva Boyer discovered copper ore on Table Mountain and several wagon trains of it were hauled to Sacramento in 1861. An arrastra was operated in New York Canyon in the early days and it is said that George Hearst had his first Nevada mining experience there. The silver-lead deposits of Cornish Camp were actively worked in the early seventies by John C. Fall and associates of Unionville; and the Linda-Jo silver-lead mine was discovered by Charles Gilbert about 1878 and produced some $30,000 from shallow workings. Antimony ore was shipped from Fenstonemaker Canyon on the W. side of the range opposite the Linda-Jo mine in the eighties. The nickel and cobalt deposits of Cottonwood Canyon were discovered by George Lovelock and Charles Bell about 1882. The Nickel Mine was worked from its discovery until 1890 when it was shut down on account of litigation; reopened in 1904, and again closed down in 1908. The Lovelock Mine shipped some 500 tons of high-grade nickel-cobalt ore to Swansea for treatment in the early days, but has been idle for many years. The Mines Dev. Co. of Nevada operated a copper-nickel-cobalt property under bond and produced several carloads of sorted high-grade copper ore containing but a trace of nickel while from the same workings they extracted ore containing 29% nickel. Gold was discovered in the district by F. L. Mason in 1922.*

Geology. The lower part of Cottonwood Canyon is in shale, limestone and quartzite, according to Ransome, and the upper part in andesite and andesite breccia. Both sedimentary and volcanic rocks are probably of Triassic age and are cut by dikes, masses, and tongues of diorite which are probably Cretaceous. The diorite and andesite are cut by white feldspathic dikes which are probably also Cretaceous in age. Resting upon these rocks at the head of the canyon is a thick series of Tertiary eruptive rocks.

*Some of the data used in this description was kindly furnished by J. T. Reid.

Nickel and Cobalt. The ore at the nickel mine occupies narrow, irregular fissures in the Triassic andesite at a diorite contact. It consists of a bright green hydrous nickel arsenate, believed to be annebergite with cores of nickel sulpharsenide, thought to be gersdorffite, while chloanthite and other nickel minerals are also probably present. At the Lovelock Mine the ore occurs in erratic veinlets in the altered andesite as at the Nickel Mine, but no diorite is exposed and the ore is much more complex, containing copper and cobalt in addition to nickel. The minerals observed by Ransome were tetrahedrite, erythrite, azurite and green crusts which were perhaps a mixture of annabergite and brochantite.

Copper and Gold. The ore minerals of the Table Mountain copper deposits are mainly pyrite and chalcopyrite carrying a little gold and a trace of silver. They occur irregularly disseminated through a bed of andesitic tuff near the base of the Tertiary eruptive series. In the richer portions of the deposit, Carpenter noted chalcocite, bornite, tenorite, cuprite, malachite, and azurite. The recent gold strike is said to have been made in a rhyolite-limestone contact vein.

Silver-Lead. The silver-lead ores of Cornish Camp and the Linda-Jo Mine occur in white quartz veins in limestone. At the latter locality, the vein is near a porphyry contact, and small values in gold are also present..

Kaolin. The Adamson-Dickson kaolin deposit lies at the mouth of New York Canyon, in the foothills of the range, sloping gently W. and N. W. towards the valley. The deposit has a thickness of from 75 ft. to 100 ft., according to Buwalda, and is a member of the Paleozoic sedimentary series. The material is a white, lean, gritty clay consisting of abundant quartz and considerable kaolin with sparse grains of rutile and occasional limonite stains. Burned at 1,150 degrees C., the absorption was 32.4%, porosity 48.9%, and color white; while at 1,300 degrees C absorption was 28.9%, porosity 45.2%, and color light cream white.

Oil Shale. Oil shale occurs in Cottonwood Canyon, according to Reid, and the surface material yields 6 gals. oil per ton on distillation.

Gypsum. A layer of gypsum occurs in the Triassic sediments two-thirds of the way up the W. flank of Table Mountain, according to Louderback.

Mines. The Boyer Copper Ms. Co. owns 49 lode claims aggregating 1,000 acres, a 10-acre millsite, and 160 acres of placer ground. A. H. Carpenter, 210 Noble Ave., Crafton, Pa., is Cons. Eng. for this company which has recently been conducting exploration work on its property. The Mines Dev. Co. of Nevada, described under the Juniper Range District, Pershing Co., has an option on a copper-nickel-cobalt prospect in the Table Mountain District upon which it has done some work. The Gibraltar-Federal Gold M. Co. is prospecting the recent gold strike. G. F. Talbot of Reno is Pres.; L. L. Hudson, V. P.; and F. L. Mason, Supt.

Bibliography.

B1867	319	MR1883-4	539,545	MR1896	329-330
R1873	212-3	MR1885	299,361	MR1901	243
		MR1886	171	MR1908 I	470
SMN1871-2	54	MR1887	109,620		
SMN1877-8	65-6	MR1893	170		

USGS Carson Sink topographic map (S. part of district only)

Buwalda, J. P., "Nevada" in "High-Grade Clays of the Eastern U. S., with Notes on Some Western Clays", USGS B 709 (1922) 122-4.

Carpenter, A. H., "Boyer Copper Deposits, Nevada", **M&SP 103 (1911) 804-5.**

Hill 507 199.
Louderbrack, G. D., "Basin Range Structure of the Humboldt Region", GSA B 15 (1904) 334,
Ransome 414 55-8.
Stuart NMR 121, 122-3.
Weed MH 1152 Boyer Copper Ms. Co.
1257 Mines Dev. Co. of Nevada.

TOY (Browns)
Tungsten

Location. The Toy District is located in N. W. Churchill Co. on the Pershing Co. border. The Toy Mine is 2 m. S. W. of Toy section house, formerly known as Browns, which is on the S. P. R. R. and the Toy Mill is at Fanning siding a short distance S. of Toy on the same railroad.

History. The Toy Mine was purchased by the St. Anthony Ms. Co., a subsidiary of the Atolia M. Co.. in 1908, and developed slowly up to 1915 when a mill was erected at Fanning to treat its ores. The mill was operated by the company in 1916 and 1917 and by lessees in 1918. The Bonanza King Group 2 m. E. of the Toy Mine has been prospected.

Geology. The country consists of shales and quartzites in which is a thin-bedded limestone, according to Hess and Larsen. These sedimentary rocks, which are perhaps of Jurassic age, have been intruded by granodiorite, which has altered the limestone and produced the ore. The ore consists of garnet, diopside, quartz, hornblende, clinozoisite, scheelite, calcite, pyrite, chloropal, and iron oxide formed from the oxidation of pyrite. The scheelite is in fine particles and is rarely visible to the naked eye.

Bibliography. MR1908 I 724 MR1910 I 739 MR1916 I 792
MR1909 I 580 MR1913 I 356 MR1918 I 979
MR1915 I 825
USGS Carson Sink topographic map.
Hess, F. L., & Larsen, E. S., "Contact-Metamorphic Tungsten Deposits of the U. S..", USGS B 725D (1921) 287-9.
Hill 507 212.
Weed MH 1324 St Anthony Ms. Co.

WESTGATE
Silver, Lead, Gold

Location. The Westgate District is at Westgate in S. E. Churchill Co., and adjoins the Eastgate District on the W. Fallon on the S. P. R. R. is 54 m. W. N. W. of Westgate.

History. Ore was produced in this district in 1915.

Bibliography. MR1905 266. USGS Carcon Sink topographic map. Hill 507 200.

WHITE CLOUD
Copper, (Zinc, Lead, Silver)

Location. The White Cloud District is at Coppereid in White Cloud Canyon on the W. slope of the Stillwater Range in N. Churchill Co. The town of Lovelock on the S. P. R. R. is 35 m. N. W. of Coppereid. The Copper Kettle District adjoins the White Cloud District on the N., and the Shady Run District on the S. W.

History. The district was discovered in 1869 and a small copper smelter was in operation there about 1893. The Nevada United M. Co. under the management of J. T. Reid carried on active prospecting operations from

1893 to 1896 and shipped a little ore. For the past 10 years only assessment work has been done in the district.

Geology. The lower part of White Cloud Canyon is in granite which has been intruded by dikes of granite porphyry and diorite porphyry. Farther up the canyon the granite appears to grade into granite porhyry which is intrusive into a sedimentary series composed of limestones, calcareous shales, and a few beds of gypsum, probably of Middle Triassic age, and has caused contact metamorphism.

Ore. The copper ore occurs in a broad zone of mineralized limestone on the ridge S. of White Cloud Canyon, which runs E. from the contact and dips about 45oN. Large outcrops of specularite and soft hematite occur carrying a little malachite and in places a litle oxidized copper ore is present. In depth the contact-metamorphic veinlets, concretions, and masses contain quartz, chalcopyrite and sphalerite, according to Ransome.

Bibliography. R1869 193 SMN1871-2 19 MR1908 I 472
R1875 187 MR1909 I 394
MR1915 I 625
USGS Carson Sink topographic map.
Hill 507 200
Ransome 414 591-61.

WHITE PLAINS see DESERT and WHITE PLAINS FLAT

WHITE PLAINS FLAT
Salt

Location. White Plains Flat is at White Plains (Huxley station) in N. W. Churchill Co. and extends 4 m. S. to Parran and 4 m. N. E. to Ocala, all three places being on the S. P. R. R. The flat lies N. E. of the Desert District, S. E. of the Jessup District, and S. W. of the Lake District.

History. Walter Smith located the salt deposit about 1870, and formed the Desert Crystal Salt Co. which produced salt by solar evaporation of the brine, according to a letter from J. T. Reid. About 200 tons of salt were shipped annually, principally to the silver mines of eastern Nevada, and some table salt was produced. Production was continued on a decreasing scale up to 1915. The International Salt Co. under a lease from the Desert Crystal Salt Co. made small productions at Parran in 1911 and 1912.

Bibliography. MR1882 543-4 MR1913 II 300 MR1914 II 300-1
MR1912 II 919 MR1915 II 273
USGS Carson Sink topographic map.
Phalen, W. C., "Salt Resources of the U. S.", USGS B 669 (1919) 140.

WONDER
Silver, Gold, Copper, Zinc

Location. The Wonder District is situated at Wonder on the W. slope of the Augusta Mts. in W. Churchill Co. Wonder is 55 m. by road E. of Fallon which is on the S. P. R. R. The elevation of Wonder is 5,800 ft. and Twin Peaks rises above it to an altitude of 7,110 ft.

History. The district was discovered on March 15, 1916, by Thomas J. Stroud; the Nevada Wonder Mine which is the most important in the district being located shortly afterwards. A rush took place from the Fairview District to the new camp. The Nevada Wonder M. Co. began operations that year, and later constructed a 10-stamp cyanide mill which started on August 1, 1911 and was finally shut down for want of ore in Dec. 1919.

Production. The total production of the Wonder District from 1907 to 1921 was $5,838,765 according to Mineral Resources of the U. S. Geol. Survey. The greater part of this production was made by the Nevada Wonder Mine.

Geology. According to Burgess, the country rock of the Wonder District is a complex aggregate of Tertiary eruptives. The Wonder rhyolite is the oldest rock in the district and the ore-bearing veins occur in it, usually near small intrusive masses of a more acid rhyolite. The veins are composed chiefly of quartz and feldspar and contain small bodies of silver-gold ore, but the Nevada Wonder is the only one which has paid to mine. The Nevada Wonder vein lies partly at the contact of the Wonder rhyolite with an intrusive body of dacite and partiy within the Wonder rhyolite, striking N25 degrees W and dipping 75 degrees E. The ore consists mainly of quartz rudely banded with feldspar and containing occasional fluorite. It is usually stained with limonite and contains a little dendritic manganese oxide. The silver is in the form of argentite and the halogen salts,—embolite, iodobromite, and bromyrite; while gold occurs both native and combined with argentite. Oxidation extends to a depth of 1,300 ft. Copper and lead are found only in traces and there is no zinc above the 1,300 ft. level.

Mines. The Nevada Wonder M. Co. was incorporated in Delaware in 1906 with a capital stock of 1,500,000 shares of $1 par value, of which 91,513 remain in the treasury. The company paid semi-annual dividends of from 5% to 10% from May, 1913, to May, 1919, amounting in all to $1,549,005. The home office of the company is at 572 Bullitt Bldg., Philadelphia. C. R. Miller is Pres., J. H. Whiteman, V. P.-Gen. Counsel, with S. Bell, Jr., M. B. Cutter, B. Hoopes, Jr., J. W. Murray, and W. M. Potts, Directors. P. S. Bickmore is Sec.-Treas. and H. C. Carlisle, Gen. Supt. The company is now seeking new properties of merit. G. A. Manning of Campbell, N.Y., is Pres. of 2 other companies which have recently been active in the Wonder District;—the Nevada Silver Cons. Co., Inc., and the New York Oro Corp.

Bibliography.

General

MR1905 116,266	MR1910 I 507	MR1916 I 469-470
MR1906 293	MR1911 I 667-8	MR1917 I 265-6
MR1907 I 345	MR1912 I 785	MR1918 I 230
MR1908 I 471-2	MR1913 I 817	MR1919 I 385-6
MR1909 I 394	MR1914 I 670-1	MR1920 I 319
	MR1915 I 625-6	MR1921 I 380

Carpenter, E. E., "Operation of the Nevada Wonder Mining Company During Four Years of War", M&SP 120 (1920) 537-8.

Weed MH 1284 Nevada Silver Cons. Co. 1289 Nevada Wonder M. Co. 1293 New York Oro Corp.

Zalinski, E. R., "Mining in the Wonder District, Nevada", E&MJ 83 (1907) 763-5.

————"Recent Developments at Wonder", E&MJ 84 (1907) 357-8.

Geology

Burgess, J. A., "The Halogen Salts at Wonder, Nevada", EG 12 (1917) 589-593

Discussions as follows:—

Young, J. W., EG 13 (1918) 224-5.

Lindgren, W., EG 13 (1918) 225-6.

Burgess, J. A., EG 13 (1918) 546-9.

Knopf, A., EG 13 (1918) 622-4.
Young, J. W., EG 14 (1919) 427-430.
Hill 507 200-1.
Ritter, E. A., "Ore Formation in the Wonder District, Nevada", E&MJ 87 (1909) 289-292.

Mining

Smither, T. M., "Stoping Methods at the Nevada Wonder Mine", M&SP 110 (1915) 757-9.

Milling

Carpenter, E. E., "Cyanide Practice of Churchill Milling Co., Wonder, Nevada", T AIME 52 (1915) 123-137.
——— Discussion, "Water in Stamp Milling", M&SP 116 (1918) 288.
Daman, A. C., "The Nevada Wonder Mill", E&MJ 101 (1916) 927-8
Megraw, H. A., "Cyaniding at the Nevada Wonder Mill", E&MJ 95 (1913) 693-5.
Van Saun, P. E., "New Mill at Nevada Wonder Mine", E&MJ 91 (1911) 520-2.
———"The Nevada Wonder Mining Company's New Mill", M&EW 35 (1911) 959-962.

Miscellaneous

Burgess, J. A., "The Nevada Wonder Pipe Line", M&SP 112 (1916) 435-8.
M&SP Editorial, "Employer and Employee", 112 (1916) 926-7.
USGS Carson Sink topographic map.

CLARK COUNTY

ALUNITE (Railroad Pass)

Gold, (Potash)

Location. The Alunite District is located at the S. E. end of Las Vegas Valley in Central Clark Co. on a cross range between the Spring Mt. and Eldorado Ranges at a low gap known as Railroad Pass. It is 22 m. by road S. E. of Las Vegas, which is on the U. P. R. R. and 12 m. E. of Erie which is on the same railroad.

History. Gold was discovered in the mountains to the W. of Railroad Pass prior to 1908. Robert T. Hill discovered alunite and gold in the pass in 1908. This discovery was prospected by the Alunite M. Co. and abandoned shortly afetrwards. Interest in the district as a possible source of potash from alunite was aroused in 1915, but the proportion of potash present was found to be too low to pay to work. The Quo Vadis gold mine began operations in 1915 and produced a little high grade ore but is at present shut down.

Geology. The mountains to the W. of Railroad Pass are composed of biotite-monzonite and those to the E. of andesite and latite, according to Hill. Within the area of the pass, between these mountains of unaltered igneous rocks, the biotite-monzonite, andesite, and latite have been alunitized and silicified and small amounts of precious metals have been deposited in them. According to Gale, the general run of rock as it would have to be mined in this locality would probably not average over 2½% of potash, or, at the most, 3½%. At the Quo Vadis Mine in the biotite-monzonite to the W. of Railroad Pass, rich stringers of gold ore occur. A letter from E. S. Giles states that the gold is free in quartz, and very little silver is present. The gold-bearing quartz is frozen to the wall on one side and to calcite colored by manganese on the other side, the calcite being frozen to the other wall of the vein.

Mines. The Quo Vadis M. Co. was incorporated in Nevada in 1915. P. Buol of Las Vegas is Pres.; C C. Ronnow, V. P.; E. W. Clark. Sec.-Treas.; with F. A. Clark and J. B. Anderson additional Directors.

Bigliography. MR1915 II 111-2 Potash. USGS Camp Mohave topo. map.
Hill 507 202.
Hill, R. T., "A Scientific Search for a New Goldfield",
E&MJ 86 (1908) 1157-1160.
———————"Camp Alunite, A New Nevada Gold District",
E&MJ 86 (1908) 1203-6.
Weed MH 1309 Quo Vadis Gold M. Co.

ARDEN

Gypsum

Location. The Arden Gypsum Mine is located in central Clark Co. in the foothills of the Spring Mountains 5 m. W. or Arden which is on the U.P.R.R. The mine is connected with the mill at Arden by a narrow gauge railroad

History. The property was purchased by the United States Gypsum Company from the Arden Plaster Company in 1919.

Geology. The gypsum deposit underlies an oval hill about 500 ft. in height and ¾ m. in length with its longer axis strending N. W. This hill is composed of Carboniferous strata which dip gently toward the E. These strata consist of an upper bed of massive, gray, cherty limestone about 125 ft. in thickness; below which comes a bed of red and green gypsiferous shale containing thin beds of gray limestone in places, which bed has a thickness of about 85 ft.; and rests upon a bed of massive gray limestone of undertermined thickness.

Gypsum. The gypsum ocurs as a bed from 25 to 80 ft. in thickness included in the gypsiferous shale which lies between the two limestones. It is pure, massive, crystalline and of medium grain. It is translucent with a reddish tinge near the surface which becomes bluish with depth. Little anhydrite ocurs in the gypsum but occasional masses up to 100 tons in size have been found and the gypsum is believed to have been derived from anhydrite by hydration. The upper surface of the gypsum is very irregular, the irregularities is some instances being filled by the shale and in others being left open as cavities covered by shale.

Mining. The United States Gypsum Co., 205 W. Monroe St., Chicago, is operating the Arden mine and mill as one of its 19 plants scattered throughout the country. The gypsum is quarried where the overburden is not excessive, and where the cover is too heavy, tunnels are driven and stopes opened. Pillars are left to support the roof when necessary, the loss of gypsum in pillars being reduced by leaving anhydrite and shale as pillars whenever possible. Steam shovels are used for handling the overburden and for loading the gypsum. The gypsum is crushed at the top of the hill and lowered in balanced skips on an incline to a bin at the foot of the hill. From this bin it is loaded into cars which are drawn by steam locomotives over the 5-mile narrow gauge railroad to the mill at Arden, where the usual kettle process is employed. The products manufactured are:—stucco, plaster of Paris, finishing plaster, moulding plaster, casting plaster, and wall plaster. The production varies with the demands of the building trade in Southern California where the principal markets are located.

Bibliography. MR1910 II 727 MR1916 II 260 MR1918 II 290
MR1912 II 648 MR1917 II 89 MR1919 II 104-5,113
MR1915 II 155 MR1920 II 63
USGS Las Vegas topographic map.

Gilbert, G. K., "Report on the Geology of Portions of Nevada, Utah, California, and Arizona Examined in the Years 1871 and 1872", USGSW100th 3 (1875) 166.

Jones, J. C., and Stone, R. W., "Gypsum Deposits of the U. S., Southern Nevada", USGS B 697 (1920) 155-8.

Louderback, G. D., "Gypsum Deposits in Nevada", USGS B223 (1904) 112.

Stone, R. W., "Gypsum Products", USBM TP 155 (1917) 48, 56.

BUNKERVILLE see COPPER KING

COLORADO see ELDORADO

COPPER KING (Bunkerville)

Platinum, Nickel, Copper, Cobalt

Location. The Copper King District is located 15 m. S. of Bunkerville in the foothills of the Virgin Range in N. E. Clark Co. By road it is some 15 m. N. E. of St. Thomas which is the terminus of the St. Thomas Branch of the U. P. R. R. The prospects are at elevations of 3,600 ft. to 4,200 ft., while Virgin Peak, the highest point in the range, reaches a height of 7,500 ft.

History. The occurance of nickel in the district was known as early as 1901, and the Key West property was worked about 1903. In 1906, the presence of platinum in the ore was known, and the Great Eastern property, as well as the Key West, was being prospected. A carload of ore was shipped from the Key West Mine by the Nevada Copper, Platinum, and Nickel Co. in 1908. The district has been inactive in recent years.

Geology. The country rock of the Copper King District according to Bancroft, is gneiss which is apparently of granite origin and probobly or pre-Cambrian age. This gneiss is cut by intrusions of various basic rocks which have undergone metamorphism with the gneiss, and by dikes of aplite and pegmatite which are later than this metamorphism. The ore-bodies are the most basic of the dikes, having the composition of enstatite-mica picrite, a variety of peridotite. These peridotite dikes constitute the only known example of pre-Cambrian ore deposits in Nevada.

Ore. The chief minerals in the peridotite dikes, and therefore the chief minerals in the ore, are augite, olivine, biotite, and enstatite. With these occur appreciable amounts of magnetite, pyrite, chalcopyrite, and pyrrhotite, the latter probably nickeliferous. A trace of platinum was also found in a hornblendite dike which showed no metallic content even under the microscope. The carload of ore from the Key West peridotite dike contained 2.30% copper, 1.79% nickel, 0.08% cobalt, 0.13 oz. per ton platinum, and traces of gold and silver.

Bibliography. MR1901 243 MR1906 560 MR1908 I 472, 736, 783
MR1905 270 MR1907 I 365 MR1909 I 395
USGS St. Thomes topographic map.

Bancroft, H., "Platinum in Southeastern Nevada", USGS B 430 (1910) 192-9.

E&MJ Special Correspondence "Nickel Ore in Nevada", 86 (1908) 23.

Hill 507 30, 201-2.
Spurr 208 131-3 Virgin Range.
Thompson, A. M., "Nickel-Copper-Platinum Ore in Nevada", E&MJ 86 (1908) 72.

CRESCENT

Silver, Gold, Lead, Copper, Turquoise, (Molybdenum, Vanadium)

Location. The Crescent District is at Crescent in S. Clark Co. on the California border. Nipton, California, on the U. P. R. R. is 6 m. W. of Crescent. The Sunset District adjoins the Crescent District on the N. W.

History. An Indian known as Prospector Johnnie discovered turquoise 3 m. S. E. of Crescent in 1894. The mine was purchased by J. R. Wood of New York in 1896 and operations were begun in the name of the Toltec Gem Co. the following year. Prehistoric workings were found at this mine and ancient stone hammers lay scattered about. Two other turquoise properties were discovered in the district later;—the Smithson and Phillips Mine 1 m. E. of the Toltec Mine, and the Morgan Mine 1 m. S. of Crescent. According to Schrader, wulfenite and vanadinite occur 4 m. E. of Crescent. Ransome states that the Crescent District was first exploited for precious metals about 1895 and that a revival of interest occurred in 1905. Small intermittent productions were made by the district from 1906 to 1917. A 1-Nissen stamp amalgamating mill wes erected at the Golden Calf Mine in 1911. The camp is at present inactive.

Geology. The prevailing country rock is gneiss, according to Ransome, with which are associated fine-grained schists and quartzites. These metamorphic orcks are probably pre-Cambrian. They are cut by granitic intrusions and basic dikes; and at the Morgan turquoise mine the granitic rock is cut by rhyolite. The turquoise occurs in seams and veinlets in the granitic intrusions and in the rhyolite. The silver and gold occur in shattered masses of quartzite which have been recemented by quartz.

Bibliography. MR1905 270 MR1908 I 472 MR1913 II 697-9 Turq.
MR1906 296-7 MR1909 I 395 MR1915 I 627
MR1907 I 365-6 MR1911 I 668 MR1917 I 266
MR1912 I 782
USGS Ivanpah topographic map.
Hill 507 201.
Ransome, F. L., "Preliminary Account of Goldfield, Bullfrog, and Other Mining Districts in Southern Nevada", USGS B 303 (1907) 79-80.
Schrader 624 199 Vanadinite; 200 Wulfenite.

ELDORADO (COLORADO)

Gold, Silver, Lead, Copper

Location. The Eldorado District is on Eldorado Canyon W. of the Colorado River is S. Clark Co. on the Arizona border. It occupies the N. end of the Opal Mts., the Searchlight District lying at the S. end; and is 24 m. N. of the town of Searchlight which is on the B. & S. R. R.

History. The district was discovered about 1857. The town of Eldorado was founded and the district organized as the Colorado District in 1861. The Techatticup Mine is the principal mine and was opened in 1863. A 15 stamp mill was erected at the mine that year, and other mills have been built in the district since.

Production. Ransome estimates the production of the district up to 1907 at from $2,000,000 to $5,000,000. The Techatticup Mine is credited with a production of $1,700,000. From 1907 to 1921, the district produced 84,010 tons of ore containing $662,892 in gold, 724,568 ozs. of silver, 2,059 lbs. of copper, and 23,483 lbs. lead, valued in all at $1,256,120, according to Mineral Resources of the U. S. Geol. Survey.

Geology. The Opal Mts. are composed of a N.-S. belt of pre-Cambrian schists and gneisses intruded by quartz-monzonite, according to Ransome. Areas of Tertiary eruptive flank this belt in places. At Nob Hill to the S. of Eldorado Canyon, narrow fissure veins in gneiss and schist contain auriferous and argentiferous pyrite galena and sphalerite in a gangue of vein quartz and altered country rock. At the upper end of Eldorado Canyon, orebodies occur in mineralized fissure zones in the quartz-monzonite. The Techatticup Mine is in this section and its ore consists of pyrite, galena, and sphalerite finely disseminated in shattered quartz-monzonite which has been cemented by stringers of quartz and calcite.

Mines. Two companies have been active in the district recently;—the Ben Ezra Copper & Gold M. Co. and the Eldorado-Flagstaff M. & M. Co. G. A. Duncan is Mgr. of the Eldorado-Flagstaff at Nelson. The officers of the Ben Ezra are H. P. Balcom, Pres.; C. W. Wright, Sec.-Treas., and W. Mackenzie, Supt.; all of Santa Paula, Cal. H. O. Russell is Res. Agt. at Searchlight.

Bibliography.

B1867 429	MR1906 297	MR1914 I 672
R1872 186	MR1907 I 366	MR1915 I 627
SMN1867-8 85	MR1908 I 473	MR1916 I 470
SMN1869-70 103	MR1909 I 395	MR1917 I 266
SMN1871-2 96	MR1910 I 508	MR1918 I 231
SMN1875-6 91	MR1911 I 669	MR1919 I 386
	MR1912 I 786	MR1920 I 320
MR1905 270	MR1913 I 818	MR1921 I 380

USGS Camp Mohave topographic map.

Davis 2, 928, 929.

Hill 507 201.

M&SP, Discusion, "Water Laws of Nevada".
Duncan, G. A., 117 (1918) 821-2; 118 (1919) 314-6.
Walker, R. T., 118 (1919) 111-3; 451.
Brooks, E. W., 118 (1919) 209-210.

Ransome, F. L., "Preliminary Account of the Goldfield, Bullfrog, and Other Mining Districts in Southern Nevada", USGS B 303 (1907) 63-76.

Stuart NMR 134-5.

Weed MH 1142 Ben Ezra Copper & Gold M. Co
1190 Eldorado-Flagstaff M. & M. Co.
Eldorado Gold Star M. Co.
1253 M. & D. M. Co.
1335 Silver Standard M. & Extraction Co.
1349 Techatticup Mine. 1370 Waterville M. & M. Co.

GASS PEAK

Zinc, Silver, Gold

Location. The Gass Peak District is situated on Gass Peak at the S. end of the Las Vegas Range in central Clark Co. The town of Las Vegas on the U. P. R. R. is 18 m. to the S.

History. The principal mine is that of the June Bug Dev. Co., Inc., of which

L. S. Scoville of Ogden, Utah, is Pres. Nearly $38,000 worth of ore was shipped from this mine in 1916 and 1917.

Geology. Zinc silicate ore containing values in silver and gold is said to occur in shoots in veins in Paleozoic limestone.

Bibliography. MR1916 I 471 MR1917 I 266
USGS Las Vegas topographic map.
Spurr 208 Las Vegas Range. Weed MH 1233 June Bug Dev. Co., Inc.

GOLD BUTTE

Copper, Silver, Zinc, Mica, (Gold)

Location. The Gold Butte Districe is at Gold Butte in the S. part of the Virgin Range in E. Clark Co. near the Arizona border. Gold Butte with an altitud of 4,300 ft. is the highest peak in the district. St. Thomas on the U. P. R. R. is 28 m. N. W. of the camp at Gold Butte.

History. Daniel Bonelli discovered mica in the S. part of the district about 1873; and in 1893 and 1894, he made trial shipments of uncut mica amounting to 1,800 lbs. In 1905, the metal mining portion of the district to the N. was discovered by Bonella, Burgess, Syphus, and Gentry. From 1912 to 1918, small shipments of argentiferous copper ore and one of zinc ore were made from the district.

Geology. Gold Butte and the area to the S. are underlain by pre-Cambrian granite, gneiss, and schist, according to Hill, while N. of the peak Paleozoic limestones and sandstones appear. Small replacement bodies of oxide copper ore consisting of hematite, limonite and oxidized copper minerals occur in the limestone; and chalcocite secondary after chalcopyrite and bornite was observed at the Lincoln Mine. The granite and granitic gneiss are intersected by narrow quartz veins in which sulphides occur sparingly, the principal one being pyrite, with lesser amounts of chalcopyrite, galena, and sphalerite. A little free gold is found at the surface in the more heavily mineralized portions of these veins.

Bibliography.

SMN1875-6 90-91 Mica	MR1905 270	MR1913 I 818
MR1883-4 911 Mica	MR1906 297	MR1914 I 672
MR1893 753-4 Mica	MR1907 I 367	MR1915 I 627
MR1902 986 Mica	MR1908 I 472-3	MR1916 I 471
MR1912 II 1084-5 Mica	MR1909 I 395	MR1917 I 267
MR1914 II 73 Mica	MR1912 I 786	MR1918 I 231

USGS St. Thomas topographic map.
Hill 507 201.
Hill 648 42-53.
Spurr 208 131-3 Virgin Range.
Weed MH 1189 Eastern C. Co.

GOODSPRINGS see YELLOW PINE

LAS VEGAS

Manganese

Location. The Las Vegas manganese district is situated on the S. side of the Las Vegas Wash, in low hills which border the Colorado R. and may be considered either as outliers of the Opal Range to the S. or of the Muddy Range to the N. The town of Las Vegas on the U. P. R. R. is 16 m. N. W. The deposits are at an elevation of about 2,000 ft.

History. The Three Kids Mine was discovered by Bob Edwards early in 1917, and other discoveries were made in the same neighborhood later. The Manganese Association took over the Three Kids Mine and began production in November, 1917. Thos. Thorkildson, Trust & Saving Bank Bldg., Los Angeles, purchased the Three Kids Mine and adjoining Las Vegas group in April, 1918. The original name was retained and the production greatly increased. The property was closed down in March, 1919, owing to the unfavorable condition of the manganese market. None of the other properties in the district produced.

Production. The Three Kids Mine produced 12,000 tons of 40% manganese ore up to July 1, 1918, at which time it was producing at the rate of 60 tons per day.

Geology. The hills of the Las Vegas manganese district are composed of volcanic flows and tuffs which are regarded as of Tertiary age by Pardee and Jones. Tuff, sand, clay, gypsum, and gravel beds, probably derived largely from the disintegration of the volcanic rocks and deposited in a Miocene lake, flank the hills and underlie the mesa extending to Las Vegas Wash. Rhyolite and andesite tuffs and breccias are the most abundant volcanic rocks, and associated with them are a few basaltic flows.

Ore Deposits. The ore deposits are replacements of tuffs and of sand and clay beds derived from them. In most cases they are clearly associated with faults or fissures through which the manganese-bearing solutions rose. The principal ore mineral is wad containing occasional streaks and grains of pyrolusite and hard manganese minerals were only observed by Pardee and Jones in one deposit. The principal impurities in the ore are the unreplaced materials of the sand or tuff bed, but some impurities, like lead and copper oxides, were probably introduced with the manganese. In addition to these impurities in the ore itself, gypsum and calcite veinlets cut the deposits and gypsum efflorescence frepuently caps them.

Bibliography. MR1917 I 684-5 MR1918 II 644

USGS St. Thomas topographic map.

Hale, F. A., Jr., "Manganese Deposts of Clark County, Nevada", E&MJ 105 (1918) 775-7.

M&SP "Manganese in the Colorado River Desert Region", 117 (1918) 755-8.

Pardee, J. T., & Jones, E. L., Jr., "Deposits of Manganese Ore in Nevada", USGS B710 (1920) 222-233.

Weed MH 1228 International Properties Syndicate.
1243 Lowney Manganese Assoc.
1247 Manganese Assoc., Ins.

LYONS see SUNSET

MOAPA

Gypsum

Location. The Moapa gypsum deposits are situated in the N. part of the Muddy Range to the S. E. of Moapa which is on the U. P. R. R. Moapa is the starting point of the St. Thomas Branch of the U. P. R. R. and the gypsum deposits are near this branch and have their shipping points upon it.

White Star Plaster Co. The White Star Plaster Co. is the only company operating in the Moapa District at the present time.* Its mail and express office is Moapa; freight office and plaster works, Arrowhead; and California office in the W. I. Hollingsworth Bldg., Los Angeles. W. B. Lenhart is Supt. Up to 1921, the White Star Plaster Co. mined and cal-

*Data concerning the White Star Plaster Co., was kindly furnished by W. B. Lenhart.

cined gypsite, but the product not proving entirely satisfactory, a gypsum deposit was developed that year and the mill has been working on gypsum since. The gypsum deposit is on the St. Thomas Branch of the U. P. R. R. about 3½ m. E. of Moapa. Two beds of gypsum occur in limestone, only the lower one being at present worked. This bed is from 30 ft. to 40 ft. thick, dips at about 30 degrees, and is exposed for a length of 3,500 ft. As the mill was located at the gypsite deposit, the gypsum has to be hauled 1½ m. to the mill. The mill originally contained two 10 ft. oil-fired calcining kettles, but was enlarged to a 3-kettle mill in 1922. During the year1922, the shipments of plaster by the White Star Plaster Co. to Los Angeles averaged about 3,000 tons per month.

Rex Plaster Co. The Rex Plaster Co. formerly shipped gypsum to a factory in Los Angeles but has not been active in recent years. According to Jones & Stone, the Rex deposit is situated 5 m. from the nearest shipping point which is 5 m. from Moapa on the St. Thomas Branch of the U. P. R. R. It is a massive body of pure white gypsum from 100 ft. to 300 ft. in thickness, dipping E. at from 40 to 70 degrees, and underlain by a gray limestone and overlain by a red sandstone.

Bibliography. MR1919 II 104.

Jones, J. C., and Stone, R. W., "Gypsum Deposits of the U. S., Southern Nevada", USGS B 697 (1920) 159.

MUDDY MOUNTAINS

Borates

Location. The Muddy Mts. lie in E. central Clark Co., N. of the Colorado R., E. of Las Vegas Valley, and . of the Virgin River District. The automobile highway between Las Vegas and St. Thomas, which is part of the Arrowhead Trail from Salt Lake City to Los Angeles, crosses the northern part of the Muddy Mts.; but the borate deposits are in the southern part of the mountains and are reached by the old road from Las Vegas to St. Thomas. The borates occur in two localities 12 m. apart,—one in White Basin in the S. E. part of the Muddy Mts., and one near Callville Wash, a tributary of the Colorado R., in the S. W. part of the mountains. The gypsum which occurs in the N. part of the Muddy Mts. is described under the Moapa District.

History. John Perkins of St. Thomas located borates in White Basin late in 1920, and his claims were purchased by the Pacific Coast Borax Co. Other claims were subsequently located in the same neighborhood, some of which have been acquired by the American Borax Co. F. M. Lovell and George Hartman studied the occurrence of borates at White Basin in December, 1920, and then prospected southwestward till they located the great deposit near Callville Wash late in January, 1921. F. M. Smith immediately purchased this property for the West End Chemical Co. The West End Chemical Co. erected a mill at the property in 1922 and began shipping the concentrates to its refineries in various parts of this country and Europe.

Geology. The Muddy Mountains are composed of sedimentary rocks belonging to two main groups. The older group consists of Paleozoic and Mesozoic rocks which have been upturned and eroded and upon which rest unconformably the younger group of Tertiary rocks. The Tertiary rocks are in turn divisable into two series separated by an unconformity. The older series is made up of the Overton fanglomerate and the Horse Spring formation probably of Miocene age, upon which lies unconformably the Muddy Creek formation probably of Pliocene age. The Horse Springs formation is 2,500 ft. in thickness and consists of beds of limestone, shale, tuff,gypsum, sandstone, and conglomerate. The borate deposits occur in the limestone which forms the lower member of the Horse Spring formation.

Borate Deposits. Colemanite is the principal boron mineral present in the deposits. A little ulexite is associated with it. At White Basin the colemante occurs in irregular layers, bunches, and balls in a shaly stratum of the Horse Spring limestone. The layers of colemanite are small, varying from 1 in. to 2½ ft. in thickness, and the district has been much disturbed by folding and faulting. Near Callville Wash, the deposit of colemanite is larger and more regular. It is in the form of a lens in the shaly limestone, 3,000 ft. in length and from 10 to 18 ft. in width, on the northern rim of a syncline. It consists of bunchy layers of massive crystalline colemanite alternating with layers of shaly limestone, the colemanite making up about one-third of the deposit.

Mines. The mine near Callville Wash is the property of the West End Chemical Co. F. M. Smith is Pres. and J. W. Sherwin is Mgr. with home office at Oakland, Calif. J. W. Wilson is Mine Supt., at Las Vegas, Nevada. The company is operating a concentrating mill at the mine with an output of 30 tons per day, and hauling the concentrates 30 m. to the railroad with caterpillar tractors. The principal mine at White Basin is owned by the Pacific Coast Borax Co., a subsidary of the British borax trust, Borax Consolidated, Ltd. The American Borax Co. also has important holdings in that locality.

Bibliography. MR1920 II 130. USGS St. Thomas topographic map.

E&MJ, "Promising Colemanite Deposit Found in Nevada", 111 (1921) 600.

Gale, Hoyt S., "The Callville Wash Colemanite Deposit", E&MJ 112 (1921) 524.

Longwall, C. R., "Geology of the Muddy Mountains, Nevada, with a Section to the Grand Wash Clixs in Western Arizona", AJS 50 (1920) 39-62.

Noble, L. F., "Colemanite in Clark County, Nevada", USGS B 735B (1922) 23-39.

Spurr 208 136-8.

POTOSI see YELLOW PINE

RAILROAD PASS see ALUNITE

ST. THOMAS see VIRGIN RIVER

SEARCHLIGHT

Gold, Silver, Copper, Lead, (Turquoise)

Location. The town of Searchlight is the terminus of the B. & S. R. R. in S. Clark Co. The principal mines of the Searchlight District are near the town, but the district is usually considered as extending E. to the Colorado R. at the Arizona boundary, N.E. to Dupont, and S.E. to camp Thurman. The district occupies the S. end of the Opal Mts., the Eldorado District covering the N. end, and its elevation ranges from 2,000 to 3,000 ft.

History. The Searchlight District was discovered in 1897. Ore was found on the Searchlight claim of what is now the Duplex Mine in 1898, and the Quartette Mine was discovered and the district organized the same year. The Quartette Mine built a mill on the Colorado R. in 1902, connecting it with the mine by narrow gauge; but water being encountered in the mine the following year, a mill was erected there instead. The B. & S. R. R. was completed in 1907. In 1908, a turquoise weighing 320 carats and valued at $2,600 was said to have been found in the district. During recent years the camp has been largely in the hands of leasers, although several companies have been carrying on prospecting and development work.

Production. Ransome estimated the production to the end of 1905 at from $1,-750,000 to $2,000,000.

RECENT PRODUCTION OF THE SEARCHLIGHT DISTRICT
1904 - 1921
(According to Mineral Resources of the United States, U. S. Geol. Survey)

Year	Mines	Tons	Gold $	Silver Ozs.	Copper Lbs.	Lead Lbs.	Total $
1904	3	16,750	380,386	13,246			388,068
1905	6	38,069	399,575	28,528	22,808	12,064	420,931
1906	5	45,668	519,785	11,543	11,182	9,655	530,227
1907	7	45,921	484,568	7,494	37,063	38,113	498,947
1908	7	52,193	271,573	10,883	14,954	44,857	281,199
1909	8	68,931	335,645	13,544	22,916	36,209	347,224
1910	13	27,331	215,097	11,489	93,848	45,847	235,237
1911	10	1,850	19,979	2,136	12,095	7,659	22,968
1912	9	4,158	46,244	2,562	10,126	10,032	49,942
1913	15	5,989	137,547	5,903	53,971	61,006	152,162
1914	5	3,057	82,657	2,855	5,556	21,919	85,830
1915	13	7,766	103,042	3,546	20,970	35,562	110,182
1916	15	6,322	80,677	4,340	24,392	9,951	90,220
1917	14	12,866	71,213	9,073	98,923	59,453	110,808
1918	14	1,700	54,761	8,395	44,477	77,863	79,670
1919	9	6,980	59,005	5,502	18,074	39,307	70,612
1920	7	2,131	70,953	5,275	44,941	91,436	92,287
1921	9	1,182	74,432	9,877	31,128	98,048	92,737
Total 1904-21		348,864	3,407,139	156,193	567,424	688,981	3,659,251

Geology. The main mass of the Opal Mountains is made up of a N.- S. belt of pre-Cambrian schists and gneisses which are much disturbed and are cut by intrusions of quartz-monzonite. These rocks are flanked by Tertiary eruptives. Near the town of Searchlight, the principal rock is biotite-schist, while to the N. the mountains consist largely of quartz-monzonite, and to the E. and W. lie areas of Tertiary volcanic rocks. The orebodies occupy mineralized fault zones near the contact between the quartz-monzonite and the gneiss and also near the Tertiary eruptives. They have a general N. W. strike and S. W. dip. The Quartette lode extends from the Tertiary andesite into the pre-Cambrian complex and has a strike of from N65 degrees W to N68 degrees W and a southwesterly dip of from 20 degrees to 70 degrees. The ore commonly consists of crushed country rock colored by chrysocolla and iron oxide, together with some quartz. The less frequent constituents are cerussite, wulfenite, cuprodescloizite, azurite malachite, specular hematite, calcite, chalcocite, galena, and pyrite. The gold in the oxide ores is free.

Mines. The Quartette Mine was for a long time the principal mine of the camp. According to figures furnished by H. O. Russell, it has probably produced more than $3,500,000; while the actual output from September 30, 1903, to October 1, 1909, was $1,997,888.34. The Duplex Mine has been the largest producer in recent years. According to G. R. Colton, the mine produced about $500,000 by company operation up to 1915 and about $250,000 under a leasing system since that date. The Duplex M. Co. was incorporated in Nevada in 1915 with a capital stock of 500,000 shares of $1 par value. M. M. Riley, 500 P. E. Bldg., Los Angeles, is Pres., and G. R. Colton, Searchlight, is Sec.-Mgr. The company owns 7 claims aggregating 102 acres. Its main working shaft has a depth of 700 ft. on an incline of 45 degrees. The ore mined was formerly treated on the property by amalgamation, but below the 400 ft. level the ore became so base that it had to be shipped to the smelters instead. Other mines recently active in the Searchlight District are as follows:—

Big Casino Mining Company, W. W. Wishon in charge.
Chief of the Hills Gold Mining Co., Leased to Riverview.
M. & M. Co., C. M. Crowley, Mgr.
Cyrus Noble Group of Mines, J. O. Thompson, Lessee.
Rand Mining Company, John S. Sartain, Supt.
Searchlight Mining & Milling Co., Walter W. Wells, Lessee.
States Mutual Cons. Mining Co., Mrs. V. A. Brewer, Supt.
St. Louis Mine, W. H. Barton, Owner.

The Dupont Copper Ms. Co. which owns 23 claims aggregating 410 acres in the Dupont section of the Searchlight District, 18 m. by road N. E. of Searchlight, in contemplating developing its property in depth. N. H. Wheeler of 10 E. 43rd St., New York City, is Gen. Mgr. The company is incorporated in Arizona with a capital stock of 2,500,000 shares of $1 par value, of which 1,500,000 have been issued.

Bibliography

MR1898 579-580 Turquoise
MR1903 183
MR1904 146
MR1905 116,270
MR1906 122,296-7
MR1907 I 368-9
MR1908 I 473-4
II 846 Turquoise
MR1909 I 395-6
MR1910 I 508
MR1911 I 669
MR1912 I 786
MR1913 I 818
MR1914 I 672
MR1915 I 627
MR1916 I 471
MR1917 I 267
MR1918 I 231
MR1919 I 387
MR1920 I 320
MR1921 I 380

USGS Camp Mohave topographic map.

Corn, G. B., "The Searchlight Review", Chamber of Commerce, Searchlight, (1917).

Hill 507 202.

Horton, F. W., "Molybdenum, Its Ores and Their Concentration", USBM B 111 (1916) 89.

M&SP, "Searchlight District, Nevada", 90 (1905) 172.

———"Cyanide Plant at the Quartette Mine, Nevada", 106 (1913) 953.

Ransome, F. L., "Preliminary Account of Goldfield, Bullfrog, and Other Mining Districts in Southern Nevada", USGS B 303 (1907) 63-76.

Stuart NMR 132-3.

Weed MH 1145 Big Casino Leasing Co., Big Casino M. Co.
1187 Duplex M. Co., Dupont Copper Ms. Co.
1302 Preseverance Ms. Co., Ltd.

1325 Searchlight Mercantile Co.
1345 States Mutual Cons. Ms. Co.

SLOAN

Limestone, Dolomite, (Radium)

Location. The Sloan limestone and dolomite deposit is located in central Clark Co., at Sloan on the U. P. R. R.

Geology. Massive gray limestone occurs as a bed 140 ft. thick with Carboniferous dolomites both underlying and overlying it. The limestone is pure, having the following analysis:—

Silica	0.81%
Alumina & Ferric Oxide	0.37%
Calcium Carbonate	98.66%
Magnesium Carbonate	0.17%
Moisture	0.06%
	100.07%

Mining. The deposit belongs to the Nevada Lime & Rock Co. P. Buol of Las Vegas is Pres.- Gen. Mgr., and P. M. Shelby is Supt. at Sloan. The company has an office and warehouse at 864 Commercial St., Los Angeles. The limestone is quarried and hand sorted. From 40,000 to 50,000 tons of this limestone are shipped annually to the sugar refineries of southern California. The Nevada Lime & Rock Co. also own a lime plant of 60 tons daily capacity consisting of a 100x7 rotary kiln, a 50x7 rotary cooler; and a hydrating plant with a capacity of 3 tons per hour. The greater portion of the product of the lime plant is shipped to Los Angeles for use in the building trades, but some is sent to Tonopah for use in the cyanide mills. In addition to limestone and lime, the Nevada Lime & Rock Co. ships dolomite, some 3,000 tons of this going to steel plants annually for open hearth furnace lining.*

Radium. Traces of carnotite were discovered in a railway cut 2 m. S. of Sloan by N. E. Williams in 1905. It forms thin coatings on the walls of joints in a Tertiary rhyolite flow. Calcite, celestite, and manganese oxide also occur in the joints and sodium sulphate coats the rhyolite near by.

Bibliography. USGS Ivanpah topographic map.
Hewett, D. F., "Carnotite in Southern Nevada", E&MJ 115 (1923) 232-5.

SUNSET (Lyons)

Gold

Location. The Sunset District is in S. Clark Co. adjoining the Crescent District on the N. and a short distance S. E. of the Yellow Pine District. It occupies a low desertrange 15 . S. E. of Jean which is on the S.P.R.R., and a short distance E. of Lyons, California, on the same railroad.

History. The only producer in the district has been the Lucy Gray Mine. This mine is situated 3 m. E. of Lyons and began operations in 1897. In 1912, the Lucy Gray Mine was operating a small cyanide plant consisting of 1 Lane mill and leaching tanks. Bullion was produced as late as 1919, but the mine has been idle the past few years.

Geology. The ore deposits are gold-bearing veins in granite.

Bibliography. MR1907 I 367 MR1911 I 669-670 MR1912 I 786 MR1913 I 818 MR1915 I 627 MR1916 I 471 MR1918 I 231 MR1919 I 387

*The data on limestone and dolomite for this article was kindly furnished by P. M. Shelby

USGS Ivanpah topographic map.

Hill 507 202.

SUTOR
(Radium)

History. Traces of carnotite were discovered 2 m. W. of Sutor which is on the U. P. R. R. by J. Is .chsen in 1921. Shortly afterwards other discoveries were made 1 m. to the E. and 1 m. to the N. E. of the first discovery.

Geology. The carnotite occurs as patches with manganese oxide on fractures and joint cracks in reddish sandstone and sandy limestone and in buff sandstone. These sandstones underlie the lowest Permian limestone. The quantity of carnotite discovered is too small and the grade too low for profitable exploitation.

Bibliography. Hewett, D. F., "Carnotite in Southern Nevada", E&MJ 115 (1923) 232-5.

VIRGIN RIVER (St. Thomas)
Salt, (Gypsum, Magnesite, Potash)

Location. The Virgin River salt district extends from St. Thomas S. along the Virgin River to its junction with the Colorado at the Arizona border. St. Thomas in the terminus of the St. Thomas Branch of the U. P. R. R.

History. The Virgin River salt beds were known as early as 1866, but have only been worked to a small extent and for local use on account of their comparative inaccessibility. In the seventies the mines of Eldorado Canyon and of the Mineral Park District in Arizona obtained their supply of salt from these beds. Magnesite was discovered on the Muddy River in the N. part of the district in 1915.

Geology. Salt ledges are exposed at several places along both sides of the Virgin R. for a distance of at least 12 m., according to H. S. Gale. The salt is associated with a great thickness of gypsum and gypsiferous shales and is included in a series of red shales and sandstones which are usually referred to as "Red Beds" of Triassic or Permian age, but in this instance are believed to be the time equivalent of the Tertiary Greggs Breccia. These beds have been much faulted and folded, and the salt has been leached out in most places where openly exposed. The largest ledge presents an almost vertical face of solid crystalline salt 100 ft. in thickness, with the bottom concealed so that it may be even thicker. This salt was tested for potash by Gale who found it to contain only a fraction of a percent of that substance. The mabnesite deposit on the Muddy R. is said to be of great size but to contain high percentages of lime and silica.

Bibliography.

B1866 96 MR1886 638
SMN1869-70 98-9 MR1915 II 108-110 Potash; 1024 Magnesite
SMN1875-6 90 MR1916 II 399 Magnesite
MR1882 543 MR1918 II 152 Magnesite

Jones, J. C., & Stone, R. W., "Gypsum Deposits of the U. S., Southern Nevada", USGS B 697 (1920) 160.

Phalen, W. C., "Salt Resources of the U. S.", USGS B 669 (1919 146-8.

Spurr208 136-8 Muddy Mt. Range.

Thompson & West 492.

YELLOW PINE (Goodsprings, Potosi)

Zinc, Lead, Silver, Gold, Copper, Platinum, Palladium, Vanadium, Cobalt, (Barite, Gypsum, Mercury, Radium)

Location. The Yellow Pine District occupies the S. end of the Spring Mountain Range in S. W. Clark Co. It extends from the California border northward for 24 m. and has a maximum width of 16 m. The gypsum deposits of Arden lie a short distance N. E., the Sloan limestone quarry a short distance E., and the Sunset District a short distance S. E. The principal town in the district is Goodsprings situated 8 m. N. W. of Jean which in on the U. P. R. R. Jean is the main shipping point for the district, but Arden to the N. and Roach to the S. are also important.

History. Aside from the early mines of the Indians and of the Franciscan monks, the Potosi Mine in the Yellow Pine Mining District is the oldest lode mine in Nevada. It was discovered by a party of Mormons returning from San Bernardino over the old Spanish Trail in 1855. Slade was appointed superintendent by the church authorities that same year, and an attempt was made to smelt the lead ore at the mine. This trial proving unsuccessful, the ore was sent to the Las Vegas way station where Dudd Leavitt and Isaac Grundy constructed a furnace in a fireplace and produced 5 tons of lead, thus inaugurating lead smelting in Nevada. Notwithstanding its early discovery, little attention was paid to the Yellow Pine District on account of its inaccessibility until the rich Keystone Gold Mine was discovered in the early nineties. The Keystone Mine shut down in 1905, and there has been no important gold production in the district since. In that same year, however, mining activities in the district were greatly stimulated by the completion of the San Pedro, Los Angeles & Salt Lake R. R. and by the recognition of the presence of zinc in the Potosi ore by C. T. Brown. In 1905, the Potosi Mine became the largest producer of zinc in Nevada; while in 1906 the zinc production of Nevada was more than quadrupled over that of 1905, the Potosi Mine continuing to be the chief producer and other mines in the Yellow Pine District beginning to produce. The largest of the new mines was the Yellow Pine, The Yelow Pine M. Co. built a 100-ton zinc-lead separating mill at Goodsprings in 1910 and the following year constructed a narrow-gauge railway from Goodsprings to the main line at Jean. Platinum and palladium were discovered in the Yellow Pine District in 1914 associated with gold in the ore of the Boss Mine. Barite of good quality was found in 1915. Vanadium ore was found and mined at the Christmas Consolidated Mine in 1917; and recently some small shipments of cobalt ore have been made from the district. Traces of carnotite were discovered 2 m. N. of Goodsprings by Nelson Nunn in 1921.

RECENT PRODUCTION OF YELLOW PINE MINING DISTRICT
(According to Mineral Resources of the United States, U. S. Geol. Survey)

Year	Mines	Tons	Gold $	Silver Ozs.	Copper Lbs.	Lead Lbs.	Zinc Lbs.	Total $
1902	1	2,078	44,174					44,174
1903	4	2,092	53,396	753	21,800	28,000		55,765
1904	1	1,929	47,878	146				47,961
1905	2	2,394	15,000	3,707		290,063	685,659	71,335
1906	5	9,481	9,150	1,573	67,341	625,175	2,885,246	234,550
1907	6	4,615	9,743	2,994	93,090	187,310	1,879,432	151,156
1908	11	4,627	7,791	10,247	42,144	720,285	1,115,851	101,482
1909	10	8,664	5,740	18,461	392	406,353	3,013,352	195,585
1910	16	5,878	1,219	16,826	122,925	1,263,837	2,707,071	227,707
1911	17	8,677	2,424	47,072	173,719	1,617,224	3,548,032	324,100
1912	22	28,386	34	223,013	103,398	6,544,917	13,254,860	1,363,354
1913	25	29,060	1,268	192,339	283,592	6,204,065	14,363,709	1,239,081
1914	19	24,537	8,034	122,703	156,389	4,185,208	11,862,149	864,882
1915	35	38,361	833	100,146	262,600	4,620,243	21,061,182	2,926,300
1916	54	71,155	14,333	56,492	494,604	8,349,850	28,889,282	4,886,281
1917	79	64,260	10,656	219,789	764,733	9,298,706	20,535,768	3,294,870
1918	38	40,614	4,869	140,211	400,792	6,459,836	15,444,731	2,108,193
1919	26	19,273	11,650	139,656	130,395	4,217,739	5,974,219	851,976
1920	24	18,611	3,273	96,557	76,567	3,927,280	9,381,503	1,196,692
1921	5	287	22	2,195	119	186,169	69,397	14,080
Total 1902-21		385,009	251,487	1,394,880	3,194,600	59,132,260	156,691,443	19,999,424

Geology. The Spring Mountain Range consists mainly of Paleozoic sediments which have undergone intense folding accompanied by faulting. Igneous rocks are scarce and are represented chiefly by quartz-monzonite porphyry dikes and sills. At the S. end of the range where the Yellow Pine District is located, lies a series or Carboniferous sedimentaries consisting largely of siliceous limestones, but including strata of pure crystalline limestone and of dolomite, and with occasional intercalated beds of fine-grained sandstone. These strata have a general W. to S. W. dip of from 15 to 45 degrees which is occasionally disturbed by local folds. The quartz-monzonite porphyry is intruded into these beds and is of post-Jurassic age, perhaps Teritary.

Ore-bodies. Two distinct types of ore-bodies occur in the district, replacement deposits of zinc-lead and of copper in the Carboniferous limestone, and gold deposits in altered quartz-monzonite porphyry. The Keystone Mine furnished an example of the latter type, which is unimportant at the present time. The limestone replacement deposits occur in fractured areas in the neighborhood of quartz-monzonite porphyry intrusions. They may be divided into two groups, irregular, oxidized, zinc-lead-silver replacements in limestone and dolomite, and still more irregular copper-gold replacements.

Ores. The principal ores of the district are the oxidized zinc-lead-silver ores. They contain smithsonite, cerussite, and galena, with smaller amounts of calamine, hydrozincite, anglesite, pyromorphite and sphalerite and a variable silver content. Vanadium has been found associated with these ores on some 14 properties and vanadium ore has been shipped from the Christmas Consolidated, Bill Nye, and Campbell Mines. Next in importance are the oxidized copper-gold ores. They contain cuprite, chryscolla, and malachite, with occasional chalcocite and liebethenite, and a gold content which is usually low. Platinum and palladium have been found associated with these ores on 4 properties and shipments of gold-platinum-palladium ores have been made from the Boss Mine. Stringers of high grade cobalt ore have been found on several properties, and shipments have recently been made from the Copper Chief, High Line, and Columbia Mines. Barite and gypsum have bee discovered in the district but no shipments have been made. At the Red Cloud Gold Mine 5 m. N. W. of Goodsprings, cinnabar occurs in sufficient quantities to be of possible commercial importance as an ore of mercury, according to a letter from Frank Williams. The carnotite N. of Goodsprings occurs as sporadic patches of thin film on reddish and yellowish sandstones of Permian age, according to Hewitt.

Mines. The principal mine is that of the Yellow Pine M. Co. J. F. Kent is Pres. at Goodsprings. The capitalization is $1,000,000 in shares of $1 par value, all issued. The stock is listed on the Los Angeles Stock Exchange and the transfer office is at 441 Security Bldg., Los Angeles. The company owns 240 acres of patented claims, a half interest in the Charleston M. Co., mine plant, 120-ton wet concentrating mill, and narrow-gauge railroad to the main line at Jean. The other properties which are present active in the district are the Bullion M. Co., A. F. Johnson, Mgr.; the Copperside Mine, O. F. Schwartz, Mgr.; the Dawn M. Co., F. Williams, Pres.; the Root Zinc M. Co., S. C. Root, Pres., and the Whale Group, F. L. Miller, Owner.

BIBLIOGRAPHY of YELLOW PINE DISTRICT

General

R1870 168-175
SMN1869-70 103
SMN1871-2 95-6
SMN1875-6 92
MR1905 270,391
MR1906 296,478
MR1907 I 369
MR1908 I 474
MR1909 I 396-7
MR1910 I 509
MR1911 I 669
MR1912 I 786
MR1913 I 818; 448-450, Platinum.
MR1914 I 672-3; 333, 339-340, Platinum.
MR1915 I 627-8; 144,726 Platinum; II 175, Barite
MR1916 I 471-3; 10, Platinum.
MR1917 I 257-8; 20, Platinum, 956, Vanadium.
MR1918 I 232; 206, Platinum.
MR1919 I 387-8; 15, Platinum.
MR1920 I 320-1; 45, Platinum; II 59, Gypsum.
MR1921 I 380.
USGS Goodsprings, Ivanpah, and Las Vegas topographic maps.

Davis, 1,217-8.

Gregory, N. B., "The Yellow Pine Mining District of Nevada", E&MJ 90 (1910) 1308-9.

Hale, F. A., Jr., "Yellow Pine District, Nevada", E&MJ 91 (1911) 803-4.

———"Developments in the Yellow Pine District, Nevada", M&EW (1911) 915.

———"Yellow Pine Mining District, Nevada", Mexican Mining Jour. 15 (1912) 29.

Hill 507 202.

Hillen, A. G.,"Zinc in Nevada", Los Angeles Min. Rev. Sept. 24 (1910) 72.

M&SP, "The Yellow Pine Mine at Goodsprings, Nevada", 121 (1920) 239-240.

Nevius, J. N., "Reconnaissance of Goodsprings District, Nevada", M&EW 42 (1915) 897-9.

Palmer, L. A.,"The Yellow Pine District, Nevada", E&MJ 102 (1916) 123-5

Scott, W. A., "Activity in Goodsprings District, Nevada", M&EW 45 (1916) 1069-1071.

Stuart NMR 31, 136-7.

Weed MH 1139 Azalia M. Co., Azurite M. Co.
1146 Bill Nye M. Co. 1150 Boss Gold M. Co.
1150 Boss Extension M. Co.
1164 Charleston M. Co., Christmas Cons. Ms. Co.
1179 Copper Peak M. Co. 1216 Goodsprings Anchor Co.
1216 Gould & Curry M. Co. 1218 Green Monster Mine.
1227 Ingomar Mine. 1229 Ironside M. Co.
1281 Nevada Lincoln M. Co.
1298 Oro Amigo Platino M. Co. 1306 Potosi Mine.
Prairie Flower M. Co. 1310 Red Streak M. Co.
1324 Saint Anthony M. Co. 1346 Sultan Mine.
1376 White Pine M. Co. 1378 Yellow Pine Ex. Co.

White, D., "Zinc Mines in Southern Nevada", PAMC, Goldfield, Nevada, (1909) 401-411.

Geology

Bain, H. F., "A Nevada Zinc Deposit", USGS B 285 (1906) 166-9.

Cramption, F. A., "Platinum at the Boss Mine, Goodsprings, Nevada", M&SP 112 (1916) 479-482.

Hale, F. A., Jr., "Platinum Ore in Southern Nevada", E&MJ 98 (1914) 641-2.

———"Ore Deposits of the Yellow Pine Mining District, Clark County, Nevada", T AIME 59 (1918) 93-111; B AIME 134 (1918), 535-553; E&MJ 105 (1918) 4551460.

Hewitt, D. F., "Carnotite in Southern Nevada", E&MJ 115 (1923) 232-5.

Hill, J. M., "The Zinc-Lead Deposits of the Yellow Pine District", (Abstract), Wash. Acad. Sci. J 3 (1913) 238-9.

———"The Yellow Pine District, Clark County, Nevada", USGS B 540F (1914) 223-274.

Kennedy, J. C., "Occurrence of Platinum at Boss Mine, Nevada", M&EW 42 (1915) 939-940.

Knopf, A., "Platinum-Gold Lode Deposit in Southern Nevada", (Abstract), B GSA 26 (1915) 85.

———"Plumbojarosite and Other Basic Lead-Ferric Sulphates from the Yellow Pine District, Nevada", Wash. Acad. Sci. J 5 (1915), 497-503.

———"A Gold-Platinum-Palladium Lode in Southern Nevada", USGS B 620A (1915) 1-18; Abstract, M&SP 110 (1915), 876-879; Wash. Acad. Sci. J 5 (1915) 370.

M&SP, "Palladium; Its Characteristice, Uses, and Discovery in the Boss Mine", 109 (1914) 990.

Mudd, S. W., "The Boss Mine, Goodsprings, Nevada", M&SP 110, (1915), 297-8.

Spurr, J. E., "Origin and Structure of the Basin Ranges", GSA B 12, (1901) 235-6.

Spurr 208 164-180.

USGSW 100th 3 (1875) 124, 166, 179, 180.

Metallurgy

Hale, F. A., Jr., "Yellow Pine Mill, Goodsprings, Nevada", E&MJ 94, (1912) 68.

———"The Yellow Pine Mill", SLMR, July 30, 1912, 33.

Miscellaneous

Waring, G. A., "Ground Water in Pahrump, Mesquite and Ivanpah Valleys, Nevada and California", USGS WSP 450 (1920) 51-81.

DOUGLAS COUNTY

BUCKSKIN

Copper, (Iron, Gypsum)

Location. The Buckskin District lies on the E. border of Douglas Co. adjoining the Yerington District in Lyon Co. on the W. The Mount Siegel District adjoins it on the S. W.

History. The Minnesota-Nevada C. Ms. Co., of which Otto Taubert of Yerington is in charge, produced copper ore in 1913, 1916, and 1918.

Geology. The geology of the Buckskin District is similar to that of the Yerington District, of which it may be considered the W. extension. At the Minnesota-Nevada Mine, a deposit some 20 ft. in width at a limestone-granite contact crries sulphide ore with a copper content of about 3% according to Mines Handbook. There is also a large iron ore deposit on this property. A bed of gypsum in Triassic limestone occurs just W. of Buckskin, according to Jones.

Bibliography. MR1907 I 346 MR1913 I 819 MR1916 I 473
MR1908 I 475 MR1918 I 232
USGS Wellington topographic map.

Jones, J. C., "Nevada" in "Gypsum Deposits of the U. S.", USGS B 697 (1920) 154,

Stuart NMR, 112.

Weed MH 1128 Albany Copper Co.
1132 Antelope Group.
1235 Kennedy Cons. M. Co.
1258 Minnesota-Nevada Copper Ms. Co.

EAGLE see GARDNERVILLE

GARDNERVILLE (Eagle)

Copper, Gold, Silver

Location. The Gardnerville District is situated on the W. slope of the Pine Nut Range in central Douglas Co. It lies to the E. of Carter's Stage Station which is on the state highway 14 m. S. E. of Minden, the terminus of the South Branch of the V. & T. R. R. The town of Gardnerville for which the district is named is 1 m. S. E. of Minden and that much nearer the district. The Gardnerville District adjoins the Red Canyon District on the W., the Silver Glance District on the N. W., and the Mt. Siegel placer district on the S. W.

History. The Peck and Mammoth lodes were opened in 1860 and explored for several years. In 1881, the only mining work in progress in Douglas Co., was the extension of the Mammoth tunnel to cut the ledge. The Pine Nut Mine built a 5-stamp mill on its property in 1907. The Ruby Hill Mine shipped copper ore in 1912 and erected a 30-ton copper leaching plant in

1915. The Veta Grande M. Co. took over the old Mammoth Mine in 1921 and began active exploraion work.

Geology. The principal country rock of the Gardnerville District is augite-andesite of the Tertiary or Triassic age. The ore deposit at the Pine Nut Mine is a silicified fracture zone in the andesite, according to Hill. The ore consists of altered andesite fragments with some quartz and calcite and contains gold associated with pyrite. At the Ruby Hill Mine, the ore deposit is similar, but the rock has been altered to sericite, calcite, and pyrite instead of being silicified. Chalcopyrite is the primary ore mineral and has been altered near the surface to cuprite and copper carbonates. The Veta Grande ore deposit is a large quartz vein between an altered augite-andesite foot-wall and a lamprophyric dike separating it from an intrusive hornblende-andesite on the hanging wall, according to information furnished by W. P. Catlin. The silver occurs in streaks in the form of argentite and stephanite carrying minor amounts of gold.

Mines. The Ruby Hill and Veta Grande Mines are active at present. R. C. Walker of Gardnerville is Mgr. of the Ruby Hill Copper & Gold M. Co. The Veta Grande M. Co. is incorporated in Nevada with a capital stock of 1,500, 000 shares of $1 par value. E. J. Roberts, 1040 Mills Bldg., San Francisco is Pres.,and J. H. Smith is Supt. at Minden. The property consists of 10 claims developed by 2 tunnels and drifts aggregating 1200 ft. and equipped with a compressor.

Bibliography. B1867 323
SMN1873-4 18
1866:20
MR1905 266
MR1907 I 346
MR1908 I 475
MR1909 I 347
MR1910 I 509-510
MR1911 I 670
MR1912 I 789
MR1913 I 819
MR1914 I 819
MR1915 I 628-9
MR1917 I 269
MR1919 I 388
USGS Markleeville topographic map.
Hill 507 202-3.
Hill 594 18, 21, 23, 25, 28, 51-64.
Spurr 208 120-5 Pine Nut Range.
Stuart NMR 111.
Thompson & West 375.
Weed MH 1367 Veta Grande M. Co.

GENOA

Copper, Silver, Gold

Location. The Genoa District lies upon the E. slope of the Sierra Nevada Mts. in W. Douglas Co. just W. of the town of Genoa.

History. The Genoa District was organized in 1860. A considerable amount of work was done without encouraging results, and work was suspended about 1865. Recently the old properties have been taken over by new companies which are carrying on prospecting operations.

Geology The country rock of the Genoa District consists of Triassic sediments which have been intruded by Cretaceous granite, according to reports by J. C. Jones and E. J. Schrader. The sedimentary rocks include limestones, schists, and quartzites which strike somewhat E. of N. and dip at variable angles toward the W. The orebodies occur within zones of fracture which are mainly in limestone and are in the form of irregular lenses. The lenses consist of more or less altered country rock cut by many cross and parallel fissures. The lode minerals, which occur both as fissure fillings and replacement of the country rocks include:—quartz, epidote, garnet, tourmaline, actinolite, chlorite, chalcopyrite, and pyrite, together with a small amount of silver and gold. The ore occurs as bunches and stringers of small size within the lenses. Secondary bornite, azurite, and malachite have been formed at the surface, and the silver there is probably in the form of chloride.

Mines. The Genoa Comstock Ms. Co. owns 4 claims in the S. part of the district and is running a tunnel to cut its main lode in depth. The company was incorporated in Nevada in 1921 with a capital stock of 1,500,000 shares of 10 cents par value. H. S. Anderson, P. O. Box 581, Reno, is Pres.; John Murphy is V. P., and L. B. Berriman, Sec.-Treas. The Simon Ex. M. Co. is conducting prospecting work in the northern part of the district.

Bibliography. B1867 323 MR1916 I 473

USGS Carson topographic map.

Davis, 813.

Lawson, A. C., "The Recent Fault Scarps at Genoa, Nevada", Seism. Soc. Am. B 2 (1912) 193-200.

Thompson & West, 375.

HOLBROOK see MOUNTAIN HOUSE

MOUNT SIEGEL
Placer Gold

Location. The Mount Siegel District lies N. of Mt. Siegel on the upper part of Buckeye Creek in E. Douglas Co. It is 20 m. E. of Minden, which is the terminus of the South Branch of the V. & T. R. R. The elevation above sea-level is 7,100 ft. The Mount Siegel District adjoins the Gardnerville District on the N. E., the Red Canyon District on the N. W., and the Buckskin District in Lyon Co. on the S. W.

History. The Buckeye placers produced during prospecting work in 1911. In 1914, and intermittently since then, the Ancient Gold Placer in the Mount Siegel District has made small productions. G. W. Slater of Minden is Pres. of this company.

Geology. The ore deposits of the Mount Siegel District are gold placers, The gold mined there in 1911 had a fineness of 0.880.

Bibliography.

MR1911 I 670	MR1914 I 674	MR1917 I 269
MR1912 I 789	MR1915 I 629	MR1918 I 232
MR1913 I 819	MR1916 I 473	MR1919 I 388

USGS Markleeville topographic map.

MOUNTAIN HOUSE (Holbrook)
Gold, Silver

Location. The Mountain House District is situated at Mountain House or Holbrook in the Pine Nut Range in S. Douglas Co. on the California border. The Orpheus Mine was recently acquired by the Carson Valley M. Co. of which F. C. Everett is Gen. Mgr. The 2-stamp mill on the property is being enlarged by the addition of a 5 stamps and concentrating tables; and a compressor is being installed to operate the machine drills in the mine, according to the Nevada Mining Press. The Carson Valley M. Co. also owns the Ventura Mine 5 m. S. E.

RED CANYON (Silver Lake)
Gold, Silver, Lead

Location. The Red Canyon District is located in the S. part of the Pine Nut Range in E. Douglas Co. It includes Red Canyon which extends down the E. slope of the Pine Nut Range into Smith Valley; the upper part of Mill Canyon which extends down the W. slope into Carson Valley; and the region N. of Red Canyon, where the Winters Mine is located, which is sometimes called the Silver Lake District. Hudson Station on the N. C. B. R. R. is 6 m. E. of the mouth of Red Canyon; and Minden, the terminus of the South Branch of the V. & T. R. R., lies 18 m. N. W. in Carson Valley. The Gardnerville District adjoins the Red Canyon District on the W., the Silver Glance District on the S., and the Mount Siegel placer district on the N. W.

History. The first ore mined in the Red Canyon District was taken from the Longfellow Mine now owned by the Detroit G. Mg. Co., and hauled to Virginia City for treatment in 1863, according to notes furnished by E. W. Carman. The early production amounted to some $50,000 and was derived from workings not over 50 ft. in depth. The Winters Mine was discovered in 1872 and produced about $8,000 up to 1881. A custom mill was built in Mill Canyon to treat the ores of the Red Canyon District about 1872. In 1905, the Winters Mine erected a 25-ton cyanide mill to treat its ores.

Production. According to Mineral Resources of the U. S. Geol. Survey, the production of the Red Canyon District from 1903 to 1911 amounted to $22,114 in silver, gold, and lead; and there has been a small intermittent production since that time.

Geology. In the Red Canyon District, sedimentary rocks probably of Triassic age have been intruded by quartz-monzonite probably of Cretaceous age. According to Hill, the sedimentary section at Red Canyon, beginning at the bottom, is as follows:—

Thick-bedded white crystalline limestone	100-300 ft.
Thin-bedded dark gray to black argillites with some lenses of white quartzite	800-1,000 ft.
Massive blue-gray limestone	1,500-2,000 ft.

Ore Deposits. Three types of ore deposits occur in the Red Canyon District; quartz veins in the quartz-monzonite, contact-metamorphic lodes in the Triassic sediments, and replacement lodes in the Triassic beds. An example of the first class of deposits occurs on the property of the Detroit G. M. Co. Vertical veins in monzonite consist of white quartz with minor amounts of pyrite and specular hematite. Galena and chalcopyrite occur with the pyrite in the rich lenses. The oxidized ore extends to a depth of 400 ft. and consists of rusty quartz with occasionoal yellow stains of lead carbonate and blue and green stains of copper carbonate. The values are principally in gold with a little silver. The Red Canyon Mine at the mouth of Red Canyon is an example of the contact-metamorphic lodes. A 10-ft. dike of quartz-monzonite in the Triassic limestone has altered that rock for 100 ft., producing a mass of epidote, quartz, and calcite, containing pyrite, chalcopyrite, and pyrrhotite. The Winters Mine represents deposits of the third class. The lode is a replacement in slightly calcareous argillites which occur in the form of a lens included in the quartz-monzonite. It contains galena and some stibnite with occasional pyrite and chalcopyrite and has been completely oxidized to a depth of 150 ft.

Mines. The only mines at present operating in the district are those owned by the Detroit Gold M. Co. E. W. Carman of Gardnerville is in charge of these properties which are equipped with power hoist, air compressor, air drills, and a 50-ton amalgamating and concentrating mill.

Bibliography. MR1905 266; MR1907 I 346; MR1908 I 475; MR1910 I 510; MR1912 I 789; MR1914 I 647; MR1916 I 473; MR1917 I 269; MR1918 I 232; MR1919 I 388

USGE Markleeville and Wellington topographic maps.

Hill 507 203.

Hill 594 51-64.

Spurr 208 120-125 Pine Nut Range.

Stuart NMR 111-2.

Thompson & West 374.

Weed MH 1241 Longfellow Gold M. Co.

SILVER GLANCE (Wellington)
Gold, Silver

Location. The Silver Glance District is in the S. E. part of the Pine Nut Mts. in E. Douglas Co. It lies to the N. W. of the town of Wellington, which is 31 m. S. S. E. of Minden, the terminus of the South Branch of the V. & T. R. R.; and 11 m. S. W. of Hudson, a station on the N. C. B. R. R. The district adjoins the Red Canyon District on the S., and the Gardnerville District on the S. E.

History. The Yankee Girl Mine was worked in the eighties, the ore being treated in arrastras on the West Walker R. near Wellington. There was formerly a 2-stamp mill on the property of the South Camp Mine; and in 1908 there was a 5-stamp mill on that of the Poco Tiempo M. Co. The Yankee Girl Mine treated ore in a 5-stamp mill on its property in 1915.

Geology. The ore deposits of the Silver Glance District are veins in quartz-monzonite which is probably of Cretaceous age. These veins are narrow and are not highly mineralized but consist mainly of white quartz with minor amounts of pyrite and specular hematite. In soome instances they contain galena and in others chalcopyrite.

Bibliography. MR1907 346; MR1908 I 475; MR1909 I 397; MR1912 I 790; MR1914 674; MR1915 I 629; MR1918 I 232

USGS Wellington topographic map.

Hill 507 203. Hill 594 51-64. Spurr 208 120-5 Pine Nut Range.

SILVER LAKE see RED CANYON

WELLINGTON see SILVER GLANCE

ELKO COUNTY

ALLEGHENY see FERGUSON SPRING

AURA (Bull Run, Centennial, Columbia)
Silver, Gold, (Lead Zinc)

Location. The Aura District is at Aura in the central part of the Bull Run or

Centennial Range in N. Elko Co. Aura is 79 m. N. of Elko which is on both the S. P. R. R. and the W. P. R. R. The old town of Columbia, after which the district is sometimes called, is a short distance N. The Aura District lies on the E. slope of the Bull Run Mts., and the Edgemont District adjoins it on the W. slope; while the Lime Mountain District adjoins the Aura District on the S. W., and the Van Duzer placer district on the N. E.

History. The Aura District was discovered by Jesse Cope and party in 1869, and by the following summer there were 10 producing mines in the camp. A 20-stamp mill was built at the Blue Jacket Mine in the early seventies, and in 1875 Edward Stokes built a custom mill at Columbia. Mining declined in the eighties and did not revive until 1906 when Aura was founded. Small shipments were made from the district up to 1919.

Geology. Paleozoic sediments make up the main mass of the Bull Run Mts., according to Emmons, and have been much faulted and intruded by granitic rock, The ore deposits are fissure veins and sheeted zones in limestone; and the silver and gold are associated with galena, pyrite, sphalerite, chalcopyrite, tetrahedrite, iron oxides, copper carbonates, and pyromorphite.

Bibliography.

R1870 144,147-9	MR1906 293	MR1913 I 819
R1872 158	MR1907 I 246	MR1915 I 629
SMN1869-70 61-2	MR1909 I 398	MR1916 I 474
SMN1871-2 22-3	MR1911 I 671	MR1917 I 270
SMN1873-4 29	MR1912 I 790	MR1919 I 388
SMN1875-6 19-20		

Emmons 408 67-74. Hill 507 203. Thompson & West 392

BULL RUN see AURA

BULLION see RAILROAD

BURNER
Lead, Silver

Location. The Burner District is situated in the Burner Hills about 10 m. W. of Good Hope in W. Elko Co.

History. The Mint Mine is the principal property. It was operated in the early eighties and some $30,000 worth of lead-silver ore shipped from it. Active operations were suspended in 1893 and little work has been done in the district since then.

Geology. The Mint lode is a nearly vertical fissure vein in andesite, with sulphides and quartz occuring as ribbons paralled to the walls and as masses impregnating the adjacent andesite, which is somewhat altered. The ore consists of galena, sphalerite, pyrite arsenopyrite, and chalcopyrite in a gangue of quartz and calcite. Near the surface lead carbonate and iron oxide are present.

Bibliography. Emmons 408 44,66-7.

CARLIN
Diatomaceous Earth, (Oil Shale, Coal)

Location. The Carlin District is at Carlin in S. W. Elko Co. Carlin is on the S. P. R. R. and the W. P. R. R., and Vivian siding where the diatomaceous earth plant is located is 3 m. E. of Carlin on the S. P. R. R.

History. Coal was discovered in the Carlin District by John. Q. A. More in 1859, and by 1874 considerable prospecting work had been done upon the coal beds. In 1895, the Humboldt Coal Co. was formed to work the coal beds but made no production. The Trip-O-Lite Products Co. erected a diatomaceous earth plant at Vivian in 1919 and after considerable experimentation made productions in 1921 and 1922.

Diatomaceous Earth. According to C. T. Hurd, Gen. Mgr. of the Trip-O-Lite Products Co. at Carlin; the company owns an extensive deposit of diatomaceous earth 1½ m. N. of Vivian. The composition of this earth is:—

Silica	84.00%
Alumina	4.00%
Ferric Oxide	0.57%
Magnesia	0.43%
Lime	1.00%
Potash	1.00%
Soda	0.76%
Loss on Ignition	9.00%
	100.76%

It is said to weigh from 15 lbs. to 24 lbs. per cu. ft., to have an average absorptive power of 129%, and to contain less than 2% of broken diatoms. The mill at Vivian has a capacity of 12 tons per day and makes several grades of product, each adapted to certain special uses.

Oil Shale. The oil shale at Carlin occur as vertical beds up to 300 ft. in thickness, according to Alderson, and has an asphalt base.

Bibliography. SMN1873-4 29-30 Coal; MR1896 556 Coal; MR1895 458 Coal; MR1897 462 Coal

Alderson, V. C., "The Oil Shale Industry", New York, (1920) 32,46.

CENTENNIAL see AURA and EDGEMONT

CHARLESTON

Placer Gold, Gold, Silver, Copper, Lead, Antimony, (Nitrates, Oil Shale)

Location The Charleston District lies N. of Charleston on 76 Creek and in the neighborhood of Copper Mt. in N. Elko Co. It adjoins the Jarbidge District on the S. W. and the Island Mountain District on the E.

History. Gold placers were discovered on 76 Creek, presumably in 1876, and the town of Charleston grew up 4 m. to the S. The Prunty Mine on 76 Creek is equipped with a 5-stamp mill which has operated intermittently since 1905. It has produced silver, gold, copper, and antimony. The Graham Mine a short distance down the creek is a copper-gold mine which has been developed recently.

Geology. The country rocks of the district are principally Paleozoic schist, quartzite, limestone, and shale, according to Schrader. They rest upon hornblende-granite which is thought to be intrusive and are cut by dikes of slightly younger granodiorite and capped locally by Tertiary rhyolite which is probably of Miocene age. Quartz veins occur in the Paleozoic rocks and associated intrusives, and contact-metamorphic and replacement deposits in limestone are probably also represented. Small productions of gold, silver, copper, lead, and antimony have been made. Rhodonite occurs on Copper Mountain as well as small croppings of sedimentary rocks said to contain nitrates. Oil shale float has been found near Charleston.

Bibliography. MR1908 I 476 MR1913 I 819 MR1918 I 233
Hill 507 204.
Schrader, F. C., "A Reconnaissance of the Jarbidge Contact, and Elk Mountain Districts, Elko County, Nevada", USGS B 497 (1912) 47-8.
———"The Jarbidge Minning District, Nevada, With a Note on the Charleston District", USGS B 741 (1923) 78-83.

COLUMBIA see AURA

CONTACT (Salmon River, Kit Carson, Porter)
Copper, Silver, Gold

Location. The Contact District is at Contact on the Salmon R. in N. E. Elko Co. near the Idaho border. Contact is 35 m. S. of Rogerson, Idaho, which is on the O. S. L. and 50 m. N. of Wells, Nev., which is on the S. P. R. R. and the W. P. R. R. The camp has an elevation of 5,400 ft. The three principal peaks in the district are the Ellen D. in the W. part with an elevation of 8,500 ft.; Middle Stack, 10 m. to the E. N. E. with an elevation of 8,100 ft.; and China, 8 m. to the S. E. of Ellen D. Mt. and 9 m. to the S. W. of Middle Stack Mt. The district is sometimes subdivided into the Contact or Salmon River District in the vicinity of the camp of Contact and Ellen D. Mt.; the Porter District in the neighborhood of China Mt. and Blanchard Mt. just E. of it; and the Kit Carson District near Middle Stack Mt.

History. Whitehill states that the Salmon and Kit Carson Districts were both discovered in 1872 by Hanks, Noll, and Lewis; but according to Schrader, a claim notice dated 1870 and signed by James Moran has been found in the district. The deposits of the Porter or China Mt. section were located in 1876 and were first worked by Chinamen on a commission basis. Small shipments of copper ore have been made intermittently since 1880. A 5-ton copper furnace erected in 1896-7 did not meet with success. The Delano Mine has been the principal producer. In 1921 the Interstate Commerce Commission issued a permit for the construction of a railroad from Rogerson, Idaho, to Wells, Nevada. If this line is built, Contact will probably become a big copper producer.

Geology. The Contact District lies in an area of folded and tilted sedimentary rocks which are mostly of Paleozoic age, are cut by granitic intrusives probably of Cretaceous age, and flooded by Tertiary lavas. Schrader states that the sedimentary rocks which formerly covered the entire area now form merely an outward quaquaversally-dipping belt encircling a granitic basin; and are in part underlain and intruded by the granitic rocks on the inner side and overlain by the Tertiary lavas on the outer side. The principal granitic rock is granodiorite, the intrusion of which produced the quaquaversal structure of the district. Both granodiorite and Paleozoic sediments have been intruded by syenitic and lamprophyric dikes.

Ore Deposits. The ore deposits of the Contact District are principally in the contact zone of the granodiorite and within the granodiorite in connection with the dikes. The deposits belong to three classes: contact-metamorphic deposits, fissure veins, and replacement deposits. The ores are copper ores containing small amounts of silver and gold. The contact-metamorphic deposits occur at or near the contacts of limestone and granodiorite, usually within the limestone. They consist of a mixture of carbonates massive garnet, axinite, epidote, calcite, quartz, actinolite, diopside, malachite, azurite, pyrite, chalcopyrite, bornite, copper oxides, specularite, hematite, limonite, chloropal, molybdenite, muscovite, etc., according to Schrader. Fissure veins occur parallel to the main contact, cutting across it nearly at right angles, and in the granodiorite in association with the dikes. Their filling is glassy quartz with a little chalcedony, and the ore minerals in them are chalcopyrite, bornite, malachite, azurite, chalcocite, cuprite, and shrysocolla. Molybdenite occurs in some of the veins. The country rock beside the veins has in some instances been replaced and constitutes ore.

Bibliography. MR1908 I 476 MR1913 I 819 MR1917 I 270
MR1909 I 398 MR1914 I 675 MR1918 I 233
MR1911 I 671 MR1915 I 629 MR1919 I 388
MR1912 I 790 MR1916 I 474 MR1920 I 321

Baily, J. T., "The Ore Deposits of Contact, Nevada", E&MJ 76 (1903), 612-3. (Discussion of article by Purington).

Hill 507 204.

Purington, C. W., "The Contact, Nevada, Quaquaversal", Colo. Sci. S., P 7 (1903) 127-138; Abstract E&MJ 76 (1903) 426-7.

Schrader, F. C., "A Reconnaissance of the Jarbidge, Contact and Elk Mt. Mining Districts, Elko County, Nevada,USGS B 497, (1912) 99-150.

Stuart NMR 114-5.

Weed MH 1154 Brooklyn M. Co. 1301 Panther City M. Co.
1177 Contact C. Co. 1325 Seattle Contact M. Co.
1264 Nevada Bellevue C. Co. 1326 Seattle M., M. & Power Co.
1276 Nevada Copper M., M. & Power Co.

COPE see MOUNTAIN CITY

CORNUCOPIA

Silver, Gold

Location. The Cornucopia District is is at Cornucopia in N. W. Elko Co. It is 65 m. by road N. N. W. of Elko which is on the S. P. R. R. and the W. P. R. R., and is 20 m. N. of Tuscarora.

History. The Leopard Mine was discovered by M. Durfee in 1872. A 10-stamp mill erected to treat the ores of this mine was destroyed by fire and a 20-stamp mill which replaced it met with a similar fate in 1880. The Panther was another important mine. Cornucopia attained a population of 1,000 in 1874, but soon declined, and the camp is at present deserted.

Production. The Cornucopia District is credited with an early production of more than $1,000,000 in silver and a little gold.

Geology. The ore deposits of the Cornucopia District are sheeted zones in altered Tertiary andesite. The ore is composed mainly of white quartz ribboned by a small proportion of dark sulphides, according to Emmons. Pyrite ,argentite, and gray copper are present, and ruby silver is said to have been an important ore mineral. Near the surface the ore minerals are mainly hornsilver and a yellow mineral which is probably pyromorphite. Some ore occured in the altered country rock beside the veins.

Bibliography. R1873 222 SMN1873-4 33-4
R1874 266 SMN1875-6 20-22
R1875 467 SMN1877-8 21-2

Emmons 408 44,62-5.

Hill 507 204.
Stuart NMR 114.
Thompson & West 393.

DELKER

Copper

Location. The Delker District is located on the N. E. side of Delker Hill which is the easternmost of a group of low hills E. of Ruby Lake. Currie on the N. N. R. R. is 25 m. S. W.

History. The Delker District was discovered in 1894, and copper ore was shipped from it in 1916 and 1917.

Geology. Quartz-monzonite is intrusive into limestones with interbedded quartzites which are probably of Mississippian age. Contact deposits and veins in the intrusive occur. The ore is an oxidized copper ore. The principal ore minerals are chrysocolla, malachite, and a small amount of azurite, while copper pitch ore is also present. A minor amount of chalcocite was noted by Hill in one instance.

Bibliography.

MR1914 I 676 MR1916 I 474 MR1917 I 270
Hill 648 28, 35, 37, 67-7.

DOLLY VARDEN (Mizpah, Granite Mountain)

Copper, Lead, Silver, Gold

Location. The Dolly Varden District is situated in S. E. Elko Co. in an isolated group of hills called the Wachoe Mts. by the Fortieth Parallel Survey. Mizpah Springs in the N. part of the district is 7 m. by road E. of Mizpah siding which is on the N. N. R. R., while Currie on the same railroad is 16 m. to the S. W. The district is sometimes divided into the Mizpah District on the N. and the Dolly Varden District on the S. Granite Mountain was an early name for part of the district, and is still sometimes applied to the extreme S. part.

History. The district was discovered in 1872 by Johnson and Murphy and actively worked for a few years. The discovery of gold in the N. section of the district in 1905 caused a brief excitement. Small and irregular shipments have been made from the district since 1908.

Geology. Quartz-monzonite which is probably of Cretaceous age underlies the N. part of the district and is intrusive into Carboniferous shale and limestone which constitute the country rock of the S. part. These intrusive and sedimentary rocks are flanked on the E. and S. by Tertiary lava flows.

Ore. Narrow gold-bearing quartz veins occur in the quartz-monzonite. According to Hill, they contain chalcopyrite and minor amounts of pyrite and bismuthinite and are said to show some free gold. Contact-metamorphic deposits of two types occur in the Carboniferous limestones and shales to the S.;—copper deposits and lead-silver deposits. The ore minerals of the copper deposits are chrysocolla, copper pitch ore, and malachite, with smaller amounts of chalcocite and bornite and with residual kernels of chalcopyrite. Limonite is abundant and contains a little residual pyrite. The gangue minerals are quartz, garnet, biotite, and tremolite with occasional calcite and antinolite. The principal minerals of the lead-silver deposits are argentiferous cerussite, anglesite, and residual kernels of galena.

Bibliography.

SMN1873-4 30 MR1910 I 510 MR1914 I 675
SMN1875-6 24 MR1911 I 671 MR1915 I 629
MR1908 I 476 MR1912 I 790 MR1918 I 233
MR1909 I 398 MR1913 I 819 MR1920 I 321
Hill 507 204.

Hill 648 26, 28, 30, 32, 35, 37, 38, 39, 40, 41, 76-88.
Thompson & West 394 Granite Mountain.
Weed MH 1230 Jersey Cons. C. & G. M. Co. 1266 Nevada Butte M. Co.

EDGEMONT (Centennial)
Gold, Silver, (Lead)

Location. The Edgemont District is situated in the central part of the Bull Run or Centennial Range in N. Elko Co. The town of Edgemont is 92 m. N. N. W. of Elko which is on the S. P. R. R. and the W. P. R. R. The Edgemont District is on the W. side of the Bull Run Mts., the Aura District adjoining it on the E. side of the range; while the Lime Mountain District adjoins the Edgemont District on the S.

History. The gold ores of the Edgemont District were discovered in the nineties. The Montana Gold M. Co. purchased the Lucky Girl Group in 1898 and erected a 20-stamp amalgamation mill and cyanide plant in 1902. From 1905 to 1909, this mine was the largest producer in Elko Co. A 10-stamp mill was built at the Bull Run Mine in 1902, but only ran for two years.

Production. The Montana and Bull Run Mines yielded about $1,000,000 up to 1908, chiefly in gold. The district made small productions up to 1917.

Geology. The country rock is a quartzite which is probably of Carboniferous age. The ore deposits are fissure veins which usually cut across the bedding of the quartzite and have been disturbed by normal faulting. The vein filling is chiefly quartz with a small amount oof pyrite, galena, and arsenopyrite. The gold and silver are associated with the sulphides. Sericite and pyrite have been developed to a small extent in the country rock near the veins.

Bibliography.

MR1905 256	MR1909 I 398	MR1912 I 790
MR1906 293	MR1910 I 510	MR1914 I 675
MR1907 I 346	MR1911 I 671	MR1916 I 474
MR1908 I 476		MR1917 I 270

Emmons 408 43, 67,-70, 75-80.
Davis 825-6. Hill 507 203-4. Stuart NMR 117.

ELKO
Oil Shale, (Lignite, Granite)

Location. The Elko oil shale deposits are situated at Elko in W. Elko Co. The oil shale belt is of considerable extent, but the principal work upon it has been done near the town of Elko which is on the S. P. R. R. and the W. P. R. R.

History. During the past fifty years, the lignite and oil shale beds of Elko have been prospected by the Central Pacific R. R. and others in an unsuccessful endeavor to find good fuel. Thirty years ago, R. M. Catlin acquired an oil shale property in the district which was opened by the Catlin Shale Products Co in 1914.* A 4-tube continuous screw retort plant was erected in 1916. This was superseded in 1918 by a plant consisting of 8 vertical retorts using superheated gases; which in turn gave way to the present plant which employs two 80-ton continuous vertical retorts. The Southern Pacific Co. opened an oil shale mine in 1918; and in cooperation with the U. S. Bureau of Mines, constructed a Scotch oil shale plant with Pumpherston retort. This property was shut down in 1919 and the plant sold to the Catlin Shale Products Co. in 1921. The Catlin plant has produced shale oil commercially but is still in the experimental stage.

Geology. The Elko oil shales occur in a lake deposit of Miocene age, according to Buwalda. The formation consists predominantly of shales and sandstones with which are interbedded conglomerate, limestone, chert, rhyolite, tuff, and bituminous and lignitic shales. It rests unconformably upon Pennsylvanian Weber quartzite, and is intruded by rhyolite. The region has been disturbed since the deposition of these beds and they are now tilted to angles of from a few degrees up to 70 degrees and are considerably warped and faulted. The oil shale occurs in zones which lie at successive horizons and range from a few feet to 60 ft. in thickness. There are two beds of oil shale of good quality 6 ft. thick, and others under 3 ft. The good shale produces from 50 to 75 gals. shale oil per ton upon dry distillation. Elko shale oil is said to be unusually high in paraffin.

Mines. The Catlin Shale Products Co. is controlled by its president, R. M. Catlin of Franklin, N. J. W. J. Sheeler is Supt. at Elko. The mine and plant are located 2½ m. S. W. of Elko. The mine is developed by a 400-ft. incline and several thousand feet of drifts. Electric drills are employed. The retort plant has a capacity of 300 tons of shale in 24 hrs., and there is a complete refining plant including wax plant. The usual run of petroleum products is manufactured, together with a high-grade, paraffin-base, lubricating oil which has given very satisfactory results.*

Bibliography. MR1913 II 1370 Granite.

Alderson, V. C., "The Oil Shale Industry", New York (1920) 31-2, 41, 42-4.

Buwalda, J. P., "Nevada" in "Oil Shale of the Rocky Mountain Region", USGS B 729 (1923) 91-102.

Day, D. T., "A Handbook of the Petroleum Industry", NY (1922) I 98, 838 890-2.

E&MJ, "Oil Wildcatting in Nevada", 109 (1920) 665.

Winchester, D. E., "Oil Shale in Northwestern Colorado and Adjacent Areas", USGS B 641 F (1916) 152, 156, 159, 161.

FERBER

Copper, Lead, Silver, Gold

Location. The Ferber District occupies some low foothills 3 m. E. of the main Toano Range in S. E. Elko. Co. on the Utah border. Most of the properties are in a low E.-W. pass through the foothills. The district is about 40 m. S. of Wendover, Utah, which is on the W. P. R. R.

History. The Ferber District was discovered by the Ferber Bros. about 1880, and the properties were relocated about 1890. Small and irregular shipments were made from the district from 1910 to 1917.

Geology. The low pass is cut through a small stock of quartz-monzonite intrusive into limestones which are probably of Carboniferous age, according to Hill. Most of the ore deposits are contact-metamorphic deposits in the crystalline limestone adjacent to the quartz-monzonite, but there are also a siliceous vein and a silicified shear zone in the quartz-monzonite.

Ore. The contact-metamorphic deposits are chiefly valuable for their copper which occurs mainly as chrysocolla and copper pitch ore with smaller amounts of carbonates. A little chalcocite and chalcopyrite were observed in depth. The orebodies in the quartz-monzonite contain limonite, lead carbonate, and a minor amount of oxidized copper minerals.

*The data concerning the Catlin Shale Products Co. was kindly furnished by W.L.Sheeler

Bibliography. MR1910 I 510 MR1913 I 820 MR1917 I 474
MR1912 I 790 MR1914 I 675,677 MR1917 I 270
Hill 648 28, 30-1, 32, 34, 35, 95-6, 98-102.

FERGUSON SPRING (Allegheny)

Copper, Lead

Location. The Ferguson Spring or Allegheny District is at Ferguson Spring Range at Don Don Pass in S. E. Elko Co.

Geology. The orebodies are irregular replacements deposits that parallel the bedding along fractures in the Paleozoic limestones, according to Hill. The ores are oxidized and consist mainly of limonite and barite with occasional copper carbonates. Oxidized coper and lead ores were shipped from three properties in the district in 1917.

Bibliography. MR1914 I 677 MR1917 I 270 Hill 648 34, 39, 95-6, 97-8

GOLD CIRCLE (Midas)

Gold, Silver

Location. The Gold Circle District is at Midas in the hills of the S. E. slope of the Owyhee Bluffs, near the édge of Squaw Valley, in W. Elko Co. Midas is 50 m. by road N. E. of Golconda which is on the S. P. R. R., 50 m. N. of Battle Mt. on the same railroad, and 33 m. N. E. of Redhouse, a station on the W. P. R. R.

History. Gold was discovered in the district in 1907, and a rush ensued. The town of Midas was laid out and its population at one time reached 2,000. A number of small mills were built, the Rex mill running for several years. The Elko Prince Mine, located in 1907, was taken over by the Elko Prince M. Co. in 1908. By agreement with the Dorr Co., a mill was erected at the Elko Prince Mine in 1915 and operated until 1922 when it burned down.

PRODUCTION OF GOLD CIRCLE DISTRICT
(According to Mineral Resources of the United States, U. S. Geol. Survey)

Year	Mines	Tons	Gold $	Silver Ozs.	Copper Lbs.	Total $
1908	6	245	36,561	1,171		37,182
1909	10	1,815	88,635	3,717		90,568
1910	12	1,042	24,767	1,152		25,389
1911	8	7,284	58,749	3,658	13	60,694
1912	15	3,137	37,053	5,173		40,235
1913	11	2,626	33,941	1,333		34,746
1914	9	3,362	62,190	11,725	23	68,677
1915	10	2,946	42,168	13,704		49,116
1916	11	23,947	243,565	157,207		347,107
1917	16	19,523	242,198	182,844		392,861
1918	6	18,692	288,591	167,684		456,275
1919	6	15,586	195,604	132,097		343,532
1920	6	14,347	104,482	94,109		207,061
1921	9	12,733	121,485	82,961		204,447
Total 1908-1921		127,285	1,580,090	858,527	36	$2,357,990

Geology. Tertiary volcanic rocks constitute the country rock of the Gold Circle District. Rhyolite flows and flow breccias cover the greater part of the area. They have been cut by dikes of andesite and are occasionally overlain by andesite flows. In the vicinity of the ore deposits, the rhyolite has been hydrothermally altered to a chalky white rock which is stained light brown by iron oxides in some places; and the andesite has also been hydrothermally altered but in less degree.

Ore Deposits. The ore deposits are fissure veins, replacement veins, and sheeted zones in the rhyolite. They strike N. W.- S. E. and dip steeply to the N. E. The fissure filling is chiefly quartz, in which gold and silver values occur. Part of the gold is free, and part is associated with pyrite. Silver is present as dark bands of argentite and polybasite in the quartz. The secondary minerals are quartz, iron oxide, manganese oxide, native silver, and hornsilver. The Elko Prince vein is from 1 ft. to 2 ft. wide and all the ore extracted from it has come from one large ore-shoot.

Mines. The Big Chief Cons. Ms. Co. and the Elko Prince M. Co. are the only companies at present active in the district. The **Elko Prince M. Co.** has a capital stock of 1,150,000 shares of $1 par value of which 1,108,566 have been issued. Its home office is at 501 Fifth Avenue, New York City; and L. L. Savage is Pres.; R. P. Jackson, V. P.; H. E. Haws, Sec.-Treas.; J. V. N. Dorr and F. F. Sharpless, additional Directors. L. D. Dougan is Supt. at Midas. The **Big Chief Cons. Ms. Co.** was incorporated in Nevada in 1919 with a capital of $2,000,000 and is controlled by C. J. and F. Berry of 250 First St., San Francisco. J. Coughlan is Supt. at Midas.

Bibliography. MR1907 I 348 MR1911 I 671 MR1915 I 629-30 MR1919 I 388-9
MR1908 I 477 MR1912 I 790 MR1916 I 475 MR1920 I 321
MR1909 I 399 MR1913 I 820 MR1917 I 270-2 MR1921 I 381
MR1910 I 510 MR1914 I 675 MR1918 I 233

Dorr, J. V. N., & Dougan, L. D., "Elko Prince Mine and Mill", T AIME 60 (1919) 78-97; Abstract, M&SP 117 (1918) 791.

Dougan, L. D., "The Elko Prince Leasing Co's. New Mill", M&EW 43, (1915) 939.

Emmons 408 48-57, 70.

Hill 507 204.

Howell, B. P., "Midas Mill", Discussion, M&SP 110 (1915) 851,979.

Stuart NMR 115-7.

Weed MH 1140 Bamberger Ms. Co. 1207 Gold Circle Buick M. Co.
1143 Berry M. Co. Gold Circle Coalition Co.
1189 Eastern Star M. Co. 1208 Gold Circle Crown M. Co.
1190 Elko Prince Leasing Co. Gold Circle Queen M. Co.
1191 Elko Prince M. Co. 1316 Rex Ms. Co.

Young, G. .J, "Cooperation Among Small Mines",E&MJ 106 (1918) 813.

GOLD CREEK see ISLAND MOUNTAIN

GOOD HOPE

Silver

Location. The Good Hope District is at Good Hope on Chino Creek in N. W. Elko Co. The Rock Creek District adjoins it on the S. Good Hope is about 25 m. by road N. W. of Tuscarora.

History. The district was discovered in 1878. The Buckeye and Ohio Mine was worked from 1882 to 1884, and its ore treated in a 5-stamp mill employing the Reese River process. In 1903 a concentrator was built but operated only for a short time. Silver ore was shipped from the Midnight Mine in 1921.

Production. The district is said to have produced more than $100,000 in silver in the early eighties.

Geology. According to Emmons, the ore deposits are sheeted zones in Tertiary rhyolite flow breccia and probably also in andesite. The country rock is altered in the neighborhood of the deposits. The Buckeye and Ohio lode is composed of veinlets of quartz and sulphides including masses of the highly altered and partly silicified country rock. The ore consists of quartz, pyrite, arsenopyrite, freibergite, stibnite, and dark ruby silver.

Bibliography. MR1921.I.381 Emmons 408 65-6 Hill 507 204
Thompson & West 393

GRANITE MOUNTAIN see DOLLY VARDEN

ISLAND MOUNTAIN (Gold Creek)

Gold, Silver

Location. The Island Mountain District is at Island Mountain in the neighborhood of Gold Creek in N. Elko Co. Gold Creek is 75 m. N. of Elko which is on the S. P. R. R. and the W. P. R. R. The Charleston District adjoins the Island Mountain District on the E.

History. The gold placers of Island Mountain were discovered by Penrod, Rousselle, and Newton in 1873. They were actively worked for the first few years and have made occasional productions since. Lodes were discovered at about the same time as the placers but were not opened until later. In 1917, the Hammond Exploration Co. built a small amalgamating mill near Gold Creek and began to explore its lode property. This company has made small productions from test runs of its mill and is still actively engaged in exploration work.

Geology. Gold with a little silver occurs in quartz veins which also contain arsenopyrite.

Mines. The Hammond Exploration Co. is under eastern ownership. It was incorporated in Colorado in 1923 and its legal address is Care C. F. Crowley, 815 E. & C. Bldg., Denver. H. W. Hammond is in charge of operations at Gold Creek. The property is equipped with a gasoline hoist and engines, compressor, pumps, and a small testing plant including small Hardinge mill, amalgamating plates, and Wilfley table.

Bibliography. SMN1873-4 27-8 MR1909 I 399 MR1919 I 399
SMN1875-6 22-3 MR1917 I 272 MR1921 I 381
MR1918 I 233

Thompson & West 394.
Weed MH 1221 Hammond Exploration Co.

IVANHOE

Mercury

Location. The Ivanhoe District is located at Ivanhoe Springs in W. Elko Co. Ivanhoe Springs is 90 m. by road N. W. of Elko which is on the S.P.R. R. and the W.P.R.R.

History. The Ivanhoe District was prospected in 1916 and produced a few flasks of quicksilver in 1917.

Geology. Cinnabar occurs as irregular stringers and as streaky disseminations in a much-altered, glassy rhyolite flow breccia, according to Ransome, and is widely distributed, covering an area of 5 or 6 sq. m. The rhyolite is altered to a cryptocrystalline white rock, and to a lesser degree to opal, chalky material, and yellowish siliceous sponge.

Bibliography. MR1916 I 767 MR1917 I 416 MR1918 I 164 MR1919 I 172

JARBIDGE
Gold, Silver

Location. The Jarbidge District lies on the upper N. slope of the divide between the Snake River Valley on the N. and the Great Basin Region on the S. It is in a region of rugged mountains with elevations ranging from 5,500 ft. to 11,000 ft. The name "Jarbidge" is derived from "Ja-ha-bich", an Indian word meaning "the devil"; and was probably applied to the district on account of its ruggedness and the presence of hot springs. The camp of Jarbidge has an elevation of 6,200 ft. and is about 65 m. by road S.W. of Rogerson, Idaho, which is on the O.S.L. The Charleston District adjoins the Jarbidge District on the S.W.

History. The Jarbidge District was prospected to a slight extent by Jack Sinclair and others in the early days, and in 1907 by C. M. Howard, but the discovery that led to the founding of the present camp was made by D. A. Bourne in 1909. A stampede ensued in 1910, in the course of which the population rose above 1,500; decreasing to 300 later in the year. A number of properties were developed and small annual shipments made up to 1918, when the mill of the Elkoro Ms. Co. went into operation and the production of the camp became large. With the decline of Goldfield, Jarbidge became the principal gold camp in Nevada, and the Elkoro Ms. Co. the largest producer. In 1919, 1920 and 1921, Jarbidge produced more gold from gold ores than any other Nevada camp.

Production. The total production of the Jarbidge District from 1910 to 1921, as recorded by Mineral Resources of the U. S. Geol. Survey, was 175,804 tons containing $1,915,787 in gold and 154,546 ozs. of silver with a total value of $2,079,376.

Geology. The Jarbidge District lies in an area of folded and tilted Paleozoic sedimentary rocks, cut by granitic intrusives which are probably of Cretaceous age, and flooded by Tertiary rhyolites, according to Schrader. The rhyolites were erupted in two distinct periods, perhaps Miocene and Pliocene. The older rhyolites were erupted along a fissure or series of craters coinciding with axis of the Crater Range and have a maximum thickness of 6,000 ft. The Crater Range is the most striking feature in the Jarbidge District, consisting as it does of seven peaks about a mile apart and with altitudes in excess of that of the Great Basin Divide.

Ore. The ore deposits of the district are gold-bearing quartz fissure veins in the older rhyolite. There are some 40 of these veins occuring in two belts, a western and more important belt on the lower W. slope of the Crater Range, and and an eastern belt at the crest of the range. The gangue of these veins is chiefly quartz and adularia pseudomorphic after calcite. The gold occurs native but is mostly in a finely divided state not readily visible to the naked eye. The gold is associated with silver which occurs mainly as argentite and as an alloy with the gold. Other minerals observed by Schrader in the Jarbidge ores are apatite, calcite, chalcedony, chlorite, epidote (?), fluorite, hematite, hyalite, kaolin, limonite, manganese oxide, marcasite, muscovite, opaline silica, orthoclase, pyrite, sericite, silver, and talc. Some veins also contain fault breccia consisting of fragments of silicified rhyolite country rock and of underlying Paleozoic shales, slates and quartzites.

Elkoro Mines. The Elkoro Ms. Co. owns the most important properties in the district. This company is a subsidiary of the Yukon Gold Co. which owns 60% of its stock, the Yukon Gold Co. being in turn controlled by the

Yukon-Alaska Trust. The Elkoro Ms. Co. was incorporated in Delaware in 1906 with a capital stock of 200,000 shares of $5 par value, all of which have been issued. It has offices at 120 Broadway, New York City and 582 Market St., San Francisco. W. Loeb, Jr., is Pres.; W. E. Bennett, V. P.- Sec.; O. B. Perry, Gen. Mgr.; with C. Earl, R. W. Strauss, E. L. Newhouse, E. E. McCarthy, F. R. Foraker, and L. Sloss, Directors. D. Steel is Res. Mgr. at Jarbidge. The Elkoro Ms. Co. owns 75 claims including the Starlight, Long Hike, and O. K. Groups. The property is developed by 11 adit tunnels aggregating 12,000 ft. and giving backs of 400 ft. to 800 ft. and is equipped with electric hoist, 2 compressors, pumps, 2 aerial trams, and a 100-ton mill employing the counter-current decantation system of cyanidation. This company is the largest producer of gold from gold ore in Nevada.

Other Mines. Five other properties are at present active in the district. The **Alpha Mine** consists of 5 claims developed by 4 tunnels and workings aggregating 3500 ft. and equipped with compressor, power line, 2 aerial tramways, and a 5-stamp mill. It is owned by O. B. Olson and associates of Tacoma, Wash., and is being operated under lease by O. B. Olson. F. Erno is Supt. at Jarbidge. The **Bluster Cons. Gold-Silver Ms. Co.** owns 16 claims and is at present conducting exploration on its 400-ft. level. T. B. Beadle is Pres.; F. Benane, V. P.; G. Winkler, Sec.; J. D. Goodwin, Treas.; with J. M. Prunty, W. J. McVicker, and J. C. Cook, additional Directors. The equipment includes a 3500-ft. aerial tram, 4000-ft. pipe line, sawmill, and 10-stamp mill. Excavation is now in progress for a 100-ton cyanide plant which is to be added to the mill. The **Flaxie M. Co.** owns 3 claims which are now being operated under lease by the Elkoro Ms. Co. W. Shopke is Res. Agt. The **Legitmiate Ms. Co.** is also under lease to the Elkoro Co. It owns 7 claims equipped with a 5-stamp test mill which has made a small production. W. van Alder is Res. Agt. The **Jarbidge-Buhl M. Co.** was incorporated in Nevada in 1917 with a capital stock of $1,000,000 shares of $1 par value. Its principal office is at Buhl, Idaho; local office in charge of W. H. Hudson, Res. Agt., at Jarbidge; and secretary's office at 424 International Life Bldg., St. Louis. B. R. Tillery is Pres. and J. C. Finch, Sec.- Treas.; with V. R. Laird, N. O. Thompson, H. C. Coleman, F. W. Fisher, W. H. McGraw, and G. C. Beach, additional Directors. J. Deemer is Gen. Mgr. The property consists of 12 claims including the Altitude Group developed by 1300 ft. of drifts and crosscuts and equipped with aerial tramway, compressor, and power line.

Bibliography.

MR1909 I 399
MR1910 I 510-1
MR1911 I 671
MR1912 I 791
MR1913 I 820
MR1914 I 675
MR1915 I 630
MR1916 I 475
MR1917 I 272
MR1918 I 233
MR1919 I 389
MR1920 I 321
MR1921 I 381

Buckley, E. R., "Geology of the Jarbidge Mining District, Nevada", M&EW 45 (1911) 1209-1210.

Butler, B. S., "Potash in Certain Copper and Gold Ores", USGS B 620J, (1915) 235.

Hill 507 205.

Schrader, F. C., "A Reconnaissance of the Jarbidge, Contact, Elk Mountain Mining District, Elko County, Nevada," USGS B 497, (1912) 11-98; Abstract, Wash. Acad. Sci. J 2 (1912) 430-440.

———"The Jarbidge Mining District, Nevada", USGS B 741 (1923).
Scott, W. A., "Jarbidge, Nevada", M&SP 100 (1910) 613-5.
Sweetser, N. W., "Geology of the Jarbidge Mining District Nevada", M&SP 101 (1910) 871-2.
Weed MH 140 Yukon-Alaska Trust. 141-2 Yukon Gold Co.
1150 Bluster Cons. M Co. Bonanza G. M. Co.
1190 Elko M. Co. 1191-2 Elkoro Ms. Co.
1229 Jarbidge-Buhl M. Co. Jarbidge Central Ms. Co.
1239 Legitimate Ms. Co.

KINSLEY

Silver, Copper, Lead

Location. The Kinsley District is at Kinsley Spring at the S. end of Kinsley Mt. between Elko and White PiPne Co's. Kinsley Spring is 28 m. S. E. of Currie which is on the N. N. R. R.

History. The Kinsley District was first discovered by Felix O'Neil in 1862. He was driven away by the Mormons, and the district was rediscovered about 1865 by George Kingsley, a soldier, and afterwards reorganized as the Kinsley District. By 1872, the camp was practically abandoned. In later years, interest revived, and in 1909, the Kinsley Development Co. erected 3 hoists and a concentrating mill.

Geology. Kinsley Mountain is a long, narrow, N.-S. ridge rising 2,000 ft. above the level of Antelope Valley. According to Hill, this ridge is composed mainly of dark-blue, cherty limestones, shaly at the base, which are probably of Cambrian age. At the S. end of the ridge there is an intrusive stock of quartz-monzonite, probably of Cretaceous age, which has metamorphosed the dolomitic limestones on all sides of it to a white crystalline limestone containing narrow belts of lime-silicate rock in which tremolite and wollastonite are common, green biotite occurs, and more rarely a gray garnet is present. The ore deposits are of contact-metamorphic origin and contain, in addition to the contact minerals noted, chrysocolla, malachite, azurite, limonite, remnants of original pyrite and chalcopyrite, and smaller amounts of copper pitch ore, chalcocite, and bornite. Silver chloride has been observed in the ore of the Morning Star Mine, and this mine has shipped oxidized lead ore. Quartz veinlets in the quartz-monzonite carry pyrite, chalcopyrite, and a small amount of galena, together with the alteration products of the sulphides.

Bibliography. B1867 411-2 SMN1869-70 63 SMN1875-6 25
SMN1866-98 103 SMN1873-4 31 MR1917 I 272
Hill 648 25, 28, 35, 37, 39, 40, 88-95.
Thompson & West 394.
Weed MH 1237 Kinsley Development Co.

KIT CARSON see CONTACT

LORAY

Silver, Lead, Copper

Location. The Loray District is at the N. end of the Toano Range in E. Elko Co., 4½ m. S. E. of Loray, a siding on the S. P. R. R.

History. Small shipments were made from the district in 1917, 1919, and 1921.

Geology. The ore is said to occur in white crystalline limestone underlain by dark limestone. According to Hill, specimens from a prospect consisted of crystalline limestone, quartz, chrysocolla, malachite, and some azurite surrounding, and in veinlets through, pitch copper ore containing lead and iron.

Bibliography. MR1914 I 677 MR1917 I 272 MR1919 I 389 MR1921 I 381 Hill 648 40, 95-7

LIME MOUNTAIN
Copper, (Gold, Silver)

Location. Lime Mountain is a ridge about 6 m. in length extending from Bull Run Creek to Deep Creek in N. Elko Co. and forming part of the Bull Run Range. It lies S. W. of the Aura District, and adjoins the Edgemont District on the S.

Geology. Lime Mt. is composed of dark-gray, Paleozoic limestone which has been intruded by quartz porphyry, andesite, and diabase; and has been locally metamorphosed to a course-grained marble with but little development of garnet or hornblende. At the Eldorado Mine there is a copper deposit carrying gold and silver values which is believed to be of contact-metamorphic origin. The ore consists of pyrite, chalcopyrite, and pyrite, intergrown with white and black mica, calcite, and quartz. Post-mineral fracturing has occurred, and there has been some secondary chalcocite enrichment.

Bibliography. Emmons 408 41,70-1.

LONE MOUNTAIN see MERRIMAC

MERRIMAC (Lone Mountain)
Lead, Silver, Copper, Gold

Location. The Merrimac District is situated on Lone Mountain, 28 m. by road N. W. of Elko which is on the S. P. R. R. and the W. P. R. R. Lone Mt. is a striking peak which rises above the main axis of the Seetoya Range to an elevation of 9,046 ft.

History. The Merrimac District was organized in 1870. About 1,000 tons of ore were produced to 1908 with a total value of over $30,000. The Rip Van Winkle Mine is said to have produced a total of $42,000 to 1916, when it was bonded to the Lone Mt. M. Co. by the Alaska Improvement Co. Silver-lead ore was produced by this property in 1921.

Geology. Lone Mt. consists mainly of dark-blue or gray massive beds of Carboniferous limestone, according to Emmons. This limestone has been intruded by quartz-monzonite and quartz-monzonite porphyry which form the crest of the mountain and have metamorphosed the limestone locally into garnet-calcite rock, green actinolite rock, hard cherty hornstone, or coarse-grained marble. The orebodies are deposits of lead and copper ore in marbleized limestone and contact-metamorphic deposits of copper ore in garnetized limestone. The ore minerals are galena, pyrite, chalcopyrite and their oxidation products.

Bibliography. R1871 219-220 R1872 158 SMN1869-70 58 SMN1875-6 25 MR1907 I 347 MR1912 I 791 MR1913 I 820 MR1916 I 475 MR1917 I 272 MR1918 I 234 MR1919 I 389 MR1920 I 323 MR1921 I 381

Emmons 408 84-6.

Weed MH 1128 Alaska Improvement Co. 1141 Lone Mt. M. Co.

MIZPAH see DOLLY VARDEN

MOUNTAIN CITY (Cope, Van Duzer)
Silver, Gold, (Lead, Zinc)

Location. The Mountain City District is located at Mountain City in the N.E. part of the Bull Run Range in N. Elko Co. Mountain City is on the Owyhee River, 1½ m. E. of the Duck Valley Indian Reservation, and 102 m. N. of Elko which is on the S. P. R. R. and the W. P. R. R. The Van Duzer placer mining district adjoins the Mountain City District on the S. and is sometimes considered a section of the Mountain City District.

History. The Mountain City District was discovered by Jesse Cope and others, in 1869, when it was organized as the Cope District, and the town of Mountain City was founded. A rush ensued and the following summer the new town had a population of about 1,000; with one mill in operation and two under construction. Over $1,000,000 in silver is said to have been recovered from shallow workings prior to 1881, but the production since that date has been small and intermittent, although three mills were built in the district in the nineteen hundreds. A 50-ton flotation plant was constructed and operated at the Nelson Mine in 1921.

Geology. The ore deposits are small fissure veins in granite and metamorphosed limestone, according to Emmons. The granite wall rock has been seciricitized and pyritized near the veins, but the limestone has not been appreciably altered by the ore-forming solutions. The principal vein filling is quartz, and the following minerals occur with it in the sulphide ore; pyrite, galena, sphalerite, tetrahedrite, arsenopyrite, chalcopyrite, argentite, pyrargyrite, stephanite, and gold. In the oxide ore, the following minerals occur with the quartz; chalcedony, iron oxides, pyromorphite, lead carbonate, copper carbonate and silicate, hornsilver, silver bromide, native silver, and native gold. Fracturing and faulting have reduced the ore locally to a white sand.

Mines. The only property at present active in the district is the Nelson Group which is being worked by Keeley and Hopkins. This property is equipped with a 9-stamp amalgamating and concentrating mill.

Bibliography.

R1869 186-7	SMN1869-70 59-61	MR1906 293
R1870 144-5	SMN1871-2 23-4	MR1907 I 346
R1871 219	SMN1873-4 31	MR1908 I 476
R1872 158	SMN1875-6 24-5	MR1915 I 630
		MR1921 I 381,382

Emmons 408 15, 26, 28, 30, 42, 43, 67,-70, 80-84.
Hill 507 205.
Thompson & West 392-3.
Weed MH 1264 Nelson Group.

MUD SPRINGS (Medicine Springs)
Lead, Zinc

Location. The Mud Springs District is located at the N. end of the Ruby Hills on the S. E. side of Ruby Valley between Medicine Spring on the N. W. and Mud Spring on the S. E. The prospects are 20 m. by road S. E. of Ruby Valley post office and 40 m. W. S. W. of Currie which is on the N. N. R. R.

History. The district was discovered in 1910 by Backman, Bardness and Martin. The Deadhorse claim was the principal prospect and the Nevada Dividend M. Co. did considerable prospecting work upon it in 1913. A little ore was shipped from the camp late in 1922.

Geology. The country rock of the Mud Springs District is Permian limestone with interbedded calcareous shale and quartzite. The deposites are replacements of the limestone along small brecciated fractures. The mineralized material consists of limonite, lead and zinc carbonates, and barite.

Bibliography. MR1914 I *676* Hill 684 *28, 34, 39, 42, 64-6.*

PORTER see CONTACT

PROCTOR

Silver

Location. The Proctor District is at Proctor on the W.P.R.R. in E. Elko Co.

History. Small shipments of silver ore were made from the Nick Del Duke and Silver Hoard properties in 1917.

Bibliography. MR1917 I *272.*

RAILROAD (Bullion)

Copper, Lead, Silver, Gold, Zinc

Location. The Railroad District is at Bullion on Bunker Hill in the Inskip or Pinyon Range in S. W. Elko Co. near the Eureka Co. boundary line. Bullion lies on the E. slope of the range at an elevation of 6,600 ft. and is 28 m. by road S. W. of Elko which is on the S. P. R. R. and the W. P. R. R., and 10 m. E. of Raines siding which is on the E. N. R. R.

History. The district was discovered in 1869 and worked during the seventies and eighties when two small smelters were in operation. Interest in the district revived in 1905, the mines were reopened in 1906, and since that time it has been a regular producer.

Production. The district is credited with a production of $3,000,000 in silver, lead, copper and gold up to 1884. From 1907 to 1921, the Railroad District produced 17,644 tons of ore having a total value of $898,544, according to Mineral Resources of the U. S. Geol. Survey.

Geology. Bunker Hill is composed mainly of gray, marbleized, Ordovician limestone, according to Emmons. This limestone has been intruded by granodiorite and by quartz porphyry which are probably of Cretaceous age. The most important orebodies are replacement deposits of lead, silver, and copper ore in marbleized limestone, and copper deposits of contact-metamorphic origin. Deposits of minor importance are copper impregnations in quartz porphyry and auriferous quartz veins in granodiorite.

Ore. The principal ore minerals in the replacement deposits are lead carbonate, hornsilver, pyromorphite, malachite, azurite, chrysocolla, cuprite, pyrite, chalcopyrite, galena, bornite, copper glance, and gray copper. The gangue is composed of quartz and calcite stained red and brown by iron oxides and showing a little manganese oxide At the contact between the granite porphyry and the Ordovician limestone at the summit of the ridge, the limestone has been altered to a rock composed of garnet, actinolite, calcite, epidote, quartz, tremolite, zoisite, and pyroxene, and at many places copper-bearing sulphide and zinc blende are intergrown with the iron-bearing silicates. The principal ore minerals are pyrite, chalcopyrite, bornite, chalcocite, galena, and zinc blende, which are locally weathered to malachite, azurite, copper oxides, iron oxides, and chrysocolla.

Mines. The Nevada Bunker Hill M. Co. is the only company active in the Railroad District at the present time. J. A. McBride is Pres.; W. W. Booher, V. P.; F. Davis, Sec.-Mgr.; J. Henderson, Treas.; with O. T. Williams,

S. E Davis, and G. S. Brown, Directors. The address of the company is Box 477, Elko; and F. Davis is Mgr. at Bullion. The company was incorporated in Nevada in 1905 with a capital stock of 2,000,000 shares of $1 par value of which 800,911 shares remain in the treasury. The property consists of 22 claims, developed by tunnels and shafts, including a deep tunnel 3,475 ft. long. The Nevada Bunker Hill claims lie on the E. flank of Bunker Hill Mt. with the Sylvania Group and Sweepstake Mine adjoining them on the W. and the Copper Belle and Bald Mt. Chief on the N.; while directly across the mountain to the S. W. in Lee Canyon are the Helen M., Gray Eagle, Ennor, and Independence. All of these properties produced in 1916 and in 1917.

Bibliography.

R1869 186
R1870 152
R1871 220-3
R1872 158-160
R1873 219-222
R1874 265-6, 420-4
SMN1869-70 57-8
SMN1871-2 28-30
SMN1873-4 34
SMN1875-6 22
SMN1877-8 22

MR1906 293
MR1907 I 378
MR1908 I 477
MR1909 I 399-400
MR1910 I 511
MR1911 I 671
MR1912 I 791
MR1913 I 820
MR1914 I 676
MR1915 I 630
MR1916 I 475
MR1917 I 272
MR1918 I 235
MR1919 I 389
MR1920 I 321
MR1921 I 381

Emmons 408 15, 25, 28, 29, 41, 88-95.
Hahn, O. H., "A Campaign in Railroad District, Nevada", T AIME 3 (1875) 176-8.
Hill 507 203.
Weed, W. H., "The Copper Mines of the U. S. in 1905", USGS B 285 (1906) 116.
Weed MH 1265 Nevada-Bullion Ms. Co., Nevada Bunker Hill M. Co.
1300 Palisade Copper Co.

ROCK CREEK

Silver

Location. The Rock Creek District is situated at the head of Rock Creek 12 m. by road W. of Tuscarora. The Good Hope District adjoins it on the N.

History. The district was discovered in 1876 and the Falcon Mine was worked from 1879 to 1881, the ore being hauled to Tuscarora for treatment. A mill was built at Rock Creek in 1884 but was never operated.

Geology. The ore deposit is a vertical fissure vein in Tertiary andesite. The vein is from 2 ft. to 5 ft. in width and the andesite in its vicinity is altered. The vein filling is mainly quartz which shows comb and ribbon structure and is banded by a small proportion of pyrite and other dark sulphides, according to Emmons. The values are in ruby silver.

Bibliography. SMN1875-6 26. Emmons 408 44, 62. Thompson & West 395.

RUBY VALLEY

Zinc, Lead, Silver

Location. The Ruby Valley District is situated on Smith Creek on the E. slope of the Ruby Range 6 m. N. of Ruby Valley post office in S. Elko Co. The shipping point is Tobar on the W. P. R. R., 45 m. to the N. E.

History. The district was discovered by Short Bros. in 1903, and the Friday

Group was located in 1906. The Michigan claim of the Friday Group produced lead ore in 1908, and was leased to the Arthur Zinc M. Co. in 1915. The Michigan Mine produced zinc and lead ore together with zinc and lead concentrates from a small concentrating mill in 1915 and in 1917.

Geology. The country rock is a white crystalline limestone probably of Paleozoic age which has been intruded by a biotite-granite which may be of Cretaceous age. The intrusion of this granite has metamorphosed the limestone at a number of places into lenses consisting largely of diopside, tremolite, and a little quartz. Some of these lenses contain galena, sphalerite, and more rarely chalcopyrite, forming small, irregular orebodies.

Bibliography. MR1908 I 477 MR1914 I 676-7 MR1915 I 630 MR1917 I 273 Hill 648 35, 41, 42, 54-9, 60-2.

SALMON RIVER see CONTACT

SPRUCE MOUNTAIN

Silver, Lear, Copper, Gold, (Manganese)

Location. The Spruce Mountain District is on Spruce Mt., the S. W. peak of the Gosiute Range in S. Elko Co. Spruce Mt. has an elevation of 10,400 ft. and the mines are on the N. and W. flanks. The district is 27 m. by road S. of Tobar, the shipping point on the W. P. R. R.; and 58 m. S. of Wells, the supply point on the S. P. R. R.

History. The Latham Mine was discovered in 1869, and in 1871 the Latham, Johnson, and Steptoe Districts were combined to form the Spruce Mountain District. The Ingot M. Co. acquired the Latham Mine and several others in 1871, built a smelter and commenced smelting, but shut down in 1872. The Starr King built a smelter shortly afterwords which also appears to have been unsuccessful. The Black Forest M. & S. Co. owns a 30-ton lead smelter which was operated intermittently up to 1909. The Bullshead M. & S. Co. built a 50-ton smelter in 1919 which made a trial run that fall but has not been operated since.

Production. The early production of the camp is not accurately known but is estimated at $700,000. The Spruce Monarch Cons. M. Co. shipped ore with a gross value of $250,000 in 1920.

RECENT PRODUCTOIN OF THE SPRUCE MOUNTAIN DISTRICT
(According to Mineral Resources of the United States, U. S. Geol. Survey)

Year	Producers	Tons	Gold $	Silver Ozs.	Copper Lbs.	Zinc Lbs.	Lead Lbs.	Total $
1902		3,600		189,072			1,272,600	145,440
1903		1,300	48	22,602	6,000		451,000	27,103
1904		346	127	8,770			122,869	9,954
1905		481	24	10,722			185,257	16,177
1906		150		3,648			59,965	5,862
1907		399	71	1,806	116,592			24,590
1908		53		75	10,977			1,489
1909		750	115	10,668	40,615		201,861	19,622
1910 No Production								
1911 No Production								
1912		18		14			13,277	605
1913		51	16	640			32,808	1,846
1914 No Production								
1915	4	71	7	975	15,415		11,821	3,755
1916	4	288	38	2,558	14,151	17,898	84,436	13,426
1917	3	627	19	4,831	42,658		252,704	37,377
1918	3	480	55	12,454	19,929		89,265	23,769
1919	2	510	8	9,757	648		277,775	25,785
1920	4	7,378		154,535	18,210		2,704,573	388,160
1921	3	1,897	39	53,099	5,065		916,264	95,023
Total 1902-1921		16,502	528	433,127	279,231	17,898	5,760,211	744,960

Geology. The country rock of the Spruce Mountain District is Mississippian limestone with some interbedded shales and quartzites which has been intruded by Cretaceous granite porphyry and by diorite, according to Hill. Spruce Mountain was formed as an anticlinal fold but has been greatly modified in the vicinity of the mines by two parallel, normal, major faults and by the intrusion of igneous rocks.

Ore. The principal ore deposits are replacements in the limestone associated with fractures and faults and with the intrusive igneous rocks, and are of both bedded and fissure types. Contact-metamorphic copper deposits also occur. The lead-silver ores contain limonite, cerussite, anglesite, and oxidized antimony mineral, and residual kernels of galena. Manganiferous silver ore occurs at the Bullshead Mine. It consists of pyrolusite, psilomelane, limonite and quartz with silver values which make it a silver ore rather than a manganese ore, according to Pardee & Jones. The minerals of the copper ores are malachite, chrysocolla, and chalcopyrite, with minor amounts of bornite and chalcocite and considerable pitch ore at the surface.

Mines. The Bullshead M. & S. Co. and the Spruce Monarch Cons. M. Co. are the only properties at present active in the Spruce Mountain District. The Bullshead M. & S. Co. was incorporated in Nevada in 1918 with a capital of 1,000,000 assessable shares of 1 cent par value of which 613,675 have been issued. The home office is at 221 Nevada State Life Bldg., Reno, and C. E. Mack is Pres. L. E. Johnson is Supt. at Wells. The company owns 10 claims, 4 of which are patented. These claims are developed by a 1,900-ft. tunnel and 4,200 ft. of workings and equipped with a 50-ton smelter. The Spruce Monarch Cons. M. Co. owns an old property whose early production is said to have been $200,000. H. Badt of Wells is Pres. The property is developed by a 2,000-ft. tunnel.

Bibliography. R1870 152 | SMN1871-2 24-6
R1872 158, 160-2 | SMN1873-4 30
R1873 217-222 | SMN1875-6 26-7
R1874 266

MR1905 266 | MR1909 I 400 | MR1917 I 273
MR1906 293 | MR1913 I 820 | MR1918 I 234
MR1907 I 347 | MR1914 I 677 | MR1919 I 390
MR1908 I 477 | MR1915 I 630 | MR1920 I 321
MR1916 I 475 | MR1921 I 382

Hill 507 205.
Hill 648 26, 29, 30, 34, 35, 39, 40, 41, 42, 67-76.
Pardee, J. T., & Jones, E. L., Jr., "Deposits of Manganese Ore in Nevada", USGS B 710F (1920) 221-2.
Stuart NMR 114.
Thompson & West 395.
Weed MH 1127 Ada H. M. Co.
1147 Black Forest M. & S. Co.
1157 Bullshead M. & S. Co.
1343 Spruce Monarch Cons. M. Co.

TECOMA

Lead, Silver, Gold, Copper

Location. The Tecoma District in the S. E. part of the Goose Creek Hills in N. E. Elko Co. on the Utah border. Tecoma on the S. P. R. R. is 10 m. S.

History. The Jackson Mine was discovered in 1906 and is the principal property.

Production. The production of the Jackson Mine to the end of 1922 was 1,500,000 lbs. lead, 27,000 ozs. silver, 44 ozs. gold, and 4,000 lbs. copper; valued in all at $103,000; according to a letter from S. F. Hunt.

Geology. The ore deposits are small irregular replacements in limestones of Devonian or Mississippian age, according to Hill, and are localized along fissures. The ore is argentiferous cerussite. At the Jackson Mine a bed of quartzite is the upper limit of the ore, and anglesite occurs in depth. Small dikes and sheets of granite prophyry have invaded the sedimentary rocks upon some of the other properties.

Mines. The Jackson Mine consisting of 6 claims is under bond to S. F. Hunt, Montello, Nev.

Bibliography. MR1908 I 477 | MR1912 I 791 | MR1916 I 476
MR1909 I 400 | MR1913 I 820 | MR1917 I 273
MR1910 I 511 | MR1914 I 677 | MR1918 I 234
MR1911 I 672 | MR1915 I 630 | MR1920 I 322

Hill 648 26, 29, 34, 39, 41, 102-5. Weed MH 1229 Jackson Mine.

TUSCARORA

Silver, Gold, Placer Gold

Location. The Tuscarora District is at Tuscarora on the S. E. slope of Mt. Blitzen in the Tuscarora Range. Tuscarora has an altitude of 6,200 ft.

Elko on the S. P. R. R. and the W. P. R. R. is 50 m. by road to the S. E.

History. The gold placers were discovered by the Beard Bros. and worked for a number of years. W. O. Weed made the first silver lode discovery in 1871 and the district was very active from 1872 to 1876 during which period 6 silver mills employing the Reese River process were in operation. The Dexter gold mine was discovered after the silver mines and operated until 1898, reopening for a brief period in 1912. It was equipped with a 40-stamp amalgamating mill later changed to a cyanide plant. The McKenzie 100-ton cyanide plant began retreating the tailings of the camp in 1908 and worked for several years. The Tuscarora-Nevada Ms. Co. engaged in an unsuccessful attempt to revive the camp from 1907 to 1915. Its property is now under lease and bond to the Stewart-Tuscarora M. Co., of which J. E. Harrington, 1465 Broadway, New York City, is Pres. The Stewart Co. erected a mill in 1909 which was destroyed by fire shortly after completion.

Production. Estimates of production range from $25,000,000 to $40,000,000, according to Emmons, the larger part of this being in silver.

Geology. The country rock of the Tuscarora District consists of Tertiary rhyolite and andesite porphyry, according to Emmons, frequently covered by a thin layer of Quaternary gravel. The ore deposits are silver lodes in andesite, gold stockworks in rhyolite, and gold placers. The country rock is highly altered in the vicinity of the lodes.

Ore. The ore minerals of the silver lodes are ruby silver, enargite, and other sulpharsenic and sulphantimony minerals, silver glance, galena, pyrite, arsenopyrite, and a little chalcopyrite, bornite, and malachite; while near the surface there is much hornsilver and native silver. Shoots of gold ore occur in the silver lodes and gold is present in most of the silver ore. The stockwork at the Dexter gold mine is composed of a network of quartz veinlets in the rhyolite. Adularia occurs in these veinlets and gold is present both in the veinlets and in the adjoining rhyolite which has been impregnated with pyrite.

Bibliography.

B1867 429-430
R1875 189
MR1905 266
MR1906 122, 293
MR1907 I 347
MR1908 I 478
MR1909 I 400
MR1910 I 510-1
SMN1871-2 24
SMN1875-6 17-19
SMN1877-8 17-21
MR1911 I 672
MR1912 I 791
MR1913 I 820
MR1914 I 677
MR1915 I 630
MR1916 I 476
MR1917 I 273
MR1918 I 234
MR1919 I 390
MR1920 I 322

Emmons 408 14-15, 25, 28, 44-5, 57-62.
Hill 507 205.
Magenau, W., "Cyaniding Stamp-Mill Tailing at Tuscarora, Nevada", M&M 21 (1900-1901) 299-300.
Stuart NMR 32, 113-4.
Thompson & West 395.
Weed MH 807 Stewart M. Co. 1292 New Tuscarora M. Co. 1346 Stewart-Tuscarora M. Co. 1362 Tuscarora-Nev. Ms. Co.

VAN DUZER (Mountain City)
Placer Gold

Location. The Van Duzer District is located on Van Duzer Creek, a tributary of the north fork of the Owyhee R., about 6 m. S. of Mountain City in N. Elko Co. It is 96 m. N. of Elko which is on the S. P. R. R. and the W. P. R. R. The Van Duzer District adjoins the Mountain City District on the S., and is some times considered a section of that district. The Aura District adjoins the Van Duzer District on the S. W.

History. The Van Duzer District was discovered in 1893 by R. M. Woodward and has been worked intermittently since. In 1908 there were two monitors on the creek about ¾ m. apart.

Geology. The depth of the Van Duzer gold placer deposit is under 15 ft. The ground consists of fine gravel with a few small baulders. The gold ranges in size from fine dust to nuggets of 6 ozs. and is worth $17 per oz.

Bibliography. MR1904 200 MR1913 I 820 MR1915 I 630
MR1905 266 MR1914 I 677 MR1918 I 234
Emmons 408 45, 84.

WARM CREEK
Zinc, Lead

Location. The Warm Creek District is located at the S. end of the East Humboldt Range in central Elko Co. The Polar Star Mine, which is the principal mine in the district, lies 5 m. S. W. of Warm Creek Ranch and 15 m. S. S. W. of Tobar, its shipping point on the W. P. R. R.

History. The Polar Star was located by Franks, Reed, and Wolverton in 1912 and is the property of the Nevada Zinc M. Co. of which G. Baglin of Tobar is Sec. Zinc and lead carbonates were shipped from the mine in 1915, and zinc ore in 1916, 1917, and 1918.

Geology. Lead and zinc carbonates occur as small, irregular, replacements along fractures in limestone which is probably of Permian age, according to Hill.

Bibliography. MR1914 I 676-7 MR1916 I 476 MR1917 I 273
MR1915 I 630 MR1918 I 234
Hill 648 35, 42, 54-9, 60.
Weed MH 1289-90 Nevada Zinc Co.

WHITE HORSE
(Copper, Lead)

Location. The White Horse District is located at White Horse Springs on the S. W. flank of the Mt. Pisgah in the Toano Range in S. E. Elko Co.

Geology. Small veins occur in a quartz-monzonite stock 2 m. in diameter which is probably of Cretaceous age. The veins carry much limonite with a minor amount of copper and lead carbonates, and their walls are sericitized for short distances. A little residual pyrite and chalcopyrite surrounded by copper pitch ore, limonite, and copper carbonates were observed by Hill.

Bibliography. MR1914 I 677. Hill 648 28, 37, 95-6, 98.

ESMERALDA COUNTY

ALIDA VALLEY see LIDA

ALUM
Potash, Sulphur

Location. The Alum District is located at Alum, just E. of the road to Silver Peak, in central Esmeralda Co. Blair Jct., on the T. & G. R. R. is 10 m. N.

History. The deposit of sulphur and alum at Alum was know as early as 1868. It was located and relocated many times, and a little development work done upon it. In 1921, the Western Chemicals, Inc., of Tonopah erected an alum-sulphur treatment plant designed to produce 10 tons c. p. alum and 10 tons 85% sulphur concentrates per day. This plant has operated intermittently since its erection.

Geology. The country rock is a dike-like mass of rhyolite intrusive into Tertiary rhyolitic tuffs, according to Spurr, and is altered in the vicinity of the deposits. In the principal deposit, sulphur coats all cracks and crevices with a thin layer which is generally not more than a fraction of an inch thick; while alum forms veinlets up to several inches in thickness which split and ramify irregularly through the altered rhyolite. Occasional seams of gypsum occur and bright red stains believed to be cinnabar. In a smaller deposit, sulphur is present as crystals in crevices.

Bibliography. SMN1867-8 96 MR1915 II 112-3

Duncan, L., "Recovery of Potash Alum and Sulphur at Tonopah", C&ME 24 (1921) 529.

Spurr, J. E., "Ore Deposits of the Silver Peak Quadrangle, Nevada", USGS PP 55 (1906) 157-8.

Young, G. J., "Potash Salts and Other Salines in the Great Basin Region", U. S. Dept. Agr. B 61 (1914) 34.

ARGENTITE

Silver

Location. The Argentite District is in the Silver Peak Range, 10 m. by air line and 24 m. by road W. of Silver Peak. Blair Jct. on the T. & G. R. R. is 40 m. to the N. The district lies at an elevation of 7,500 ft. to 8,000 ft.

History. The Argentite District was discovered in 1920 by Sanger and Taylor who located the Francis Mine. A small production of silver ore was made from this mine in 1922.

Geology. The country rocks of the Argentite District are rhyolite and limestone, according to a letter from E. S. Giles. The ore deposits are veins carrying silver values in argentite in a gangue of barite and quartz stained with copper carbonates.

Mines. The Francis Mine is now under lease and bond to the Natural Soda Products Co. of Bishop, Cal., of which W. W. Watterson is Pres. This company is carrying on active development work under the superintendence of F. Taylor of Tonopah. The Mohawk Mine adjoins the Francis on the S. and the Inman Group on the N.

BASALT see BUENA VISTA, MINERAL COUNTY

BUENA VISTA see MINERAL COUNTY

COALDALE

Coal, Variscite, Turquoise, (Agate, Chalcedony)

Location. The Coaldale District is in N. W. Esmeralda Co. at Coaldale on the T. & G. R. R. The coal beds are in the N. end of the Silver Peak Range at elevations of 5,000 ft. to 5,500 ft. Coal has also been found to the N. E. of Coaldale at the S. end of the Monte Cristo Range. The variscite and turquoise deposits are likewise located at the S. end of this range and are at altitudes of 6,150 ft. to 6,700 ft.

History. Coal was discovered in the district by William Groezinger, who mined 150 tons and sold it to the Columbus Borax Works in 1894. In 1911, two mines were operating in the district, the Nevada Coal & Fuel Company's Mine and the Darms Mine. Small productions were made from 1911 to 1915. Variscite and turquoise were discovered about 1909 and small productions made from 1910 to 1912. The U. S. Geol. Survey received specimens of agate and chalcedony of commercial grade from near Coaldale in 1910.

Coal. In the producing section at the N. end of the Silver Peak Range, 4 m. S. S. E. of Coaldale, there are 4 coal beds. These beds are interstratified with Tertiary rhyolites, rhyolite tuffs, sandstones, gravels, and conglomerates. They have been folded and faulted and have variable dips which are sometimes high. The material of the coal beds shows all gradations from black shale to semi-bituminous coal with a brilliant luster, according to Hance. There is considerable bone and sandstone parting, especially in the two younger beds; and the ash content of the best coal is rather high. The coal in a section may aggregate 2 ft. or 3 ft. but is usually made up of a number of thin streaks. It is reported to be a fairly good steaming coal, an excellent gas coal, and to coke well.

Coal Companies. H. A. Darms is Mgr. of the Reorg. Darms Coal M. Co. which owns the old Darms Mine and holds a 2,560-acre prospecting permit. The Darms Mine is developed by 2 incline shafts of 480 ft. and 520 ft. which cut 3 coal beds, and by drifts from these shafts. The coal is of better quality and more regular structure away from the rim of the basin where it outcrops, according to a letter from H. A. Darms, and the company is now planning to do some diamond drilling on its leased ground at a distance from the rim. The Western Coal Ms. Co. has taken over the 2,560-acre prospecting permit of J. Darms and is also preparing to do some diamond drilling.

Variscite and Turquoise. The country rock of the turquoise and variscite deposits at the S. end of the Monte Cristo Range is a slaty, dark-gray rhyolite with a few small altered trachytic dikes and some large outcrops of decomposed quartz porphyry. The variscite and turquoise occur in seams, veinlets, and nodules in the rhyolite; and include some good grade gem material.

Bibliography. MR1896 458
MR1896 556
MR1897 462
MR1907 II 152
MR1909 II 788 Turquoise
MR1910 II 849-850 Agate, Chalcedony; 886 Turquoise; 890-2 Variscite
MR1911 II 155
MR1912 II 1057 Variscite
MR1914 II 700
MR1915 II 403
USGS Tonopah and Lida topographic maps.

Davis 885-6.

Hance, J. H., "The Coaldale Coal Field, Esmeralda Co., Nevada", USGS B 531 (1913) 313-322.

Knapp, S. A., "The Coal Fields of Esmeralda Co., Nevada", M&SP 74 (1897) 133.

Spurr, J. E., "Coal Deposits Between Silver Peak and Candelaria, Esmeralda Co., Nevada", USGS B 225 (1904) 289-292.

——— "Ore Deposits of the Silver Peak Quadrangle. Nevada", USGS PP 55 (1906) 165-8.

Stoneham, W. J., "A Nevada Coal Field", E&MJ 77 (1104) 1009.

COLUMBUS MARSH

Borax, Salt, (Potash)

Location. Columbus Marsh lies S. E. of the old camp of Columbus in N. W. Esmeralda Co. on the Mineral Co. border. The T. & G. R. R. skirts the E. side of the playa, which has an area of 32½ sq. m. and an altitude of 4,512 ft.

History. Columbus Marsh was located as a salt deposit by Smith and Eaton in 1864, and the town of Columbus was founded in 1865. William Troop discovered ulexite in 1871 and 4 borax companies were at work on the marsh shortly thereafter. The Pacific Borax Co. moved its plant to Fish Lake Valley in 1875, and the town of Columbus declined rapidly. Experiments carried out by the U. S. Geol. Survey in 1912 and 1913 indicate that potash is not present in commercial amount.

Geology. Columbus Marsh is a typical playa formed by the desiccation of a small lake having a maximum depth of 104 ft.

Bibliography. B1866 94-5 SMN1869-70 108 MR1882 545, 567, 569
B1867 311 SMN1873-4 24 MR1889-90 503
SMN1875-6 38 MR1911 I 858
R1868 115 MR1917 II 421-2 Potash
USGS Tonopah and Hawthorne topographic maps.

Ayers, W. O., "Borax in America", Pop. Sci. Monthly 21 (1882) 350.

Davis 351,856.

Free, E. E., "The Topographic Features of the Desert Basins of the U. S. with reference to the Possible Occurrence of Potash", U. S. Dept. Agr. B 54 (1914) 32.

Gale, H. S., "Potash Tests at Columbus Marsh, Nevada", USGS B 540, (1914) 422-7.

Hanks, H. G., "Report on the Borax Deposits of California and Nevada", Cal. S. M. Bur. 3rd AR (1883) II 45, 53, 55.

Hicks, W. B., "The Composition of the Muds of Columbus Marsh, Nevada", USGS PP 95 (1915) 11 pp.

Phalen, W. C., "Salt Resources of the U. S.", USGS B 669 (1919) 142.

Thompson &West 419-420.

Young, G. J., "Potash Salts and Other Salines in the Great Basin Region", U. S. Dept. Agr. B 61 (1914) 1, 8, 33, 39, 53-4, 64.

CROW SPRINGS

Turquoise, Variscite, (Diatomaceous Earth, Silver, Lead, Copper, Gold)

Location. The turquoise and variscite region of Esmeralda and Nye Co's. is located in the N. E. section of the Monte Cristo Range and foothills to the N. E. and S. W. of Crow Springs. Crow Springs is in Esmeralda Co. 11 m. N. W. of Millers which is on the T. & G. R. R. Diatomaceous earth occurs in the vicinity of Crow Springs, and the Carrie silver-lead mine is in the Monte Cristo Range not far distant. The Crow Springs District adjoins the Royston section of the San Antone District in Nye Co. on the W.

Turquoise and Variscite. The Royal Blue turquoise mine in Nye Co. nearly 7 m. N. E. of Crow Springs was owned and operated for a number of years by W. Petry of Los Angeles who sold it to the Himalaya M. Co. in 1907, which operated for 2 yrs. The turquoise occurs in seams, veinlets, and nodules in altered trachyte, and both blue and matrix gems are of fine quality. The Wehrend prospect is ½ m. N. of the Royal Blue Mine and is geologically similar; while the Petry prospect ¾ m. S. of Crow Springs is in rhyolite and the Myers and Bona Mine which was actively worked in 1908 is on a quartz porphyry-slate contact. Variscite and turquoise were discovered in the Monte Cristo Range about 10 m. S. W. of Crow Springs in the direction of Blair Jct. by Mrs. M. Lovejoy in 1910. The gem material occurs in rhyolite which is cut by quartz porphyry dikes and masses.

Bibliography. MR1907 II 827 Turquoise
MR1908 II 846 Turquoise
MR1909 II 624 Diatomaceous Earth
781-5,788 Turquoise.
MR1910 II 886 Turquoise
892-4 Variscite
MR1917 I 278
USGS Tonopah topographic map.
Spurr 208 105-6 Monte Cristo Range.
Weed MH 1162 Carrie Silver-Lead Ms. Co.

CUPRITE

Silica, Sulphur, Silver, Gold, Copper, Lead, (Potash)

Location. The Cuprite District is at Cuprite on the T. & T. R. R. 14 m. S. of Goldfield. It occupies a low range of hills exteuding N. E. from Mt. Jackson.

History. The copper ores which gave the Cuprite District its name were discovered in 1905. Small amounts of these ores and of lead-silver ores have been shipped. A few shipments of sulphur were made prior to 1909, and a considerable number since. Silica plants have been erected recently by the Super-Silica Corp. and the American Tripolite Products Co., Inc., and have made small productions.

Geology. According to Ball, the rocks of the Cuprite Hills in order of formation are:—Cambrian sediments of which the dominant member is limestone; a diorite porphry dike probably of Cretaceous age; and Tertiary Siebert lake beds, rhyolite flows, and basalt.

Silica and Sulphur. The deposits of silica, sulphur, and potash have been described by Ransome. One large deposit lies 1½ m. N. W. of the railroad station at the top of a hill of Siebert conglomerate. It is nearly horizontal stratum of white ashy material containing masses of sulphur. This white pulverulent material was probably originally rhyolite tuff or glass but is now altered to silica and alunite. A second deposit is ½ m. N. E. of Cuprite and consists of solfatarically altered rhyolite pumice containing alunite. Most of the silica is low alunite, and the alunite is a soda variety low in potash, so the deposits have no importance as possible sources of commercial potash.

Ore Deposits. The ore deposits of the Cuprite District are copper-silver-gold replacements in the Cambrian limestone and gold-bearing veins in Tertiary rhyolite. The minerals of the replacement deposits are chalcopyrite, pyrite, galena, calcite, and quartz. They were deposited as sporadic masses in the limestone, as seams along joints, and as lenticular bodies along shear zones. Descending waters, often assisted by post-mineral fracturing, have produced chalcocite, carbonates, and oxides, together with iron-stained chalcedony. Silver and gold are present, the former predominating in value.

Mines. The Super-Silica Corp. is working the deposit N. E. of Cuprite. This company is incorporated in Nevada with a capital stock of 10.000,000 shares of 10 cents par value of which 2,500,000 shares have been issued, and its office is in Goldfield. C. S. Sprague is Pres.;J. K. Turner and E. D. Foster are V. P's.; B. Gill is Sec.-Treas.; with M. Opp, A. J. Mc-

Pherrin, and A. W. Ward, additional Directors. The Super-Silica Corp. owns 400 acres and has a lease on 400 acres more of silica ground. It is equipped with a 60-ton silica plant. American Tripolite Products Co., Inc., is a close corporation with home office at 44 E. 23rd St., New York City, and local office at Goldfield. The company owns 160 acres of silica land N. W. of Cuprite and has erected a plant with a capacity of 3 tons per h. to 325-mesh. J. N. Theo owns a sulphur deposit from which he has shipped sulphur to Honolulu for fertilizing purposes since 1915.

Bibliography. MR1908 I 483 MR1913 I 821 MR1917 I 274
MR1912 I 792 MR1915 II 112 Potash MR1918 I 235
MR1917 II 113 Potash, 410-1 Sulphur
USGS Lida topographic map

Ball 285 59-61.
Ball 308 46, 69-71.
Hill 507 206.
Ransome, F. L., "The Geology and Ore Deposits of Goldfield, Nevada", USGS PP 66 (1909) 109-110.
Weed MH 1182 Cuprite Copper Co., Cuprite Esmeralda Sulphur Co. Cuprite Sulphur Corp.

DIVIDE (Gold Mountain)

Silver, Gold, Lead

Location. Six miles S. of Tonopah, the road to Goldfield crosses its highest summit which is on the flank of Gold Mt. and known as the "Divide". The Divide District was named after this summit which is situated at its center. The general altitude of the district is 6,000 ft. Gold Mt. has an elevation of 6,600 ft. and is one of a number of isolated hills, the highest of which is 7,000 ft. The Tonopah District in Nye Co. adjoins the Divide District on the N.

History. Runge and Rochelle discovered the gold vein of the Tonopah Divide Mine in 1901 and the Tonopah Gold Mt. M. Co. was organized to work it the following year. Wingfield and Brougher reorganized this company as the Tonopah Divide M. Co. in 1912. In 1917 Brougher struck rich ore in the silver vein while crosscutting for the gold vein, and by 1919 a mining boom was in progress which rivalled the Goldfield rush. Some 350 mining claims covering 40 sq. m. were located over an area of 100 sq. m. having Gold Mt. as its center, but the boom collapsed after a few months. Ore was found upon a few of the properties; while others, which saw no chance of success with their Divide holdings and had money in their treasuries, used their funds to explore more favorable prospects in other districts.

Production. The principal production of the district has come from the Tonopah Divide Mine. This mine produced $282,107.83 in 1919; made similar productions in 1920 and 1921, and in 1922 milled 18,885 tons with a return of $552,752.95.

Geology. The prevailing country rock is the Fraction rhyolite breccia of the Tonopah District, according to Knopf. Siebert lake bed tuffs interstratified with this breccia place its age as Upper Miocene. Some rhyolite flows are also associated with it, and it is intruded by several rhyolite stocks corresponding to the Tonopah Oddie rhyolite, and also by a large mass of Divide andesite. Latite flows of Pliocene age rest unconformably upon the Miocene rocks.

Ore Deposits. The principal orebodies are silver-bearing zones of fracturing and shearing in the Fraction rhyolite breccia. The lode filling consists of slightly mineralized rhyolite breccia containing a small amount of disseminated pyrite and cut by thin stringers of quartz. The silver is present mainly in the form of hornsilver concentrated in rich masses along irregular seams of sericite in the lodes. The rare molybdenum minerals, molybdite and powellite are found in the ore of the Tonopah Divide Mine. A few narrow gold veins occur in the Oddie rhyolite. The vein filling of these veins consist of brecciated rhyolite cemented by quartz, and an unidentified mineral resembling apatite.

Tonopah Divide Mine. The Tonopah Divide M. Co., which owns the principal mine in the district, was incorporated in Nevada in 1912 with a capital stock of 1,250,000 shares of $1 par value of which 1,014,800 have been issued. The stock is listed on the San Francisco Stock Exchange and the New York Curb. E. B. Brougher is Pres.; W. J. Douglas, V. P.; E. J. Erickson, Sec.-Treas.; with E. W. Blair and C. A. Heller, additional Directors. Wm. Watters is Supt. at Tonopah. The company owns 14 patented claims aggregating 160 acres in the center of the Divide District. The property is developed by a 1,445-ft. shaft and 5 m. of workings and equipped with motor-driven machinery including 2 hoists and 2 compressors.

Other Mines. There are 9 other active companies in the district:—The **Belcher Ex. Divide M. Co.,** of which A. Walz is Pres. and A. Kelly in charge at Tonopah, owns 9 claims situated 1 m. W. of the Tonopah Divide Mine and developed by shafts 110 ft. and 300 ft. in depth. The **Brougher Divide M. Co.,** of which E. B. Brougher is Pres. and G. Hanson is Supt. at Tonopah, owns 72 acres N. of the Tonopah Divide developed by a 500-ft. shaft and a 296-ft winze and equipped with electric hoist and compressor. The **Divide Ex. M. Co.,** of which Z. Kendall is Pres. and E. J. Bevis, Supt., owns 20 acres adjoining the Tonopah Divide on the N., developed by shafts of 425 ft. and 200 ft. and equipped with electric hoist and compressor. The **Gold Zone Divide M. Co.,** of which W. Forman, 216 State Bank Bldg., Tonopah, is Pres., and H. R. Bradley, Supt., owns 9 claims adjoining the Tonopah Divide on the S., developed by a 900-ft. shaft and 1½ m. of workings, and equipped with electric hoist and compressor. The **Tonopah Hasbrouck M. Co.,** of which A. A. Busey is Pres.-Mgr. and A. A. Codd, 421 Clay Peters Bldg., Reno, is Sec., owns the old Hasbrouck. Mine, 1 m. W. of the Tonopah Divide, developed by shaft and tunnel and equipped with hoist and compressor. The **New Alto Divide M. Co.,** of which J. Mayer is Pres.-Mgr. and B. Gill is Sec.-Treas., owns 4 claims E. of the Tonopah Divide developed by a 400-ft. shaft and equipped with hoist and compressor. The **New Sutherland Divide M. Co.,** of which C. R. Evans, Goldfield, is Pres. and J. B. Kendall, Mgr., owns 7 claims 1½ m. N. of the Tonopah Divide, developed by a 900-ft. incline shaft, and equipped with a 50 hp hoist. The **Tonopah Dividend M. Co.,** of which E. P. Cullinan, 219 State Bank Bldg., Tonopah, is Supt., owns 85 acres adjoining the Tonopah Divide on the N., developed by shafts of 850 ft. and 400 ft., and equipped with electric hoist and compresor. The **Reorg. Belcher Divide M. Co.,** of which N. M. McCormick is Pres. and S. E. Jameson, Supt., owns 4 patented claims ½ m. N. W. of the Tonopah Divide, developed by a 550-ft. shaft and equipped with electric hoist and compressor.

Bibliography. MR1919 I 392-3 MR1920 I 323 MR1921 I 382-3
USGS Tonopah and Lida topographic maps.

Ball 308 82.

Carpenter, J. A., "The Divide District" E&MJ 107 (1919) 859-861.

Hill 507 207.

Knoph, A., "The Divide Silver District. Nev.", USGS B 715K (1921), 147-170.

Lincoln, F. C., "The Divide Mining District: Its History, Its Geology, Its Mines", Nevada News Letter 11 (1919) No. 20, 14-17, 44.

M&SP Editorial, "The Geology of the Divide District", 118 (1919) 622.

Sizer, F. L., "The Divide District", M&SP 118 (1919) 631-3.

Spurr, J. E. "Ore Deposits of Tonopah and Neighboring Districts, Nev.", USGS B 213 (1903) 87.

———"Geology of the Tonopah Mining District", USGS PP 42 (1905) 99.

———"Genetic Relations of Western Nevada Ores", T AIME 36 (1906), 381.

Weed MH 1127-1378 Describes 215 Divide mining companies.

Young, G. J., "Divide Silver-Gold District of Nevada", E&MJ 109, (1920) 62-6.

DYER

Silver, Lead

Location. The Dyer District is located E. of Dyer's ranch in the foothills of the W. flank of the Silver Peak Range in W. Esmeralda Co. Coaldale on the T. & G. R. R. is 28 m. N.

History. The district was first prospected in 1863 and 1864; again and more intensively from 1884 to 1887 when a little ore was shipped; and once more at the time of the Goldfield boom, some lead-silver ore being shipped in 1912.

Geology. The ore deposits occur in Ordovician limestone a short distance from a small body of intrusive granite. They are bedding-plane deposits in irregular zones of crushed and decomposed limestone, according to Spurr, and the ore consists of erratic bunches of quartz containing black copper-silver sulphide, in places oxidized to copper carbonate, iron oxide, and silver chloride.

Bibliography. MR1912 I 792 USGS White Mountain topographic map.

Hill 507 206.

Spurr, J. E., "Ore Deposits of Silver Peak Quadrangle, Nevada", USGS B 225 (1904) 115.

———"Ore Deposits of the Silver Peak Quadrangle, Nevada", USS PP 55 (1906) 84-5.

FISH LAKE MARSH

Borax, (Petroleum)

Location. Fish Lake Marsh occupies 10,000 acres in the lower part of Fish Lake Valley in W. Esmeralda Co. The marsh has an elevation of 4,825 ft., and drains into Columbus Marsh which is 10 m. N.

History. Mott & Piper produced borax from the marsh as early as 1873; the Pacific Borax Co. moved its plant there from Columbus in 1875; and the borate deposits were worked for a number of years thereafter. In 1920, the California Excelsior Oil Co. drilled for oil on the McNett ranch and obtained artesian water. The Merger Oil Co. is now drilling at the mouth of Ice House Canyon and has attained a depth of 1,400 ft.

Geology. The marsh contains a number of tracts in which borates occur, mainly in the form of ulexite. Potash is not present in important amounts.

Bibliography. R1875 458 MR1882 567-9 MR1911 II 858
SMN1867-8 95 Salt MR1883-4 861-2 MR1917 II 422 Potash
SMN1873-4 24-5
USGS Lida topographic map.
Hanks, H. G., "Report on the Borax Deposits of California and Nevada", Cal. S. M. Bur. 3rd AR (1883) II 53, 55.
Spurr, J E., "Ore Deposits of the Silver Peak Quadrangle, Nevada", USGS PP 55 (1906) 158-161.
Young, G J., "Potash Salts and Other Salines in the Great Basin Region", U. S. Dept. Agr. B 61 (1914) 39, 42.

GOLDFIELD

Gold, Silver, Copper, Lead, (Diatomaceous Earth)

Location. The Goldfield District is situated at Goldfield in the Goldfield Hills in E. Esmeralda Co. on the Nye Co. border. Goldfield is a station on the T. & G. R. R. and the T. & T. R. R. The altitude of Goldfield is 5,684 ft. and the surrounding hills rise to elevations in excess of 6,000 ft.

History. The Goldfield District was discovered by Harry Stimler and William Marsh on December 2, 1912. After a brief excitement, most of the prospectors left the camp and the original claims were allowea to lapse. A. D. Meyers and R. C. Hart located the Combination Lode on May 24, 1903. Ore was discovered on that claim the following October, shipments began in December, and the great Goldfield stampede ensued. The railroad was extended to Goldfield in 1905, at which time the population of the town was 8,000. The Goldfield Cons. Ms. Co., the principal company in the camp, was organized by George Wingfield and Senator Nixon in 1906, and its 100-stamp mill was completed in 1908. It was in the latter year that Goldfield reached its maximum population of 20,000. From 1904 to 1918, the Goldfield District was the most important gold-producing district in Nevada. By that time, the large known orebodies had become nearly exhausted and the production fell off rapidly. Production still continues, however, and exploration work is being pushed.

PRODUCTION OF GOLDFIELD DISTRICT
(According to Mineral Resources of the United States, U. S. Geol. Survey)

Year	Tons	Gold $	Silver Ozs.	Copper Lbs.	Lead Lbs.	Total $
1903		70,670	287			70,825
1904	8,000	2,341,979	19,954			2,353,353
1905		1,882,951	8,589			1,888,139
1906	59,628	7,026,154	15,648			7,036,638
1907	101,136	8,408,396	71,710			8,455,725
1908	88,152	4,880,251	30,823	1,606		4,896,799
1909	297,199	9,383,261	33,164	52,015		9,407,268
1910	339,219	11,137,150	117,598	107,282		11,214,278
1911	390,431	10,287,075	126,406	72,998		10,363,195
1912	362,777	6,239,747	125,736	579,539	3,581	6,412,859
1913	364,785	5,019,419	153,984	1,257,319	16,063	5,308,017
1914	367,166	4,705,169	129,830	1,059,021	4,018	4,919,302
1915	418,935	4,389,385	165,306	1,679,423		4,767,094
1916	383,456	2,651,158	129,781	1,317,400	2,418	3,060,801
1917	339,488	1,900,082	78,184	728,255		2,163,320
1918	264,237	1,212,572	90,560	548,847		1,438,697
1919	16,435	740,265	39,912	125,418	1,460	808,373
1920	6,571	155,789	6,081	13,440		164,890
1921	1,903	146,782	1,761	3,689		149,019
Total 1903-1920		$82,578,255	1,345,314	7,556,252	27,540	$84,878,592

Dividends. The following dividends have been paid by Goldfield companies according to a table prepared by Miss N. E. Preece, Geol. Dept. International Smelting Co., Salt Lake City.

Florence Goldfield	$ 840,000
Frances Mohawk	546,000
Goldfield Almo	42,000
Goldfield Combination Fraction	92,111
Goldfield Consolidated	29,897,212
Jumbo Extension	735,749
Little Florence	430,000
Reorganized Booth	550,000
Total	$33,133,072

Geology. The oldest rock in the Goldfield District is Cambrian shale which has been intruded by alaskite that is probably of Cretaceous age. Upon these older rocks rest unconformably a series oof Tertiary lavas and lake sediments. Ransome gives the following sequence beginning with the oldest:—Vindicator rhyolite, Latite, Kendall tuff, Sandstorm rhyolite, Morena rhyolite, Milltown andesite, Dacite, Dacite vitrophyre with intercalated Chispa andesite, Meda rhyolite, Andesite breccia, Espina breccia, Siebert lake beds with intercalated Mira basalt, Pozo conglomerate, Spearhead rhyolite, Rabbit Spring breccia, conglomerate, and sandstone, and Malpais basalt.

Ore Deposits. The Goldfield ore deposits are irregular lodes in the fractured and highly altered country rocks. Replacement of the country rocks by quartz, kaolinite, alunite, and pyrite has occurred. The ore shoots are in the form of irregular bodies in the irregular lodes and their limits can only be determined by assays. The principal ore bodies are in dacite, though some are in rhyolite, andesite, and latite, and low-grade ore with occasional rich shoots occurs at the latite-shale contact.

Ore. The principal gangue mineral is compact quartz derived from the silicification of volcanic rock, with which are associated kaolinite and alunite. The ore minerals occur mainly in the quartz though at times in or near alunite. They consist of fine-grained pyrite and marcasite, bismuthinite, goldfieldite, arsenical famatinite, native gold and tellurides, with minor amounts of other sulphides. Concentric shells of ore minerals about altered rock fragments are characteristic of the rich ore.

Goldfield Consolidated. The Goldfield Consolidated Mine is one of the world's greatest gold mines and was for many years the most important gold producer in Nevada. The Goldfield Cons. Ms. Co. was incorporated in Wyoming in 1906 with a capital stock of 5,000,000 shares of $10 par value of which 3,559,148 shares have been issued. The stock is listed on the San Francisco and Salt Lake City Stock Exchanges and New York and Boston Curbs. The home office is in the Reno Nat. Bank Bldg., Reno. G. Wingfield is Pres.; B. J. Henley, V. P.; W. E. Zoebel, Sec. Treas.; E. A. Julian, Gen. Mgr.; with J. H. Carstairs, W. Woodburn, and S. C. Mitchell, Directors. A. H. Lawry is Gen. Supt. at Goldfield. The company began leasing in 1906 and in 1920 adopted a policy of leasing blocks of 300 ft. x 300 ft. to individuals. The property consists of 390 acres developed by 6 working shafts.

Other Mines. Seven other companies are at present active. The **Florence Goldfield M. Co.** own 67.7 acres adjoining the Goldfield Consolidated and developed by a 1,200 ft. shaft and several miles of workings. L. E. Whicher, 25 Broad St., New York City, is Pres., F. S. Schmidt is V. P.-Gen. Mgr., and R. C. McCarthy is Supt. at Goldfield. The company paid dividends from 1908 to 1911 since which time it has been worked by lessees. The **Goldfield Deep Ms. Co. of Nev.** is a consolidation of 36 claims effected to secure funds for deep mining. A. I. D'Arcy is Pres. and E. Burt is Supt. The property adjoins the Goldfield Consolidated on the N. W. and a 2,400-ft. shaft which is being sunk upon it has now reached a depth of 1,400 ft. The **Goldfield Great Bend M. Co.,** of which C. S. Sprague is Pres. and J. K. Turner V. P.-Gen. Supt., owns 112 acres developed by 2 400-ft. shafts and equipped with hoist, compressor, pump, and old mill. It is being worked by lessees. The **Red Hill Florence M. Co.,** of which M. MacKenzie is Pres., H. G. McMahon is V. P.-Gen. Mgr., and A. H. Howe, Reno, is Sec.-Treas., owns 70 acres adjoining the Florence Goldfield on the S. and W. The **Reorg. Cracker Jack M. Co.,** of which E. P. Junor is Pres., G. K. Cremer is V. P.., and H. G. MacMahon is Sec.-Treas.-Mgr., owns the Adams Group, Great Western Group, and other claims. The **Reorg. Kewanas Gold M. Co.,** of which A. I. D'Arcy is Pres., C. Barnes V. P., and H. G. MacMahon Sec.-Treas., owns 40 acres adjoining the Goldfield Consolidated on the W. and an option on the Eagle Mine 3 m. W. of Goldfield. The **Silver Pick Cons. Ms. Co.,** of which H. Zadig, 265 Russ Bldg., San Francisco, is Pres., A. S. Wollberg V. P.., and C. D. Olney, Sec., owns 60 acres developed by a 1,500-ft. and 280-ft. shaft; and has recently been producing rich ore from its Red Top lease on the Goldfield Consolidated ground.

BIBLIOGRAPHY OF THE GOLDFIELD DISTRICT

General

MR1903 183
MR1904 197
MR1905 267; 116
MR1906 293-4; 122, 495 Quicksilver
MR1907 I 349-353, 391-4
MR1908 I 479-482, 714
MR1909 I 402-5
MR1910 I 512-4
MR1910 II 887 Turquoise
MR1911 I 673-5
MR1912 I 792-4
MR1913 I 821-3
MR1914 I 678-81
MR1915 I 631-33
MR1915 II 112 Potash; 1024 Magnesite
MR1916 I 476-8
MR1916 II 410-1 **Sulphur 113 Potash; 399 Magnesite**
MR1917 I 274-6
MR1918 I 235-7
MR1918 II 544 Strontium
MR1919 I 390-2
MR1920 I 322-3
MR1921 I 382

Draper, M. D. "The District of Goldfield, Nevada", E&MJ 78 (1904) 383-4.
———"The Goldfield District, Nevada", M&SP 90 (1905) 150-2.
E&MJ, Editorial, "Goldfield Consolitated", 91 (1911) 101-2.
Hill 507 206-7.
Lamb, M. R., "Reminiscences of Goldfield, Nevada", E&MJ 87 (1909) 441-2.
M&EW, Staff Correspondence, "The Mining Situation at Goldfield", 38 (1913), 441-2.
———, "Tonopah and Goldfield and Their Rapid Development", 90 (1905) 84-5.
———, Special Contributor, "Goldfield, Nevada", 94 (1907) 721-3.
M&SP, Editorial, "Goldfield Consolidated", 102 (1911) 194-5.
———, Editorial, "Goldfield Consolidated", 120 (1920) 296.
Rickard, T. A., "Goldfield, Nevada", M&SP 96 (1908) II 664-7.
———"Goldfield Revisited", M&SP 110 (1915) I 797-9, II 829-31, III 907-9.
Stuart NMR 47-55.
Taft, H. H., "Goldfield and Tonopah", E&MJ 81 (1906) 557-8 T AIME 37, (1907) 187-191.
Tyssowski, J., "Goldfield and the Goldfield District of Nevada", E&MJ 87, (1909) 1229.
Weed MH 1170-1380, Describes 31 Goldfield companies.

Geology

Ball 308 28, 32-4, 71-6, 107
Barnes, C., & Byler, E. A., "Relation of Faulting and Mineralization in Goldfield, Nevada", M&SP 107 (1913) 59-60.
Becker, Arnold, "Depth at Goldfield", Discussion, M&SP 95 (1908) 62.
———, "Geological Possibilities at Goldfield", M&SP 96 (1908) 846.
Butler, B. S., "Potash in Certain Copper and Gold Ores", USGS B 620J (1915), 233.
Collins, E. A., "The Combination Mine, Early Development and Geologic Structure", M&SP 95 (1907) 397-9.
Cutler, H. C., "Goldfield and Its Present Boom", E&MJ 99 (1915) 221-4.
Dominian, L., "Geology of Goldfield, Nevada", Ores and Metals, 13 No. 20, (1904) 25.
———, "The Goldfield District, Nevada", E&MJ 78 (1904) 581-2.
Gage, R. B., "Determination of Alunite", E&MJ 87 (1909) 1121-4, Discussion, G. Surr, E&MJ 88 (1909) 31; C. B., 88 (1909) 743.
Hastings, J. B., "The Geology of Goldfield, Nevada", E&MJ 81 (1906) 843-4.
Hastings, J. B., & Berkey, C. P., "The Geology and Petrography of the Goldfield Mining District, Nevada", TAIME 37 (1907) 140-159

Hill, R. T., "The Goldfield, Nev., Type of Ore Occurence", E&MJ 86 (1908), 1096-9; Discussion, F. L. Ransome, "Alunitization", E&MJ 87, (1909) 177.

Knopf, A., "Some Cinnabar Deposits in Western Nev.", USGS B 620D (1915) 65.

Lewis, J. V., "Prospecting for Ores of the Goldfield Type", E&MJ 87 (1909), 1121-2.

Lincoln, F. C., "Certain Natural Associations of Gold", EG 6 (1911) 269-9.

Locke, A., "The Ore Deposits of Goldfield, Nev.", E&MJ 94 (1912) 797-802, 843-849.

Pardee, J. T., & Jones, E. L., Jr., "Deposits of Manganese Ore in Nevada", USGS B 710 (1920) 233.

Ransome, F. L., "Preliminary Account of Goldfield, Bullfrog, and Other Mining Districts in Southern Nev.", USGS B 303 (1907) 7-39; Abstract, M&SP 94 (1907) 436-9.

———, "The Association of Alunite with Gold in the Goldfield District, Nevada", EG 2 (1907) 667-692; Abstract, Science, N. S. 27 (1908) 189; Discussion, F. C. Lincoln, EG 2 (1907) 801-3.

———, "The Geology and Ore Deposits of Goldfield, Nevada", PP 66 (1909); Abstract, E&MJ 88 (1909) 1162.

———, "Geology and Ore Deposits of the Goldfield District, Nevada", EG 5, (1910) 301-311, 438-470.

———, "Wurtzite at Goldfield, Nev.", Wash. Acad. Sci. J 4 (1914) 482-5.

Rickard, T. A., "Goldfield, Nevada", M&SP 96 (1908) 738-42.

Shannon, E. V.,"Famatinite from Goldfield, Nev.",AJS 4th Ser. 44 (1917) 469-470.

Sharwood, W. J., "Gold Telluride", M&SP 94 (1907) 731.

Shaw, E. S., "Structure of Goldfield Ores", E&MJ 91 (1911) 714.

Spurr, J. C., "Genetic Relations of Western Nevada Ores", T AIME 36 (1906), 381-2.

———, "The Goldfield District, Nevada", Abstract, Franklin Inst., J 164 (1907), 155-160.

Turner, J. K., "Mining in the Shale at Goldfield", M&SP 110 (1915) 995-6.

Winchell, H. V., "Notes on Goldfield, Nev.", Am. Geol. 35 (1905) 382-5.

Spurr, J. E., "Notes on the Geology of the Goldfield District, Nevada", USGS, B 225 (1904) 118-9.

———, "The Ores of Goldfield, Nevada", USGS B 260 (1905) 132-9; Abstract, M&SP 90 (1905) 393-5.

Mining

Barbour, P. E., "Cost of Shaft Sinking at Goldfield", E&MJ 84 (1907) 1106.

———, "Details of Wooden Headframe", E&MJ 92 (1911) 344.

———, "Bailing Bucket Dumping Device", E&MJ 97 (1914) 1099.

———, "Hoist and Compressor Power Curves", E&MJ 97 (1915) 988-9.

Collins, E. A., "A Prospecting Shaft in the Goldfield District, Goldfield, Nev.", Inst. M. & M. T 15 (1906) 540-2.

E&MJ, "Ore Chute Construction", 90 (1910) 307.

———, "Stoping at Goldfield Consolidated", 89 (1910) 952.

———, "A Sanitary Underground Latrine", 91 (1911) 556-7.

———, "Shaft Sinking, Goldfield Merger Mines", 94 (1912) 402.

———, "Shaft Sinking at Goldfield", 94 (1912) 691.

———, "Small Timber Headframes", 101 (1916) 645.

Prince, E., "Electric Power for Goldfield", Eng. News, July 6, 1915; Abstract, E&MJ 80 (1905) 111.

Rice, C. T., "Goldfield, Nevada", E&MJ 82 (1906) 339-342.
———, "Leasing at Goldfield, Nevada", E&MJ 82 (1906) 482-4.
———, "Fast Driving at the Goldfield Consolidated Mines", E&MJ 90 (1910), 1246-7.
———, "Development of Goldfield Mines", E&MJ 91 (1911) 119-122.
———, "Goldfield Consolidated Fire Equipment", E&MJ 91 (1911) 311-2.
———, "Mining Methods at Goldfield", E&MJ 92 (1911) 797-802.
———, "Mining Costs at Goldfield, Nevada", E&MJ 92 (1911) 985.
———, "A System of Keeping Stoping Costs", E&MJ 92 (1911) 1032-4.

Metallurgy

Barbour, P. E., "The Goldfield Consolidated 600-Ton Mill", E&MJ 86 (1908), 467-474, 680.
———, "Foundations of the Goldfield Consolidated Mill", E&MJ 87 (1909) 1173-6.
———, "Hydraulic Cone Classifiers", E&MJ 98 (1914) 438.
Bosqui, F. L., "Milling Versus Smelting in the Treatment of Tonopah-Goldfield Ores", M&SP 92 (1906) 217; Also in "Recent Cyanide Practice", (1907) 33-6.
———, "Ore Treatment at Combination Mine, Goldfield", M&SP 93 (1906) 431-5, 451-4; Also in "Recent Cyanide Practice", 51.
Church, J. A., "Goldfield Consolidated Sampling Mill", E&MJ 87 (1909) 311-2.
E&MJ, "The Goldfield Consolidated Mill", 86 (1908) 416-7.
———, "Deister No. 3 Concentrating Table", 86 (1908) 610.
———, "Canvas Tables of the Combination Mill", 94 (1912) 207-8.
———, "Temporary Traveling Cableway for Tailing Reclamation", 100 (1915) 673.
Hutchinson, J. W., "Operation of Goldfield Consolidated Mill, Nevada", M&SP, 102 (1911) 616-620, 652-4, 686-8, 716-9, 782-4.
———, "Lime in Cyanidation", E&MJ 94 (1912) 170.
———, "Treatment of Concentrate at the Goldfield Consolidated Mill", M&SP, 106 (1913) 170-2; 204-9; Discussion, H. Haas, 386.
Lamb, M. R., "Stamp-Mill and Cyanide Plant of Combination Mines Co. at Goldfield, Nevada", E&MJ 81 (1906) 1236-8.
Leaver, E. S., "Milling Practice in Nevada-Goldfield Reduction Works", M&SP, 97 (1908) 254-5; "More Recent Cyanide Practice", (1910) 198-200.
Martin, A. H., "Goldfield Consolidated Mill", MS Feb. 11, 1909.
———, "Milling Conditions in Goldfield District, Nev.", M&EW, Mar. 13, 1909.
———, "The Flotation Process at Goldfield, Nevada", M&EW 44 (1916) 1041-2.
M&SP, "Operating Costs at the Goldfield Consolidated Mill", 104 (1912) 137-140
MS, "Goldfield Consolidated Mill", Many 7, 1908.
Morris, H. G., "Equipment and Practice at Florence-Goldfield Mill", E&MJ 89, (1910) 356-8.
Parsons, A. B., "Cyanide Treatment of Concentrates", E&MJ 91 (1911) 368-370.
Rice, C. T., "Milling at Florence-Goldfield", E&MJ 91 (1911) 761-2.
Rickard, T. A., "Goldfield, Nev., V. Metallurgical Development", M&SP 96, (1908) 840-3.
Shaw, E. S., "The Goldfield Consolidated Bullion Refinery", E&MJ 91 (1911), 799.
———, "A Large Shipment of Gold", E&MJ 91 (1911) 903.
Sinclair, J., "Tailings Reclaimed by Cableway at Goldfield, Nevada", M&EW, 43 (1915) 643-4.
Smith, A. M., "Vacuum Slime Filters at Goldfield", M&SP 99 (1909) 65.
Tyssowski, J., "Goldfield Consolidated Mill Operations", E&MJ 89 (1910) 1230.

Miscellaneous

Barbour, P. E., Correspondence, "The Labor Troubles at Goldfield", E&MJ, 85 (1908) 124-5.
E&MJ, "Map of the Goldfield District", 87 (1909) 1230-1.
———, "Flow Sheet of Reports", 89 (1910) 1217.
———, Editorial, "Goldfield Consolidated", 90 (1910) 2; "Safety in Mines and Mills", 11-12.
———, "Listing of Goldfield Consolidated", 90 (1910) 256.
———, "Inspection Department of the Goldfield Consolidated", 91 (1911) 408.
———, "Goldfield Consolidated Cost Curves", 96 (1913) 124-5, 132, 150-1.
Farrell, M. A., "The Goldfield School of Mines", M&SP 119 (1919) 821-2.
Locke, A., "The Cost of the Goldfield Mining Boom", M&SP 101 (1910) 541-3.
M&SP, Editorial, "Gambling in Mining Shares", 93 (1906) 700-1; Discussion, F. S. Harris, "Goldfield Mines", 94 (1907) 16.
———, Editorial, "Rights of Shareholders", 98 (1909) 449; Discussion, J. H. MacKenzie, "Goldfield Consolidated", 98 (1909) 513-4.
———, "Apex Litigation at Goldfield", 110 (1915) 996-7.
Rickard, T. A.,"Goldfield, Nevada, IV—Rich Ore and its Moral Effects", M&SP. 96 (1908) 774-7.
USGS Lida and Goldfield Special topographic maps.

GOLD MOUNTAIN see DIVIDE and TOKOP

GOOD HOPE

Silver

Location. The Good Hope District is 7 m. S. of Piper Peak on the W. flank of the Silver Peak Range in S. W. Esmeralda Co. The Palmetto District adjoins it on the S. E.

History. Ore from this district was formerly worked for silver at a little furnace on Furnace Creek, Cal.

Geology. The country rock consists of slates which are probably of Ordivician age, according to Turner, and contain an interbedded layer of quartzite. At the junction of this quartzite with the underlying slate, silver ore occurs in the form of quartz containing galena and showing copper stains.

Bibliography. USGS Lida and Silver Peak topographic maps.

Spurr, J. E., "Ore Deposits of the Silver Peak Quadrangle, Nev.", USGS, PP 55 (1906) 85.

GREEN MOUNTAIN see SYLVANIA

HORNSILVER (Lime Point)

Silver, Gold, Lead, Zinc, Copper

Location. The Hornsilver District is located at Hornsilver in S. Esmeralda Co. Stonewall on the T. & T. R. R. is 16 m. N. E. and Goldfield on the T. & T. R. R. and T. & G. R. R. is 30 m. N. The district was formerly known as Lime PPoint. It adjoins the Tokop District on the N. and was considered a part of that district in the early days. Hornsilver is 5,900 ft. above sea-level, and the hills to the S. of the town rise from 500 ft. to 1,000 ft. higher.

History. The Tokop District, of which the Hornsilver District formerly constituted a part, was discovered by Thomas Shaw in 1866. In the eighties, ore was hauled from Lime Point to a mill near Lida, but the camp was later abandoned. The camp was rejuvenated at the time of the Goldfield boom, and work on the Great Western Mine began in 1905. The Orleans

Mine was opened by th Champ d'Or M. Co., a French company, about 1912, and is now being operated under lease and bond by the Orleans Hornsilver M. Co.

Production. From 1907 to 1921, the Hornsilver District produced 31,702 tons of ore containing $203,654 in gold, 411,788 ozs. silver, 7,146 lbs. copper, 37,495 lbs. lead, and 17,680 lbs. zinc, valued in all at $514,942, according to Minearl Resources of the U. S. Geol. Survey.

Geology. The rocks of the district are Cambrian limestones and calcareous shales intruded in places and more or less metamorphosed by masses of granite, according to Ransome. The principal veins constitute a parallel system cutting the bedding of the Cambrian sedimentary rocks and mainly in the shale. They are parallel to some fine-grained diorite dikes. The quartz of the veins is frequently crushed and recemented by limonite or chalcedonic quartz stained by iron oxide, according to Turner, while manganese oxide is present in appreciable amount. Silver values predominate in the upper portion of the veins, and gold in depth. Silver occurs as hornsilver, and sparingly as bromyrite, and gold is native. Lead is present as galena and cerussite.

Great Western and Orleans Mines. The principal mines are the Great Western and the Orleans, which are credited with productions of $500,000 and $450,000 respectively. The Great Western was purchased at a receiver's sale in 1922 by C. A. Stoneham of New York. The property consists of 17 claims developed by a 900-ft. shaft with 3,000 ft. of lateral workings and equipped with an 8½ m. pipe line and 150-ton cyanide plant. The Orleans Mine is being worked by the Orleans Hornsilver M. Co. of which A. I. D'Arcy is Pres., R. H. Downer, V. P., and H. D. King, Tonopah, Sec.-Treas. The property consists of 5 claims developed by a 200-ft. shaft and a 580-ft. incline.

Other Mines. Six other companies have recently been active. The **Goldsmith Divide M. Co.** has acquired 5 claims which it is working. The **Hornsilver American Co.,** controlled by E. G. Reinert, Opera House Bldg., Denver, owns 8 claims developed by shafts 200 ft. and 140 ft. deep. The **Hornsilver-Mizpah M. Co.,** controlled by J. Pincolini of Reno and A. Borcherding of Goldfield, owns 3 claims. The **Lucky Boy Divide M. Co.,** of which H. McNamara, Tonopah, is Pres., E. Marks, V. P., and H. E. Dugan is Sec., owns 2 claims adjoining the Great Western on the E. and opened by 3 prospect shafts. D. F. Murphy is Pres.-Mgr. of the **Reorg Rosetta Divide M. Co.** which is operating its property consisting of 5 claims. The **Reorg. West Divide M. Co.,** of which W. E. Sirbeck, Goldfield, is Pres., and J. K. Turner is Cons. Eng., owns 2 claims developed by 50-ft. and 70-ft. shafts.

Bibliography.

MR1907 I 354 | MR1912 I 794 | MR1916 I 478
MR1908 I 483 | MR1913 I 823 | MR1917 I 276
MR1909 I 405 | MR1914 I 681 | MR1918 I 238
MR1910 I 514 | MR1915 I 633 | MR1919 I 393
MR1911 I 675 | | MR1921 I 383

USGS Lida topographic map.

Ball 285 62.

Hill 507 207.

Ransome, F. L., "The Hornsilver District, Nev.", USGS B 380 (1909), 41-3.

Turner, J. K., "The Hornsilver Mining District", M&SP 124 (1922) 93-4.

Weed MH 1213 Gold Mt. M. & M. Co. 1215 Goldsmith Divide M. Co.
1225 Hornsilver American Co., Hornsilver Apex M. Co., Hornsilver May Co., Hornsilver New Orleans M. Co., Hornsilver Red Top M. Co.
1243 Lucky Boy Divide M. Co.
1298 Orleans Hornsilver M. Co.
1314 Reorg. Rosetta Divide M. Co.
1333 Silver Mines Corp. 1342 Southwestern Ms. Co.

KLONDYKE (Southern Klondyke)
Silver, Lead, Gold, Copper, Turquoise

Location. The Klondike District lies in the Southern Klondyke Hills in E. Esmeralda Co. on the Nye Co. border. It adjoins the Divide District on the S. E. and is sometimes considered a section of that district. The Klondyke District is 14 m. S. of Tonopah which is on the T. & G. R. R. and a few miles E. of Klondyke, a station on the same railroad. The altitude of the camp is about 5,500 ft. and the hills rise but a few hundred feet above it.

History. The district was discovered by Court and Bell in 1899. While on his way to this camp in 1900, Jim Butler discovered Tonopah.

Production. Some $50,000 was produced from one property in the district up to 1905, and the district has made annual productions since. From 1908 to 1921 it produced 11,236 tons of ore containing $36,467 in gold, 277,466 ozs. silver, 3,472 lbs. copper, and 217,147 lbs. lead, valued in all at $263,700 according to Mineral Resources of the U. S. Geol. Survey.

Ore. The country rocks of the Klondyke District are Cambrian sediments which have been intruded by granite that is probably of Cretaceous age and are capped in places by Tertiary volcanic rocks. According to Spurr, the Bell and Court vein cuts at a small angle across the sedimentary rocks and is parallel to a granite dike. Silver, chiefly in the form of chloride, predominates, with minor values in gold. The vein filling is mainly quartz containing small amounts of a copper-antimony-silver compound resembling stetefeldtite, galena, and pyrite; together with their oxidation products. Other minerals present in small amounts are siderite, calcite, hematite, and wad. Pyrite-bearing quartz veins of irregular character occur at the granite-limestone contact, particularly in the granite; and the rhyolite-limestone contact is also characterized by erratic quartz veins.

Turquoise. The Smith turquoise mine is in the Cambrian sediments ½ m. S. of Klondyke. The turquoise occurs in seams and veinlets filling joints and fractures in black slaty jasperiod near overlying rhyolite. Black matrix turquoise was shipped by the California Gem Co. from this deposit in 1909.

Mines. The **Original Klondyke Divide M. Co.** owns the mine which has made the largest production in the camp. It is incorporated in Nevada with a capital stock of 1,500,000 shares of 10 cents par value of which 650,000 shares remain in the treasury. The property consists of 12 claims developed by 25 tunnels and shafts no one which reaches a depth in excess of 200 ft. H. McNamara, Tonopah, is Pres., E. Marks is V. P. and H. E. Dugan is Sec. The **Knox Divide M Co.,** of which C. E. Knox, Tonopah, is Pres., controls 4 claims belonging to the Golden State Divide M. Co. adjoining the Original Klondyke, besides owning the Knox Group of 5 claims in the Divide District. The Klondyke property has made a considerable production and is developed by a 130-ft. shaft.

Bibliography. MR1905 268 | MR1906 294 | MR1908 I 483 | MR1909 I 405 | II 786-7 Turquoise | MR1910 I 523 | MR1912 I 804 | MR1913 I 832 | MR1914 I 696 | MR1915 I 643 | MR1916 I 487 | MR1917 I 276 | MR1918 I 238 | MR1919 I 393 | MR1920 I 323 | MR1921 I 383

USGS Lida topographic map.

Ball 285 58.

Ball 308 28, 32, 33, 44-9, 77-81.

Hill 507 210.

Suprr, J. E., "Ore Deposits of Tonopah and Neighboring Districts, Nevada", USGS B 213 (1903) 86-7.

———"The Southern Klondyke District, Esmeralda County, Nevada", EG, 1 (1906) 369-382.

———"Genetic Relations of Western Nevada Ores", T AIME 36 (1906), 400.

Weed MH 1183 Desert Klondyke M. Co. 1208 Golden State Divide M. Co.
1214 Gold Ridge Divide M. Co. 1236 Keystone Divide M. Co.
1237 Klondyke-Portland M's. Co.
Klondyke Standard Divide M. Co.
1237-8 Knox Divide M. Co. 1243 Lucky Boy Divide M. Co.
1249 Marinette M. & M. Co. 1335 Silver State Divide M. Co.
1297 Original Klondyke Divide M. Co.

LIDA (Tule Canyon, Alida Valley)

Silver, Lead, Gold, Copper

Location. The Lida District is at Lida in the Silver Peak Range in S. Esmeralda Co. Goldfield on the T. & G. R. R. and T. & T. R. R. is 30 m. N. E. The elevation of Lida is 6,100 ft. The district was formerly known as the Alida Valley District; and its S. section in Tule Canyon is sometimes called the Tule Canyon District. The Lida District adjoins the Railroad Springs District on the S., the Palmetto District on the E., and the Sylvania District on the N. E.

History. The district was discovered in 1871 by Scott and Black. Hiskey & Walker erected an 8-stamp mill there in 1873 and a second mill of 5-stamps was built in the district later. Interest revived in recent years and small annual productions have been made. The Ingalls Mine is now being worked under lease and bond.

Production. In 1871, 2½ tons of ore were shipped to Columbus and netted $622 in silver and gold. In 1872 and 1873, the mill treated 657 tons recovering $162,646 in bullion. From 1908 to 1921, the camp produced 7,096 tons of ore containing $13,633 in gold, 114,094 ozs. silver, 79,602 lbs. copper, and 406,129 lbs. lead, valued in all at $147,191, according to Mineral Resources of the U. S. Geol. Survey.

Geology. The country rock consists of Cambrian limestones and shales, according to Ball, which have been intruded by granite-porphyry and by dikes of diorite porphyry. The ore deposits are partly quartz veins in the Cambrian limestone, and partly impregnations of the country rock; and the oxidized ores are largely replacements of the limestone. The sulphides present are galena, sphalerite, pyrite, and chalcopyrite.

Bibliography. B1867 419 | R1871 183 | R1872 174, 175 | MR1906 294 | MR1907 I 354 | MR1908 I 483 | MR1915 I 633 | MR1916 I 479 | MR1917 I 276

R1874 281-2
SMN1871-2 41-2
SMN1873-4 22-4
SMN1875-6 36
MR1909 I 405
MR1910 I 514
MR1911 I 676
MR1912 I 794
MR1913 I 823
MR1918 I 238
MR1919 I 393
MR1920 I 323
MR1921 I 383

USGS Lida topographic map.

Ball 285 61-2.
Ball 308 25, 31, 45, 55-65.
Davis 587-8.
Hill 507 207-8.
M&SP, Editorial on Tule Canyon gold that cannot be found by assay, 120 (1920) 591.
Root, W. A., "The Lida Mining District of Nevada", M&EW 31 (1909), 123-5.
Spurr 208 184-6 Silver Peak Range.
Stuart NMR 46, 57-9.
Thompson & West 416-7.
Weed MH 1143 Benton-Radford-Smith Group.
1227 Indian Springs M. Co.
1332 Silver Hills Nevada Ms. Co.
1342 South Silver Hills M. Co.
1348 Syncline Gold-Silver-Copper M. Co.

LIME POINT see HORNSILVER

LONE MOUNTAIN (West Divide)

Silver, Lead, Copper, Gold, Zinc, Barite

Location. The Lone Mountain District is situated in the neighborhood of Lone Mt. in central Esmeralda Co. The T. & G. R. R. describes a semicircle about the district so that the S. E. part of it lies to the N. W. of Klondyke which is on that ralroad and the N. W. part S. E. of Blair Jct. Lone Mt. Peak is 9,121 ft. high, while its foothills reach a maximum elevation of 7,600 ft. The district has sometimes been called the West Divide District.

Production. The production of the Lone Mountain District from 1902 to 1921 was 6,333 tons containing $8,236 in gold, 838,123 ozs. silver, 88,981 lbs. copper, 3,223,476 lbs. lead and 15,402 lbs. zinc, valued in all at $585,250, according to Mineral Resources, U. S. Geol. Survey.

Geology. Lone Mt. is a batholith of granite which has intruded Cambrian limestones and shales, according to Ball. Both granite and sediments are cut by diorite. The ore deposits are veins and replacements in the Cambrian rocks. In the neighborhood of the Gen. Thomas Mine, diorite porphyry has been injected into these beds. The oxidized ore at this mine consists of cerussite, malachite, azurite, and chrysocolla in limestone heavily stained with limonite and associated with gypsum. Galena and pyrite were also observed. The American Barium Co. has mined barytes in the N. W. part of the district near Blair Jct.

Mines. The **Gen. Thomas Mine** which is owned by the Gen. Thomas of Tonopah Co. controlled by the John H. Dern Estate of Salt Lake City, was leased to Reno men in 1922. Three other companies are active in the district. The **Electric Gold Ms. Co.,** managed by F. E. Horton, Tonopah, owns the Weepah group on the S. W. slope of Lone Mt., developed by 4 shafts of which the deepest is 170 ft. and equipped with 15 hp hoist. The **Interstate Cons. Ms. Co.** at Camp Harding, 12 m. from Klondyke station, was incorporated in Arizona in 1909 with a capital stock of 2,500,000 shares of $1 par value. F. Harding is Pres.-Gen. Mgr.; J. L. Simmons, V. P.; and R. W. Force, Sec.-Treas. The property consists of 41 claims developed by a 127-ft. shaft and two 327-ft. tunnels and having 1 m. of workings. The **Occident Divide M. Co.** of which R. S. Wilber is Pres.; F. E. Schultz, V. P.; R. T. Armstrong, Sec.-Treas.; and H. F. Bruce, Cons. Eng.; owns 8 claims developed by a 70-ft. shaft and tunnels.

Bibliography.

MR1905 268	MR1913 I 822	MR1920 I 323
MR1906 294	MR1914 I 681	MR1921 I 383
MR1908 I 483	MR1915 I 633	MR1907 II 688 Barite
MR1909 I 405	MR1916 I 478	MR1915 II 175-6 Barite
MR1910 I 514	MR1917 I 276	MR1916 II 250 Barite
MR1911 I 676	MR1918 I 238	MR1920 II 194 Barite
MR1912 I 794	MR1919 I 393	MR1921 II 129 Barite

USGS Tonopah and Lida topographic maps.

Ball 285 57-8.

Ball 308 28, 31, 46, 51-55.

Hill 507 208.

Lake, A., "The Lone Mt. District, near Tonopah, Nevada", M&SP 88, (1904) 246-7.

Spurr, J. E., "The Ore Deposits of Silver Peak Quadrangle, Nevada", USGS B 225 (1904) 114-5.

———"Ore Deposits of the Silver Peak Quadrangle, Nevada", PP 55, (1905) 75-83.

———"Genetic Relations of Western Nevada Ores", T AIME 37 (1906), 396-7.

Stuart NMR 93.

Weed MH 1190 Electric Gold Ms. Co. 1228 Interstate Cons. Ms. Co. 1276 Nevada Croesus S. Ms. Co. 1295 Occident Divide M. Co. 1333 Silverpah Cons. Ms. Co. Silverpah Ex. Ms. Corp. 1357 Tonopah Gold Hill Dev. Co. 1271 West Divide Ex. M. Co. West Divide M. Co. 1377 Wilson Divide Ms. Cons. Co.

MONTEZUMA

Silver, Gold, Lead, Copper, (Bismuth)

Location. The Montezuma District is at Montezuma on the W. slope of Montezuma Peak in the Montezuma Hills in E. Esmeralda Co. Goldfield on the T. & G. R. R. and the T. & T. R. R. is 9 m. E.; and the Montezuma District adjoins the Goldfield District on the W. Montezuma Peak has an elevation of 8,426 ft.

History. The Montezuma District was discovered by Nagle, Carlyle, and Plunkett on May 24, 1867, and organized shortly afterwards. A 10-stamp mill was brought from Yankee Blade to Montezuma in 1870 but ouly operated about 4 mos. The district continued active until 1887, up to which time it produced about $500,000 mainly in silver but with some gold. In 1905, the district was again prospected and since then has been making small intermittent shipments.

Geology. Cambrian limestones, shales, and quartzites make up the principal mass of the Montezuma Hills. They have been intruded by granite, quartz-monzonite, and diorite; and are capped in places by Tertiary volcanic rocks and interbedded Siebert lake beds. The orebodies consist of veins in limestone and shale and of replacements in limestone which are sometimes at quartz-monzonite contacts. The gangue is chiefly quartz with a little calcite and kaolin, and in some cases altered limestone. At the surface the ore minerals are cerussite, copper, malachite, azurite, manganese dioxide, and limonite; while in depth these minerals give place to galena, chalcocite, and pyrite. The principal values are in silver which, according to a letter from E. S. Giles, is in the form of chlorides at the surface and of argentite in depth. Giles also reports the presence of jamesonite in the Arizona Mine of the Montezuma Silver Ms. Corp., and of bismuth ore on the Bessler Bros. property.

Mines. The **Harmill Divide M. Co.,** of which G. B. Hartley, Goldfield, is Pres.-Gen. Mgr., has stopped work upon its Divide property and is developing 8 claims in the Montezuma District. The **Montezuma Silver Ms. Corp.** is a Nevada corporation, with a capital stock of 1,000,000 shares of $1 par value of which 150,000 remain in the treasury. H. B. Nedham, Goldfield, is Pres., and E. S. Giles, Mine Mgr. The Company owns 9 claims including the old Arizona, Bullion, and Caracas Mines which were big producers in the early days, and is planning further active development. The **Washington Montezuma M. Co.,** of which R. H. Kiehm, Postal Telegraph Bldg., Chicago, is Pres., owns 2 claims which it has recently been developing.

Bibliography.

R1870 130
R1872 174
SMN1869-70 88
SMN1871-2 40
SMN1873-4 78
SMN1875-6 108
MR1905 267-8
MR1906 295
MR1908 I 484
MR1910 I 514
MR1912 I 794
MR1915 I 633
MR1916 I 479
MR1917 I 276
MR1918 I 238
MR1919 I 394
MR1920 I 323

USGS Lida topographic map.

Ball 285 58-9.
Ball 308 63-4.
Davis 857.
Hill 507 208.
Spurr 208 184-6.
Stretch, R. H., "The Montezuma District, Nevara", E&MJ 78 (1904) 5-6.
Thompson & West, 417.
Weed MH 1221 Harmill Divide M. Co. 1261 Montezuma Ms. Corp. 1370 Washington Montezuma M. Co.

ONEOTA see BUENA VISTA, MINERAL COUNTY

ORIENTAL WASH see TOKOP

PALMETTO (Windypah)
Silver, Gold, Lead

Location. The Palmetto District lies N. E. of Palmetto in the Palmetto Mts. section of the Silver Peak Range in S. Esmeralda Co. Goldfield on the T. & G. R. R. and T. & T. R. R. is 42 m. N. E. Palmetto is 6,158 ft. in elevation; the Palmetto Mine, 7,641 ft.; and Palmetto Peak, 8,885 ft. The Railroad Springs and Lida Districts adjoin the Palmetto District on the E.; and the Good Hope District adjoins it on the N. E. The N. W. part of the district is sometimes called the Windypah District.

History. The Palmetto District was discovered by Bunyard, Israel, and McNutt in 1865. A 12-stamp mill was built and put in operation that same year, but the ore did not last long. The McNamara Mine was located in 1880. In 1903, J. G. Fesler discovered the Windypah section of the Palmetto District, and considerable prospecting ensued. A little lead ore containing gold and silver was shipped in 1908. The Allied M. & M. Co. built a 50-ton concentrating mill on its property in 1920, and lessees have been working on the ground recently.

Production. The Palmetto Mine is credited with a production of $6,500,000 in silver, and the McNamara Mine made a considerable production of lead-silver ore in its early days.

Geology. The country rocks are Paleozoic sediments intruded by granite; and the ore occurs in veins and contact-metamorphic deposits. The McNamara lead-silver vein is said to be from 6 ft. to 30 ft. in width, traceable for more than 1 m., and to have an alaskite hanging wall and a limestone footwall.

Mines. The **Allied M. & M. Co.**, which owns the McNamara Mine, was incorporated in Montana in 1919 with a capital stock of 5,000,00 shares of 10 cts. par value, of which 2,650,000 have been issued. A. J. Heinecke, Lewiston, Mont., is Pres., E. G. Ivins, Sec., and D. Trepp, Goldfield, Mgr. The property consists of 16 claims developed by ½ m. of tunnels and workings, and equipped with a 50-ton concentrating mill. The **Palmetto Cons., Inc.**, owns the Palmetto Mine. The company was incorporated in Nevada in 1916 with a capital stock of 1,500,000 shares of which 1,150,000 have been issued. The stock is listed on the New York Curb, the home office is at 57 Post St., San Francisco. E. Cebrian, 57 Post St., San Francisco, is Pres., and L. Alegria, Sec. The property consists of 25 claims developed by shafts the deepest of which is 400 ft.

Bibliography. B1867 319, 419 SMN1866 44 MR1905 268
SMN1867-8 96 MR1906 295
SMN1869-70 104-5 MR1907 I 355
SMN1871-2 43 MR1908 I 484
USGS Lida and Silver Peak topographic maps.
Davis 857.
Hill 507 208, 210.
Spurr, J. E., "Ore Deposits of Silver Peak Quadrangle, Nev.", USGS B, 225 (1904) 115-6.
——"Ore Deposits of the Silver Peak Quadrangle, Nev.", USGS PP 55, (1916) 85-96.
Stuart NMR 25, 46.
Thompson & West 417.
Weed MH 1129 Allied M. & M. Co. 1300 Palmetto Cons.. Inc.

RAILROAD SPRINGS

Gold, Silver, Copper

Location. The Railroad Springs District is at Railroad Springs on the N. slope of the Silver Peak Range. Goldfield on the T. & G. R. R. and the T. & T. R. R. is 25 m. by road N. E. The Railroad Springs District adjoins the Palmetto District on the N. E. and the Lida District on the N., and is sometimes considered as a section of the latter district.

History. Small shipments of copper ore have been made from the Marvel Group, and in 1908 several small lots of silver-gold ore were shipped from the Silver Coin property.*

Geology. The country rocks of the district are Cambrian limestones and calcareous shales which have been intruded by diorite dikes that are probably

*The data for this article was obtained from a report by J. K. Turner on the Silver Coin property and from letters from E. S. Giles and B. Gill.

Cretaceous, and capped in places by Tertiary rhyolite. The veins on the Silver Coin property are principally in the shale, cutting its bedding, and parallel to fine-grained diorite dikes. Their filling consists of quartz and iron oxides and sulphides and the valuable constituents are silver and gold with lead and traces of copper. At the Gold Hill Mine the values are mainly in gold which occurs with limonite in a porous quartz vein in limestone and shale.

Mines. Three companies are at present active. The **Goldfield Dev. Co.,** of which C. Barnes, Goldfield, is Pres.; A. I. D'Arcy, V. P.; H. G. McMahon, Sec.-Treas.; and E. Burt, Mine Supt.; owns the Gold Hill Group of 8 claims, developed by a 500-ft. and an 1,100-ft. tunnel and equipped with small hoist and compressor. The **Goldfield Jackpot M. Co.,** owns the Park Hill Group adjoining the Gold Hill Group. The **Grandma Cons. Ms. Co.,** which has important holdings in Goldfield, acquired 3 of the 10 claims belonging to the Silver Coin M. Co. in the Railroad Springs District under 5-yr. lease and bond in 1922. C. S. Sprague is Pres.; J. K. Turner, V. P.; and B. Gill, Sec.-Treas.

Bibliography. USGS Lida and Silver Peaks topographic maps.

Ball 308 55-63 Northeastern Extension of Silver Peak Range.
Spurr 208 184-6 Silver Peak Range.
Weed MH 1211 Goldfield Dev. Co. 1312 Reorg. Cracker Jack M. Co.
1217 Grandma Cons.M.Co. 1325 Sandstorm-Kendall C.M.Co.
Grandma M. Co. 1329 Silver Coin M. Co.

RED MOUNTAIN see SILVER PEAK AND RED MOUNTAIN

SILVER PEAK AND RED MOUNTAIN

Gold, Silver, Lead

Location. The Silver Peak and Red Mountain District is located in the E. part of the Silver Peak Range W. of Silver Peak in central Esmeralda Co. Blair Jct. on the T. & G. R. R. is 20 m. N. Silver Peak has an elevation of 4,307 ft. and Red Mt. in the W. part of the district reaches an altitude of 8,940 ft. The W. section of the district was formerly called the Red Mountain District. Silver Peak Marsh adjoins the Silver Peak District on the E. and the Argentite District adjoins it on the S. W.

History. The Red Mountain section was discovered in 1863 and a 3-stamp mill was erected there in 1864. The Silver Peak section to the E. was discovered by Robinson Bros. in 1864, and a 10-stamp mill was brought there from the Reese River District and set up in 1865. The Great Salt Basin M. & M. Co. acquired these mills together with the principal mines and in 1867 erected a 30-stamp mill at Silver Peak which operated for 2 years. Subsequent to 1870, a little intermittent mining and milling was conducted in the district, but no important operations took place in the district until the Pittsburgh Silver Peak Gold M. Co. acquired the Mohawk, Alpine, Silver Peak, Drinkwater and Mary mines in 1906 and constructed a branch railroad line 17½ m. long from the T. & G. R. R. to Blair, a few miles N. of Silver Peak. In 1907, this company put in operation a 100-stamp cyanide mill at Blair; and in 1910 absorbed the Silver Peak—Valcalda Mine and added 20-stamps to the mill at Blair. This was the largest stamp mill in Nevada and the Pittsburgh Silver Peak was for years the largest producer of low-grade ores in the state. While there were other producers in the Silver Peak District, the bulk of the production came from the Pittsburgh Silver Peak, and activities in the district were practically suspended when that company shut down in 1915 owing to the lack of ore, moved its mill to California, and tore up the branch railroad to Blair. Recently, the Lucky Boy Divide M. Co. obtained a lease upon the Pittsburgh Silver Peak property and has developed ore which it plans to treat in a 100-ton cyanide mill at Silver Peak. Tom Fisherman, a Shoshone Indian, located the Nivloc Mine 8 m. W. of Silver Peak in 1907. The mine is owned by Mr. Colvin of Chicago, and is being explored under the management of F. Vollmer of Silver Peak.

Production. Spurr estimates the production of the district prior to 1906 at $1,418,000. Subsequent production by the Pittsburgh Silver Peak Gold M. Co. amounted to about $7,000,000. After that company had abandoned operations, J. B. Fanchini, followed by B. A. Rives produced more than $100,000 from leasing operations on the Mary Mine, the ore extracted being hauled to Silver Peak and there milled.

Geology. The country rock of the Silver Peak District consists of Paleozoic sedimentary rocks which have been intruded by alaskite which is probably Cretaceous. The ore deposits are veins in which the ore occurs in the form of overlapping quartz lenses. The values in the quartz are chiefly in gold, although rich silver ores were mined in the early days, and a little silver is nearly always present. The gold is finely disseminated in a native state, and also occurs in the small amount of pyrite and galena scattered through the quartz.

Bibliography.

B1866 126, 134	MR1905 268	MR1913 I 823-4
B1867 335,338-9,402,419	MR1906 294	MR1914 I 582
R1868 116	MR1907 I 354-5	MR1915 I 633
R1869 193-4	MR1908 I 486	MR1916 I 479
SMN1866 41-4	MR1909 I 407	MR1917 I 276
SMN1867-8 95-6	MR1910 I 514	MR1918 I 238-9
SMN1869-70 106-7	MR1911 I 576	MR1919 I 394
SMN1871-2 41	MR1912 I 794	MR1920 I 323
		MR1921 I 383

USGS Lida and Silver Peak topographic maps.

Davis 856.

Hanson, H., "Pittsburgh Silver Peak Mill", M&M 29 (1909) 569-573.

Hastings, J. B., "Are the Quartz Veins of Silver Peak, Nevada, the Result of Magmatic Segregation?", T AIME 36 (1906) 647-654.

Hill 507 209.

Martin, A. H., "Pittsburgh Silver Peak Mill", M&EW 33 (1910) 461.

——— "Silver Peak Mill", MS Nov. 18, 1909.

Smith, L., "Operation of the Pittsburgh-Silver Peak Mill", E&MJ 98, (1914) 595-9.

Spurr, J. E., "Ore Deposits of Tonopah and Neighboring Districts, Nevada", USGS B 213 (1902) 85-6.

——— "Ore Deposits of Silver Peak Quadrangle, Nevada", USGS B 225 (1904) 112-4.

——— "The Silver Peak Region, Nevada", E&MJ 77 (1904) 759-760.

——— "Ore Deposits of the Silver Peak Quadrangle, Nevada", USGS, PP 55 (1906) 34-74.

Spurr 208 184-6 Silver Peak Range.

Sutart NMR 46, 55-7.

Thompson & West 418.

Weed MH 1293 Nivloc M. Co., Nivloc Northern Ms. Co.
1305 Pittsburgh Silver Peak G. M. Co.

SILVER PEAK MARSH
Salt, (Borax, Potash)

Location. Silver Peak Marsh lies E. of Silver Peak in central Esmeralda Co. Blair Jct. on the T. & G. R. R. is 20 m. N. W. The marsh has an area of 32 sq. m. and is at an elevation of 4,340 ft. It occupies the lowest part of Clayton Valley which has a drainage area of 570 sq. m.

Geology. The marsh is a salt playa covered for the most part with a white crust of sodium chloride. The water level is near the surface, but the marsh is only covered with water after an exceptionally heavy rainfall. According to Dole, the formations to a depth of 50 ft. are chiefly salt clays and muds with layers of crystallized salt covered irregularly by gypsum-bearing clays. The sodium chloride is of high grade, and small amounts have been mined for local use. It is estimated that 15,000,000 tons of salt are within 40 ft. of the surface of the playa. Potash is not present in sufficient quantity or concentration to be commercially valuable. According to a letter from E. S. Giles, thin streaks of ulexite were discovered in the shales at the N. E. corner of the marsh in 1922.

Bibliography. B1866 95-6 MR1882 545
B1867 311 MR1917 II 92.
SMN1867-8 96 MR1917 II 423 Potash
USGS Lida topographic map.

Dole, R. B., "Exploration of Salines in Silver Peak Marsh, Nev.", USGS, B 530 (1913) 330-345.

Free, E. E., "The Topoographic Features of the Desert Basins of the U. S. with Reference to the Possible Occurrence of Potash", U. S. Dept. Agr. B 54 (1914) 32.

Phalen, W. C., "Salt Resources of the U. S.", USGS B 669 (1919) 142-4.

Spurr, J. E., "Ore Deposits of the Silver Peak Quadrangle, Nev.", USGS, PP 55 (1906) 158-9.

Young, G. J., "Potash Salts and Other Salines in the Great Basin Region", U. S. Dept. Agr. B 61 (1914) 6, 12, 19, 38, 39-43, 53.

SOUTHERN KLONDYKE see KLONDYKE

SYLVANIA (Green Mountain)
Silver, Lead

Location. The Sylvania District is situated in the Sylvania Mts. in S. Esmeralda Co. on the California border.

History. The district was discovered by Kincaid in 1870 and organized as the Green Mountain District in 1872, the name being changed to Sylvania the following year. A 30-ton lead smelting furnace was erected to treat the ores by Broder and Moffat in 1875 and operated for several years.

Geology. The chief country rocks are limestone and quartzite intruded by granite. The principal veins are in limestone and contain argentiferous galena and lead carbonates.

Bibliography. SMN1871-2 43 SMN1873-4 21-2 SMN1875-6 35-6
USGS Lida topographic map.

Davis 857.

Thompson & West 417-8.

TOKOP (Gold Mountain, Oriental Wash)
Gold, Silver, Lead, Copper

Location. The Tokop District is located at Tokop in S. Esmeralda Co. extending from the Nye Co. boundary on the E. to the California boundary at the head of Death Valley on the W. It embraces Gold Mt. on the E. and

Oriental Wash, sometimes considered as a separate district, on the W. The Hornsilver District which was formerly a part of the Tokop District adjoins it on the N. Bonnie Clare on the T. & T. R. R. is 15 m. E. Tokop has an elevation of 6,850 ft. and Gold Mt. of 8,150 ft.

History. The district was discovered by Thomas Shaw in 1865 but did not attract attention until he found the rich Oriental Mine in 1871. The ore of the Oriental Mine was worked in a 6 ft. arrastra. Several mills were built in the district in the early days and the concentrates hauled to Austin and Belmont. Tokop was recently rejuvenated and produced again from 1905 to 1919.

Production. The early production of the Tokop and Hornsilver Districts together is estimated at $500,000. The production of the Tokop District from 1905 to 1921 as recorded by Mineral Resources, U. S. Geol. Survey, was 9,958 tons containing $92,337 in gold, 46,898 ozs. silver, 7,280 lbs. copper, and 174,203 lbs. lead, valued in all at $133,112.

Geology. The country rocks of the Tokop District are Cambrian sediments which have been intruded and metamorphosed by biotite granite which is probably of Cretaceous age. The granite contains inclusions of quartz-monzonite and both granite and sedimentary rocks are cut by diorite porphyry dikes. In places, the remains of rhyolite and basalt flows of Tertiary age rest unconformably upon the older rocks, according too Ball. Gold ores containing subordinate silver values are found in veins traversing the granite and metamorphosed Cambrian sediments, according to Ransome. The vein material is mainly oxidized but pyrite, chalcopyrite, galena, and tetrahedrite occur in the filling. The gangue is quartz, stained with iron and manganese oxides.

Bibliography.

R1872 174, 175, 176	MR1905 267	MR1913 I 823
R1873 232-3	MR1906 295	MR1914 I 681
R1874 281, 282-3	MR1907 I 355	MR1915 I 633
	MR1908 I 482	MR1916 I 478
SMN1871-2 42-3	MR1909 I 405	MR1917 I 276
SMN1873-4 23	MR1910 I 514	MR1918 I 239
	MR1911 I 672	MR1919 I 394
	MR1912 I 792	MR1921 I 383

USGS Lida topographic map.

Ball 285 62-5.

Ball 308 45, 182-195.

Davis 856-7.

Hill 507 208, 210.

Ransome, F. L., "Preliminary Account of Goldfield, Bullfrog, and Other Mining Districts in Southern Nevada", USGS B 303 (1907), 80-83.

Stuart NMR 25, 46, 59.

Thompson & West 416.

Weed MH 1261 Mt. Vernon M. & M. Co. 1275 Nevada Co-Oper. M. Co. 1370 Washington Gold Quartz M. Co.

TULE CANYON see LIDA

WEST DIVIDE see LONE MOUNTAIN

WINDYPAH see PALMETTO

EUREKA COUNTY

ALPHA
Silver, Lead

Location. The Alpha District is situated in central Eureka Co., 5 m. E. of Alpha, a station on the E.-N. R. R.

History. A 10-stamp concentrator which was built 3 m. W. of the prospects did not prove a success.

Geology. The lodes are sheeted zones and replacement deposits in Devonian limestone, according to Emmons. The ore minerals include freibergite, galena, blende, pyrite, and copper carbonates; while barite is the principal gangue mineral and in some places constitutes more than half the ore.

Bibliography. Emmons 408 99.

BARTH see SAFFORD

BUCKHORN (Mill Canyon)
Gold, Silver, Lead

Location. Mill Canyon lies N. of Tenabo Peak on the N. W. slope of the Cortez Range, and Buckhorn is 5 m. distant on the S. E. side of the range. The Buckhorn District adjoins the Cortez District on the N. E. and is sometimes considered a section of that district. Beowawe on the S. P. R. R. and the W. P. R. R. is 30 m. N.

History. The Mill Canyon District was discovered at the same time as the Cortez District in 1863, and a mill erected there in 1864. The Mill Canyon mines did not prove very productive and the mill treated ore from the Cortez District. Buckhorn was discovered by Joe Lynn in the winter of 1908-9 and the principal claims acquired by George Wingfield for the Buckhorn Ms. Co. in 1910 This company erected an 800-hp electric power plant at Beowawe, and a 300-ton cyanide mill at Buckhorn. The mill operated from January 1914 to February 1916, when it was shut down and dismantled for lack of ore.

Geology. The Ordovician limestone of Mill Canyon has been intruded by granodiorite which forms the crest of the range, while on the E. slope at Buckhorn the older rocks are capped by Tertiary eruptives. According to Emmons, the ore bodies at Mill Canyon include fissure veins of silver ore in granodiorite, silver-lead replacement deposits in limestone, and ferruginous deposits of gold ore replacing limestone. According to Cook, the Buckhorn orebody occupies a N-S fault plane in basalt. The gold ore occurs in a shallow, kaolinized mass, oxidized at the surface and containing marcasite below but with the mineralization giving out at a depth of 250 ft.

Bibliography. MR1909 I 408 MR1913 I 824-5 MR1915 I 634
MR1910 I 515 MR1914 I 683 MR1916 I 480

Cook, P. R., "Cyaniding Clayey Ores at Buckhorn, Nevada", T AIME, 55 (1917) 437-445.

Emmons 408 106-110.

Hill 507 210.

M&SP, "Progress at the Buckhorn Mine", 107 (1913) 452.

Richards, E. R., "Refinery Practice and Flue Losses in Smelting Cyanide Precipitate at Buckhorn, Nevada", E&MJ 113 (1922) 865-7.

Siebert, F. J., In "Nevada Mining News", M&SP 108 (1914) 547.
Weed MH 1155 Buckhorn Ms. Co.

CORTEZ

Silver, Gold, Lead, Zinc, Copper, Turquoise

Location. The Cortez District is located at Cortez on the S. W. slope of Mt. Tenabo near the S. W. end of the Cortez Range at about the middle of the boundary line between Eureka and Lander Counties. Cortez is 36 m. S. of Beowawe, a station on the S. P. R. R. and the W. P. R. R. The camp has an elevation of 6,280 ft. and Mt. Tenabo is 9,240 ft. high. The Buckhorn District adjoins the Cortez District on the N. E. and is sometimes considered a part of it.

History. The Cortez District was discovered by prospectors from Austin in 1863. Simeon Wenban, one of the original locators, went into partnership with George Hearst in 1864, and the rich ores which they shipped to Austin helped to found the Hearst fortune. An 8-stamp mill was erected in Mill Canyon by the Cortez Co. in 1864 and was later enlarged to 16 stamps. Wenban bought out Hearst and acquired all the important mines in the district in 1867. Rich ore was discovered in the Garrison Mine in 1868. Ore was milled at Mill Canyon until a new mill was constructed at Cortez in 1886. Wenban's Tenabo Mill & Ms. Co. was operated by the British Bewick-Moreing Syndicate under the title of Cortez Mines, Ltd., from 1889 to 1891 but was turned back to Wenban in 1892. Wenban died in 1895, and leasers worked the Tenabo ground on a small scale up too 1919, when the Cons. Cortez Silver Ms. Co. took possession. This company erected a 100-ton concentration and cyanide mill in 1923.

Production. Emmons estimates the production of Cortez and Mill Canyon at $10,000,000, most of which came from the Garrison Mine in the Cortez District; while Burgess figures that the Cortez District alone produced not less than $15,000,000.

Geology. Tenabo Peak is composed of Paleozoic limestone and quartzite intruded by granitic rocks and porphyries, according to Emmons. On the S. W. slope where the Cortez District lies there is a massive bed of quartzite from 200 ft. to 300 ft. thick dipping E. at 23 degrees, which is probably the Ordovician Eureka quartzite. It is underlain by not less than 2,000 ft. of gray limestone, presumably the Pogonip limestone; and overlain by another gray limestone which is perhaps the Lone Mountain limestone. The sedimentary formations are cut by altered porphyry dikes with an E.-W. course.

Orebodies. The most important dike runs from the Garrison Mine on the W. to the Fitzgerald Mine on the E.; cutting the junction of the Pogonip limestone with the overlying Eureka quartzite in the Garrison Mine, and that between the Eureka quartzite and overlying Lone Mountain limestone in the Fitzgerald Mine. In the W. part of the Garrison Mine, large, irregular replacement deposits of high-grade ore were mined in the Pogonip limestone in the vicinity of the dike. E. of these, the prorphyry dike itself is replaced by ore; and further E. and higher up, at the junction of the Pogonip limestone with the Eureka quartzite, extensive mineralization occurs and ore is present mainly in the limestone but also extending into the quartzite. Still further to the E. in the Fitzgerald Mine, ore occurs in the quartzite at and near its contact with the dike, and extends into the

Lone Mountain limestone, although very little ore has been mined from this upper limestone.

Ore. Emmons states that the minerals in the replacement chambers are quartz, calcite, galena, stibnite, pyrite, sphalerite, stromeyerite, gray copper, and other minerals containing antimony and arsenic. Burgess adds chalcopyrite and pyrargyrite to this list. The oxidized ore consists of hornsilver, copper carbonates, and iron and manganese oxides.

Turquoise. Narrow veinlets of turquoise in altered rhyolite occur at the White Horse turquoise mine a few miles N. W. of Cortez, and some good gem material is said to have come from this mine.

Mines. The Cons. Cortez Silver Ms. Co. was incorporated in Nevada in 1919 with a capital stock of 2,000,000 shares of $1 par value. F. M. Manson, Reno, is Pres.-Treas.; F. C. Hunter, 80 Maiden Lane, New York City, is Sec.; and F. Solinsky and J. P. Gorman are additional Directors. C. D. Kaeding, 648 Mills Bldg., San Francisco, is Gen. Mgr.-Cons. Eng.; and W. H. Englebright is Supt. at Cortez. The property consists of a compact block of about 690 acres, including the old Garrison, Mt. Tenabo, St. Louis, Fitzgerald, and Artic Mines. It is developed by the 4,000-ft. Garrison tunnel and the 2,170-ft. Artic tunnel and equipped with compressor, hoist, and 100-ton concentration and cyanide mill.

Bibliography.

B1867 405-411	SMN1875-6 54-5	MR1914 I 683;
R1868 81-2, 96-8	MR1882 546, Salt	II 333-4, Turquoise
R1870 115-9	MR1905 269	MR1915 I 643
SMN1866 98, 101-2	MR1906 295	MR1917 I 277
SMN1867-8 44	MR1907 I 358	MR1918 I 239
SMN1869-70 45	MR1908 I 487, 492	MR1919 I 394
SMN1871-2 66-7	MR1909 I 408, 415	MR1920 I 324
SMN1873-4 47	MR1911 I 676	MR1921 I 385

USGE 40th 3 (1870) 401-7.

Bancroft 281-2.

Emmons 408 41, 100-6.

Hill 507 210-1, 215.

Weed, W. H., "The Copper Mines of the U. S. in 1905", USGS B 285, (19[illegible]) 116.

Weed MH 1173 Cons. Cortez Silver Ms. Co.

DIAMOND

Silver, Lead

Location. The Diamond District is situated on the W. slope of the Diamond Range in E. Eureka Co. Eureka on the E. N. R. R. is 25 m. S.

History. The district was discovered in 1864. A smelter was erected at the Champion Mine, which was the principal property, in 1873, and a small amount of bullion was produced, but the district was abandoned shortly afterwards. In 1922, the Eureka Silver M. Co., of which E. B. McCabe is Pres., and A. Eldridge is Sec., 212 Atlas Blk., Salt Lake City, undertook the development of a property in the district.

Geology. The ore consists of argentiferous galena and lead carbonate with iron and antimony, occuring as pockets and bunches in veins in limestone.

Bibliography. B1867 411 SMN1866 98 SMN1871-2 79
R1873 210 SMN1869-70 65 SMN1873-4 87
R1874 256-7 SMN1875-6 167
R1875 192

Davis, 840.
Spurr 208 81-4 Diamond Range.
Thompson & West 429.
Weed MH 1520 Eureka Silver M. Co.

DIAMOND MARSH (Williams Marsh)

Salt

Location. Diamond Marsh is in Diamond Valley W. of the Diamond Range in E. Eureka Co. Eureka on the E. N. R. R. is 40 m. S. and Mineral on the same railroad is 15 m. N. W. The salt covers an erea of 1,000 acres on the upper part of a flat 6 m. wide by 15 m. long.

History. Salt was mined from this deposit in the sixties and seventies to supply the silver mills of Eureka, Mineral Hill, and Hamilton. In 1881, salt was being produced with the aid of artificial heat at the rate of 5,000 lbs. per day.

Geology. The marsh is situated upon the Quarternary plains which occupy the center of Diamond Valley. In wet weather it is covered by a few inches of water and in dry by incrustations of salt. The salt obtained by gathering the incrustations contained only 60% of sodium chloride while that coming from the evaporation of the brine contains 95%.

Bibliography. SMN1871-2 80 USGE 40th (1877) 550.
Phalen, W. C., "Salt Resources of the U. S.", USGS B 669 (1919) 145.
Russell, I. C., "Geological History of Lake Lahontan", USGS M 11, (1885) 84.
Thompson & West 436.

EUREKA

(Ruby Hill, Secret Canyon, Pinto, Silverado, Spring Valley)

Silver, Gold, Lead, Copper, Zinc, (Molybdenum)

Location. The Eureka District is located at Eureka on the narrow gauge E.-N. R. R., 84 m. S. of Palisade, which is on the S. P. R. R. and the W. P. R. R. The camp of Ruby Hill is 2 m. W. of the town of Eureka. The Eureka District is sometimes subdivided into the Ruby Hill District in the immediate vicinity of Eureka and Ruby Hill, the Secret Canyon District to the S., the Pinto or Silverado District to the S. E., and the Spring Valley District to the S. W. The town of Eureka lies at an elevation of 6,500 ft. above sea-level, and the highest mountain in the district, Prospect Peak, has an altitude of 9,604 ft. Ruby Hill, after which the town of Ruby Hill was named, is a northerly spur of Prospect Mt., and the great Eureka and Richmond mines are situated upon it. The hill has an elevation of 7,291 ft. The town of Ruby Hill lies N. of the hill, and N. of the town is Adams Hill, 6,948 ft. in elevation.

History. The mines of Eureka were the first important lead-silver mines in the United States. The district was discovered in 1864, but did not become important until smelting was introduced, since the high lead content of the ore prohibited milling. In 1869, C. A. Stetefeldt constructed a furnace for Major McCoy and his associates which operated unsatisfactorily at the outset; but in the fall of the same year, Col. G. C. Robbins built a small furnace and demonstrated that the Eureka ores could be successfully smelted. In 1870, Albert Arentz introduced the siphon tap which revolutionized the method of discharging bullion. The McCoy furnace was leased to Col. David E. Buel and associates who bonded several mines and formed the Bateman Association to build a large smelter. This association was consolidated with the holdings of William Lent to form the Eureka Cons. M. Co. which became the largest producer. The second largest mine was the Richmond Consolidated which erected its own refinery at Eureka in 1874. The E.-N. R. R., formerly known as the Eureka & Palisade R. R., was completed in 1875. The first apex case, forerunner of the numerous apex suits which have since proved such a bane to the mining industry, was between the Eureka and Richmond companies, brought in 1877 and settled in faver of the Eureka Consolidated in 1881. Water was encountered in the Eureka Consolidated shaft at 765 ft. in 1881, and drowned out the shaft in 1882, so that pumping has been necessary since. As the bonanza orebodies became exhausted, mining passed into the hands of leasers who made the main production from 1885 to 1890. In 1890, the Richmond smelter was shut down, and in 1891, the Eureka smelter was abandoned and the camp entered upon a long period of inactivity. In 1905, a revival of mining interest occurred as a result of the organization of the Richmond-Eureka M. Co., to take over the interests of the old Richmond and Eureka companies. Another mining revival has occurred recently, culminating in the formation of the Eureka S. & M. Co., which has purchased the E.-N. R. R., acquired a number of mines, and plans to build a large smelter.

Production. According to Hague, the production of the Eureka District from 1869 to 1883 was approximately $40,000,000 in silver, $20,000,000 in gold, and 225,000 tons of lead. The production fell off rapidly after that, reviving in recent years as may be seen by the table of annual productions which follows.

RECENT PRODUCTION OF THE EUREKA DISTRICT
(According to Mineral Resources of the United States, U. S. Geol. Survey)

Year	Mines	Tons	Gold $	Silver Ozs.	Copper Lbs.	Lead Lbs.	Zinc Lbs.	Total $
1902		4,508	77,364	96,287		903,875		141,571
1903		11,667	80,125	122,730		576,748		147,065
1904		2,582	33,965	45,587		496,306		79,970
1905	7	1,692	13,457	36,965		416,308	9,120	55,511
1906	6	11,796	49,530	68,760	6,653	992,929		153,479
1907	9	35,092	173,036	152,873	115,557	2,936,672	3,075	452,867
1908	21	28,000	172,355	109,826	72,477	2,310,572		337,174
1909	17	87,936	419,778	179,315	90,100	4,340,768		711,388
1910	8	11,923	70,537	33,349	702	672,801		118,238
1911	7	15,484	101,715	5,584		27,570		105,915
1912	16	20,810	116,648	18,546	2,394	195,036		137,226
1913	26	1,096	13,227	24,173	3,053	287,959		40,971
1914	24	807	18,457	28,441	2,379	194,803	13,439	42,783
1915	25	1,135	22,578	34,308	5,653	393,526		59,457
1916	19	3,666	38,731	80,657	41,420	1,290,448	20,870	193,829
1917	31	3,304	20,131	39,708	31,771	693,757		121,188
1918	20	3,804	27,643	51,980	63,517	1,142,937		176,461
1919	15	3,283	20,988	66,459	40,713	1,179,723		165,520
1920	19	8,825	62,385	78,878	120,268	2,457,094		367,059
1921	14	9,644	49,280	45,886	206,948	776,943		156,824
Total 1902-21		267,154	$1,581,930	1,319,688	803,605	22,286,775	46,504	$3,754,496

Dividends. The following sums have been paid in dividends by Eureka companies according to a table prepared by Miss N. E. Preece, Geol. Dept., International Smelting Co., Salt Lake City.

Eureka Consolidated	$4,817,500
Hamburg	16,000
Jackson	10,000
Richmond Consolidated	3,042,387
Total	$7,869,887

Geology. The geologic section at Eureka comurises 30,000 ft. of Paleozoic rocks and constitutes the type section for Nevada. It is made up of 7,700 ft. of Cambrian sediments, including in order of age the Prospect Mt. quartzite, Eldorado limestone, Secret Canyon shale, Hamburg limestone, and Dunderberg shale; 5,000 ft. of Ordovician sediments, including the Pogonip limestone, Eureka quartzite, and Lone Mt. limestone; 6,000 ft. of Nevada limestone of Devonian age; and 11,300 ft. of Carboniferous sediments, including the White Pine shale, Diamond Peak quartzite, Lower Coal Measures, Weber conglomerate, and Upper Coal Measures. These sedimentary rocks were intruded, probably in late Mesozoic time, by granite porphyry, and quartz porphyry; while still later, probably in Tertiary time, andesitic intrusions occurred in rhyolitic and basaltic flows took place on the surface. The beds have been folded and faulted, the folds being open and the faults mainly normal.

Orebodies. The orebodies occur in the Paleozoic sedimentary rocks, and particularly in the Cambrian Eldorado and Hamburg limestones. They are replacement deposits of various shapes including veins, masses, and bedded deposits; associated with fissures. Oxidation of the orebodies has formed caves above the larger ones. The fall of rock into such openings has caused later fissuring, and a redistribution of the metals has sometimes ensued. The depth reached by oxidation is about 1,000 ft.

Ore. The principal ore of the Eureka District has been argentiferous and auriferous galena, but there has also been a considerable production of oxidized ferruginous gold ore. According to Curtis, the ore above water-level is principally composed of galena, anglesite, cerussite, mimetite, and wulfenite in a gangue consisting mainly of iron oxides with a little quartz and calcite. Below water-level, the ore is chiefly composed of pyrite, arsenopyrite, galena, and blende. The lead minerals carry more silver than gold, and the iron minerals more gold than silver by value. Silver occurs both as chloride and sulphlde, and gold is found in the native state. Enough wulfenite is present to be of possible commercial importance, according to Horton.

Mines. The famous old Eureka Consolidated and Richmond Consolidated Mines are now owned by the **Richmond-Eureka M. Co.**, a subsidiary of the U. S. Smelting, Refining & M. Co. The operating officials at 912 Newhouse Bldg., Salt Lake City, are G. W. Heintz, V. P. in charge; C. E. Allen, Gen. Mgr.; D. D. Muir, Jr., Mgr. of Mines; W. A. Howard, Smelter Mgr.; and J. C. Brumblay, Field Rep. There are 6 other companies now active in the district. The **Cyanide Mine** is under the management of C. F. Wittenberg; and the **Eureka-King M. Co.** under the superintendence of J. D. Peckner.

Eureka-Prince. The Eureka-Prince M. Co. is incorporated in Nevada with a capital stock of 1,000,000 shares of $1 par value, of which 667,400 shares are outstanding. The stock is listed on the San Francisco Stock Exchange. The western office is at Eureka, and the eastern office at 342 Madison Ave., New York City. R. B. Todd, Gazette Building, Reno, is Pres.; H. H. Salmon is V. P.; O. K. Newell, Sec.-Treas.; G. M. Todd, Asst. Sec.-Treas.; with C. Smith, additional Director. The company owns 100 acres on the N. side of Adams Hill, which formerly constituted the old Silver Lick mine. Plans have been made to install the first 50-ton unit of a cyanide plant at an early date and to sink a new 300-ft. shaft to tap the orebodies in depth.

Eureka Uncle Sam. The Eureka Uncle Sam Cons. M. Co. is incorporated in Nevada with a capital stock of 2,000,000 shares of $1 par value, of which 1,400,000 have been issued. The stock is listed on the San Francisco Stock Exchange. The home office is at 1069-71 Monadnock Bldg., San Francisco. F. T. Torpey is Pres.; G. Rodeik, V. P.; and L. L. Rosenshine, is Res. Mgr., at Eureka. The company owns 8 claims including the old Hamburg mine on the E. slope of Prospect Mountain in New York Canyon. The property has been under development since 1920 and has shipped ore taken out in the course of the work. It is proposed to erect a cyanide and flotation mill and to sink the shaft from 650 ft. to 1,500 ft.

Holly Consolidated. The Bullwacker Cons. Ms. Co. and the Eureka-Holly Cons. M. Co. were consolidated in 1922 to form the Holly Cons. M. Co. The capital stock is 1,500,000 shares of $1 par value. The Bullwhacker property consisted of 111 acres on Adams Hill including the old Bullwhacker mine; and the Eureka-Holly property consisted of 12 claims aggregating 206 acres, formerly constituting the old Idaho mine, on Mineral Point, a northerly extension of Adams Hill. The company proposes to construct a flotation mill.

Eureka Smelting & Mining. The Eureka Smelting & M. Co. was incorporated in Nevada in 1922 with a capital stock of 10,000,000 shares of $1 par value. The home office is at 342 Madison Ave., New York City. G. T. Wilson is Pres.; R. B. Todd, F. L. Torres, and A. E. Stillwell are V. P's.; and H. F. Thomas, Treas.; with the following additional Directors, A. Heckshire, J. F. Ballard, P. T. Brady, F. Hurdle, A. R. Whaley, T. W. Pelham and E. C. Randall. C. N. Sigison is Sec.-Asst. Treas. W. Lindgren is Cons. Geol., and A. P. Mayberry, Operating Min. Eng. The company has purchased the E.-N. R. R. and a number of mining properties at Eureka, including the Eureka-Croesus Mine, formerly known as the Dunderberg, which not only made a large production in the early days, but has also made a steady production from development workings since 1919. It is proposed to erect a custom smelter on the line of the E.-N. R. R. which will have a capacity of 600 tons and will later be enlarged to 2,500 tons.

BIBLIOGRAPHY OF THE EUREKA DISTRICT

General

B1867 405, 411, 431
R1869 177
R1870 2, 116, 119-128
R1871 141, 171-180, 181, 379, 380, 383, 385, 441, 518, 399, 400, 401, 407, 386, 392, 393, 394.
R1872 142-152
R1873 35, 242-257
R1875 179-185, 365
SMN1866 98-102
SMN1867-8 50
SMN1869-70 45-55, 63-5
SMN1871-2 67-79
SMN1873-4 35-47
SMN1875-6 41-54
SMN1877-8 28-62
MR1882 309
MR1883 84, 418-9 462-473.
MR1885 250
MR1886 143
MR1887 104
MR1888 86
MR1905 268
MR1906 295, 450
MR1907 I 357
MR1908 I 487
MR1909 I 409
MR1910 I 515
MR1911 I 676-7
MR1912 I 795
MR1913 I 824
MR1914 I 683
MR1915 I 534
MR1916 I 479-80
MR1917 I 277
MR1918 I 239
MR1919 I 394
MR1920 I 324
MR1920 II 54, Arsenic
MR1921 I 384

USGE 40th 1 (1878) 188-9, 2 (1877) 547-8, 3 (1870) 405.

Bancroft 281-5.

Campbell, H. P., "Windfall Mine and Mill", Pacific Miner, Sept., 1909.

Davis 830-6, 840-6.

Hill 507 211.

Hillen, A. G., "Review of Conditions of the Eureka Mining District, Nevada", M&EW 45 (1916) 571-4.

Ingalls, W. R., "The Silver-Lead Mines of Eureka, Nevada", E&MJ 85 (1907); 1051-1058: Discussion, H. M. Chance, "The Silver Lead Deposits of Eureka, Nevada", E&MJ 85 (1908) 123-4.

——"Lead and Zinc in the United States", New York, 1908: ix. 13, 19, 22, 31, 32, 36, 39, 40, 46, 48-52, 54, 61, 77, 79, 154, 176, 178-183, 185. 186, 189, 190, 192-194, 206, 209-211, 231, 233.

M&SP Editorial, "Eureka-Croesus Mining Co.", 118 (1919) 802-3.

—— Editorial, "Proposed Exploration at Eureka", 120 (1920) 145.

Science, "The Ruby Hill Mines, Eureka, Nevada", 4 (1884) 459-460.
Stuart NMR 71-82.
Thompson & West 425-6, 429-433.
Weed MH 1129 Alkali M. Co. 1131 American-British Corporation.
1157 Bullwhacker Cons. Ms. Co. 1181 Croesus Ex. M. Co.
1182 Cyanide Mine. 1194 Eureka-Climax M. Co.
1195 Eureka-Croesus M. Co.
1197 Eureka-Holly M. Co., Eureka-King M. Co., Eureka-Nevada M. Co., Eureka-Prairie M. Co., Eureka Prince M. Co., Eureka-Uncle Sam Cons. M. Co.
1124 Hope Cons. M. Co. 1316 Richmond-Eureka M. Co.

Geology

Barus, C., "The Electrical Activity of Ore-Bodies", T AIME 13 (1885) 417-477; "Experiments Made at the Hichmond Mine, Eureka District, Nevada", 435-475.
Becker, G. F., "Reconnoissance of the San Francisco, Eureka, and Bodie Districts", USGS AR 1 (1880) 37-47.
Blake, W. P., "The Ore Deposits of Eureka District, Eastern Nevada, T AIME, 6 (1879) 554-563.
Curtis, J. S., "Abstract of Report on the Mining Geology of the Eureka District, Nevada", USGS Fourth AR (1884) 221-251.
———"Silver-Lead Deposits of Eureka, Nevada", USGS M 7 (1884).
E&MJ, "Oil Shale", 107 (1919) 105 (Mention only).
Hague, A., "Report of Work in the Eureka District", USGS AR 1 (1880) 32-5.
———"Report of Work in the Eureka District", USGS AR 2 (1882) 21-35.
———"Abstract of the Report on the Geology of the Eureka DistrictNevada", USGS Third AR (1883) 237-290.
———"Geology of the Eureka District, Nevada", USGS M 20 (1892).
———"Geological Section of the Eureka District", 10th Census, U. S., 13, (1885) 33.
Harder, E. C., "Manganese Deposits of the United States", USGS B 380 (1909), 274 (Mention only).
Horton, F. W., "Molybedenum; Its Ores and their Concentration", USBM, B 111 (1916) 89; Wulfenite.
Iddings, J. P., "Microsopical Petrography of the Eruptive Rocks of the Eureka District", USGS M 20 Appendix B, (1892) 337-396.
Keyes, W. S., "The Eureka Lode of Eureka, Eastern Nevada", T AIME 6, (1879 344.
Penrose, R. A. F., Jr., Manganese: Its Uses, Ores and Deposits, AR Geol. Sur., Ark. 1890, 1 (1891) 477.
Spurr 208, 81-4.
Walcott, C. D., "Paleontology of the Eureka District", USGS M 8 (1884).
———"Systematic List of Fossils Found at Eureka, Nevada", USGS M 20 Appendix A, (1892) 319-333.
Whitney, J. B., "Pcstdam Fossils from near Eureka, Nevada", Cal. Acad. Sci., P 4 (1873) 200.

Metallurgy

Hahn, Eilers & Raymond, "The Smelting of Argentiferous Lead Ores in Nevada, Utah, and Montana", T AIME 1 (1873) 91.

Miscellaneous

Raymond, R. W., "The Eureka-Richmond Case", T AIME 6 (1879) 371.

LYNN

Placer Gold, Gold

Location. The Lynn District is at Goldville on a low divide at the head of Lynn Creek in the Tuscarora Range in N. Eureka Co. Goldville has an elevation of about 6,070 ft. and is about 20 m. N. W. of Carlin which is on the S. P. R. R. and W. P. R. R.

History. The placer deposits of the Lynn District were discovered by Joe Lynn in 1907, the Lynn Big Six Mine was located the same year by W. E. Barney, and a brief boom ensued. The Lynn Big Six M. Co. erected a 10-ton amalgamating mill in 1917.

Production. The Lynn Big Six Mine obtained some $6,000 in bullion from its mill and shipped over $15,000 worth of ore to Salt Lake smelters. The placer deposits of the district have been worked intermittently since their discovery and have made the bulk of the production of the camp. From 1909 to 1921, the camp produced $99,878 mostly in gold, according to Mineral Resources, U. S. Geol. Survey.

Geology. The country rock of the Lynn District is bedded rhyolite of Tertiary age which has been cut by porphyritic intrusions, according to Emmons, and the ore occurs in zones of shattering in the rhyolite. Within these zones, veinlets of quartz and silicified, iron-stained rhyolite carry high values in gold. Bismuth in both oxide and sulphide form is said to have accompanied the gold in the rich ore; and has been found in the placers. The placers are rich, but narrow and thin, and in places covered by heavy overburden. The placer gold is course, rough, and pure.

Mines. The Lynn Big Six M. Co. was incorporated in Nevada in 1912 with a capital stock of 1,000,000 shares of 25 cts. par value, of which 260,000 remain in the treasury. The home office is at 701 McIntyre Bldg., Salt Lake City. D. Keith is Pres.; L. A. Marks, V. P.; R. M. Holt, Sec.-Treas.; with F. Davidson and H. Harker, Directors. The company owns 23 claims which are developed by 1 m. of workings. Plans have been made to replace the old mill with a new one of at least 50 tons capacity.

Bibliography.

MR1908 I 476	MR1913 I 824	MR1917 I 277
MR1909 I 409	MR1914 I 683	MR1918 I 239-240
MR1910 I 515-6	MR1915 I 634	MR1919 I 395
MR1911 I 677	MR1916 I 480	MR1920 I 324
MR1912 I 795		MR1921 I 384

Emmons 408 45, 70, 87-8.
Hill 507 211.
M&SP, Editorial on "spirit ore", 122 (1921) 409-410.
O'Brien, T. F., Discussion, "The Koering Process", M&SP 122 (1921), 573-4.
Weed MH 1245 Lynn Big Six M. Co.

MAGGIE CREEK

Silver, Lead, Copper, Gold

Location. The Maggie Creek District lies S. W. of Maggie Canyon where Maggie Creek cuts through a low ridge joining Carlin Peaks in the Tuscarora Range with Maggie Peak in Independence Range. Carlin on the S. P. R. R. and W. P. R. R. is 10 m. S. E.

History. The Nevada Star Mine, 9 m. N. W. of Carlin, produced in 1906 and was under development up to 1909. The Copper King Mine, 2 m. S. W. of the Nevada Star made a production in 1917.

Geology. The Nevada Star lode is a small replacement vein in Paleozoic limestone. It contains bunches of galena and lead and copper carbonates, according to Emmons, and is said to carry good silver values. At the Copper King Mine, bunches of oxidized copper ore occur irregularly in shattered rhyolite of Tertiary age.

Bibliography. MR1906 293 MR1909 I 410 MR1916 I 480
MR1908 I 487 MR1913 I 824 MR1917 I 277
Emmons 408 87. Hill 507 212.

MILL CANYON see BUCKSKIN

MINERAL HILL

Silver, Lead, Gold, Copper, Zinc

Location. The Mineral Hill District is situated at Mineral Hill at the N. end of Mineral Hill Ridge which rises 700 ft. above Pine Valley. Mineral on the E.-N. R. R. is 5 m. N. W.

History. The Mineral Hill District was discovered by prospectors from Austin in 1868. A 15-stamp mill was erected, and the property sold to the Mineral Hill Silver M. Co., Ltd., an English corporation, which erected an additional 20-stamp mill. The operations of this company proved unsuccessful, and in 1880 it sold out to the Austin and Spencer Co. which operated until 1887. Small productions were made by a number of companies from 1912 to 1919.

Production. Emmons places the early production of the district at a little more than $6,000,000, mostly in silver. From 1905 to 1920, the district produced 11,417 tons of ore containing $5,303 in gold, 133,626 ozs. silver, 65,993 lbs. copper, 3,643,132 lbs. lead, and 75,118 lbs. zinc, valued in all at $362,041, according to Mineral Resources, U. S. Geol. Survey.

Geology. The ore deposits occur in Paleozoic limestone cut by altered dikes. The orebodies are chambers or irregular replacement deposits which cut across the bedding of the limestone, according to Emmons. The ore minerals are quartz, calcite, barite, silver chloride, argentite, gray copper, galena, zinc blende, copper carbonates, pyromorphite, lead carbonate, pyrite, iron and manganese oxides, to which Eissler adds polybasite, stephanite, bromide of silver, and molybdenite.

Bibliography. R1871 141, 180-1 MR1905 268-9 MR1916 I 480
R1872 152-3 MR1908 I 287 MR1917 I 277
R1873 210 MR1909 I 410 MR1918 I 240
R1874 35, 257 MR1912 I 796 MR1919 I 395
SMN1869-70 55-7 MR1913 I 825
SMN1871-2 22 MR1914 I 683
SMN1875-6 27 MR1915 I 634
USGE 40th 1 (1878) 191, 3 (1870) 407-8.

Eissler, M., "Treatment of Silver Ores at Mineral Hill, Nevada", "The Metallurgy of Silver", N. Y., (1898) 154-167.

Emmons 408 95-9.

Maynard, T. P., "The Mineral Hill Mining District, Nevada", M&EW 42, (1915) 1117-9.

Thompson & West 435.

Toll, R. H., "Mineral Hill, Nevada", M&SP 104 (1912) 888-9.

Weed MH 1257 Mineral Hill Cons. Ms. Co. 1363-4 Union Ms. Co.

PALISADE see SAFFORD

PINTO see EUREKA

ROBERTS

Silver, Lead, Copper

Location. The Roberts District lies 42 m. S. E. of Austin which is on the N. C. R. R.

History. It was discovered by Roberts and Tucker in 1870.

Geology. The principal mine was the O'Dair which had a 10-ft. vein between walls of syenite and limestone. This vein carried argentiferous galena at the surface which changed to copper at a depth of 25 ft.

Bibliography. SMN1869-70 445 SMN1873-4 61.

RUBY HILL see EUREKA

SAFFORD (Barth, Palisade)

Iron, Silver, Lead, Copper, Gold

Location. The Safford District is located in Safford Canyon S. of Barth on the S. P. R. R. Barth is 5 m. W. of Palisade on the S. P. R. R., W. P. R. R., and E.-N. R. R. Safford Canyon is on the W. slope of the Cortez Range S. of the Humboldt River and runs N. W. into Palisade Canyon through which the Humboldt flows.

History. The West iron mine was discovered and described by the geologists of the Fortieth Parellel Survey. In 1881, James Safford discovered the Onondaga silver mine, and this mine and the Zenoli silver mine became the principal producers. About 1907, the American Smelting & Refiining Co. leased the West iron mine from the Central Pacific R. R. and worked it for a number of years for iron flux for the company's Salt Lake smelters. The lease was relinquished and the West Mine is now flodded and abandoned. Small productions were made by the silver mines up to 1917.

Production. According to Emmons, the silver production up to 1910 came almost entirely from the Onondaga and Zenoli Mines and amounted to about $200,000. During the 6 yrs. prior to 1913, according to Jones, over 240,000 tons of iron flux were shipped from the West Mine.

Geology. The country rocks of the Safford District collected by Emmons included fine-grained diorites that carried some orthoclase and quartz, glassy vesicular andesites, and dacite porphyries. These he believed to be the eruptions or the intrusions from a center of Tertiary volcanism.

Silver Ore. Silver ore occurs in fissure veins in the Tertiary eruptive rocks. The wall rocks of the veins are somewhat altered. The country rock of the Zenoli Mine is quartz-andesite and the sulphide ore contains quartz, calcite, barite, stibnite, gray copper, galena, pyrite, and chalcopyrite. The oxidized ore contains iron oxide, copper carbonates, and hornsilver. The country rock of the West iron mine is fine-grained diorite, according to Emmons, which near the mine is in contact with an altered rock which is probably a metamorphosed limestone. The ore deposit is believed to be of contact-metamorphic origin. The iron ore consists mainly of hematite, according to Jones, with probably a little magnetite; and contains minor quantities of quartz, fiuor-apatite, and phlogopite. Where weathering has occurred, limonite has been produced and calcite has been deposited in seams. The ore is of high quality and uniform grade and while the phosphorus content is above the Bessemer limit, it is lower than that of the Alabama ores.

Bibliography. R1871 141; MR1909 I 410; MR1911 I 677; MR1912 I 796; MR1913 I 826; MR1914 I 683; MR1915 I 635; MR1917 I 277

Emmons 408 110-3

Hill 507 211-2

Jones, J. C., "The Barth Iron Ore Deposit", EG 8 (1913) 247-263; Abstract, GSA B 24 (1913) 96-7.

Stuart NMR 71.

Weed MH 1324 Safford Copper Co.

SECRET CANYON see EUREKA

SPRING VALLEY see EUREKA

WILLIAMS MARSH see DIAMOND MARSH

SILVERADO see EUREKA

HUMBOLDT COUNTY

ADELAIDE see **GOLD RUN**

AMOS (Awakening)
Gold, Silver

Location. The Amos District lies W. of Amos in the Slumbering Hills in central Humboldt Co. It is 30 m. N. W. of Winnemucca which is on the S. P. R. R. and W. P. R. R.

History. Gold ore containing a little silver was shipped from the distrist in 1912, and in the following year ore of similar character was treated in the 5-stamp amalgamating mill of the D. & C. M. Co. Lessees treated ore in this mill the succeeding two years. In 1917, the Oklahoma property was building a 30-ton amalgamation and concentration mill in the district; and in 1918 the Colorado No. 1 produced some gold bullion in an arrastra.

Bibliography. MR1912 I 797; MR1913 I 826; MR1914 I 684; MR1915 I 635; MR1917 I 278; MR1918 I 240

USGS Disaster and Paradise topographic maps.

ASHDOWN see WARM SPRINGS

AWAKENING see AMOS

BATTLE MOUNTAIN see LANDER COUNTY

COLUMBIA see VARYVILLE

DISASTER
Placer Gold

Location. The Disaster is in N. Humboldt Co., on the Oregon border. Winnemucca on the S. P. R. R. and W. P. R. R. is 100 m. S.

History. Placer gold was produced in the district in 1914.

Bibliography. MR1914 I 684. USGS Disaster topographic map. Hill 507 212.

DONNELLY see WASHOE COUNTY

GOLCONDA
Manganese

Location. The Golconda manganese deposits in S. E. Humboldt Co. lie at the foot of the Edna Mts. which rise 1,000 ft. to 1,200 ft. above them to the

E. Golconda on the S. P. R. R. is 3 m. W.

History. The manganiferous deposits of Golconda were discovered in 1885 and prospected for precious metals. The Noble Electric Steel Co. shipped manganese ore in 1917 and Louis Navarine made a shipment in 1918, but the total production did not exceed 100 tons, according to Pardee & Jones.

Geology. The Edna Mts. are composed largely of shales, limestones, and quartzites belonging to the Triassic Star Peak formation, according to Harder. Near the foot of the mountains are knolls of Quaternary tufa. The manganese deposits are of two types;—a horizontal layer of manganese oxide in the tufa knolls, and replacement deposits of manganiferous limonite along a crystalline limestone layer in the Star Peak formation a short distance up the slope from the tufa beds. A small amount of tungsten is present in the manganese ore.

Bibliography. MR1885 349 MR1889-90 134 MR1892 200
MR1886 197 MR1891 136 MR1907 I 100

Harder, E. C., "Manganese Deposits of the U. S.", USGS B 380 (1909) 270.

———"Manganese Deposits of the U. S.", USGS B 427 (1910) 153-7.

Hill 507 213.

Pardee, J. T., and Jones, E. L., Jr., "Deposits of Manganese Ore in Nevada", USGS B 710F (1920) 235-8.

Penrose, R., Jr., "Manganese: Its Uses, Ores, and Deposits", AR Geol. Sur. of Ark. for 1890 1 (1891) 64, 469-476, 551.

———"A Pleistocene Manganese Deposit near Golconda, Nevada", JG 1, (1893) 275-282.

HARMONY see SONOMA MOUNTAIN

IRON POINT

Manganese, Silver, Gold, Lead

Location. The Iron Point District is located at the N. end of the Battle Mt. Range in S. E. Humboldt Co. Iron Point on the S. P. R. R. lies to the N.

History. A few carloads of manganese ore were shipped from the Hansen & Knudsen claim 11 m. S. W. of Iron Point in 1918. The Silver Coin shipped several lots of silver ore containing a little lead and gold from its property 3 m. S. of Iron Point in 1920, and has continued shipping since. It is under lease to B. A. Goldsworthy.

Geology. The manganese deposits lie at an altitude og 5,500 ft., 1,000 ft. below the summits of the neighboring hills. Small lenses of oxidized manganese ore are associated with masses of quartz which stand out from weathered shale and quartzite of the Triassic Star Peak formation, according to Pardee and Jones, while unoxidized lenses of manganiferous chert also occur. The ore lenses are from 10 ft. to 25 ft. in diameter and from 1 ft. to 2 ft. thick and consist of psilomelane and pyrolusite associated with rhodonite and possibly rhodocrosite.

Bibliography. MR1920 I 325 MR1921 I 384.

Pardee, J. T., and Jones, E. L., Jr., "Deposits of Manganese Ore in Nevada", USGS B 710F (1920) 238-9

JACKSON CREEK

Copper, Lead, Silver

Location. The Jackson Creek District is located on Jackson Creek on the W. slope of the Jackson Mts. Sulphur on the W. P. R. R. is 35 m. S.

History. Small amounts of argentiferous copper ores and leaa ores nave been shipped from the district.

Geology. The country rock consists of granite and limestone and the ore deposits are veins and contact-metamorphic deposits. At the Nelson mine on Jackson Creek, according to a letter from J. T. Reid, the granite is cut by granodiorite dikes along which run contact veins containing copper carbonate and oxides, while in the next canyon to the S. replacement orebodies occur in a thin bed of limestone.

Bibliography. MR1912 I 797 MR1918 I 241 MR1920 I 324
USGS Disaster topographic map.
Hill 507 213.

MOUNT ROSE see PARADISE VALLEY

GOLD RUN (Adelaide)

Copper, Gold, Silver, Lead, Zinc, Placer Gold

Location. The Gold Run District is at Adelaide on Gold Run Creek on the E. slope of the Sonoma Range in S. E. Humboldt Co. Golconda on the S. P. R. R. is 10 m. N. The Golconda District adjoins the Gold Run District on the N.

History. The district was organized in 1866. In 1868 the Golconda Mine was operating an 8-stamp mill. In 1897, the Glasgow and Western Exp. Co. acquired the principal mines, constructed a 12-mile narrow-gage railroad from Golconda to Adelaide, and built a concentrating plant and smelter at the junction of the branch with the main line. This plant was operated upon ores from Adelaide and from Battle Mountain for a time, and some matte was produced, but the results were not satisfactory. In 1907, the mill was remodeled and Macquisten tubes employed for concentration, but operations were again suspended in 1910. The Yerington Mountain Copper Co. purchased the Adelaide Mine in 1916. Within recent years, a number of small mines have been in operation.

Production. From 1908 to 1920, the district produced 23,583 tons of ore, containing $75,364 in gold, 65,206 ozs. silver, 812,841 lbs. copper, 297,883 lbs. lead, and 1,700 lbs. zinc, valued in all at $349,942, according to Mineral Resources, U. S. Geol. Survey.

Geology. The country rock at the Adelaide Mine consists of slate and limestone belonging to the Star Peak Triassic formation, according to Ransome. The ore is a metasomatic replacement of a limestone layer which is from 50 ft. to 75 ft. thick. The ore minerals are pyrrhotite, chalcopyrite, sphalerite, and a little galena in a gangue of garnet, vesuvianite, diopside ,calcite, orthoclase and a little quartz. Veins carrying gold and silver occur in the Triassic Koipato formation W. of the Adelaide Mine.

Mines. The only properties at present active in the district are the Good Hope Group of which W. P. Hammon is Lessee, and the Kelsey M. Co. of which P. R. Kelsey is Mgr.

Bibliography.

R1868 124-5	SMN1869-70 28-9	MR1913 I 827
R1869 192	SMN1875-6 61-2	MR1914 I 648
R1870 133	MR1898 185	MR1915 I 635
R1872 157-8	MR1907 I 359	MR1916 I 481
R1873 214	MR1908 I 489-490	MR1917 I 278
R1874 261-2	MR1909 I 412	MR1918 I 241
R1875 188-9	MR1910 I 517	MR1919 I 395
SMN1866 54	MR1911 I 678	MR1920 I 324
SMN1867-8 41-2	MR1912 I 796-7	MR1921 I 384

USGE 40th 3 (1870) 316-7.

Hill 507 212.
Ingalls, W. R., "Concentration Upside Down", E&MJ 84 (1907) 765-770.
Ransome 414 62-4.
Thompson & West 450-1.
Weed MH 1181 Crown Mines. 1207 Golconda Gold Ledge M. Co.
1369 Warmack Gold M. Co. 1381-2 Yerington Mt. M. Co.

NATIONAL

Gold, Silver

Location. The National District is located on the W. slope of the Santa Rosa Range in N. E. Humboldt Co. The camp of National lies at an elevation of 6,100 ft. and Buckskin Peak rises behind it to an altitude of 8,800 ft. National is 74 m. N. of Winnemucca which is on the S. P. R. R. and W. P. R. R.

History. National was discovered in 1907 by J. L. Workman. He leased the property in small blocks, and in 1909 rich ore was found on the Stall Bros. lease. In 1910, the Stall Bros. sold out to the National Ms. Co. of Chicago. The high value of the ore, which was worth from $10 to $75 per pound, led to considerable ore stealing, and encouraged apex litigation.

Production. The production of National from 1909 to 1921 as recorded by Mineral Resources of the United States, U. S. Geol. Survey, was $3,444,997, mostly in gold.

Geology. The Santa Rosa Range at National is composed of Teritary flows resting upon a basement of Triassic clay slate, according to Lindgren. Basalt is the predominant rock and probably represents the S. margin of the Columbia River Miocene basalt. Latite is the principal country rock of the National Mine, and there is a closely related trachyte. Rhyolite dikes are numerous and a flow of rhyolite forms the summit oof Buckskin Peak.

Ore Deposits. The ore deposits of National are narrow veins in latite, rhyolite, basalt, basalt tuff, and trachyte. The country rock near the veins is altered with the development of pyrite, calcite, and a little sericite and adularia. The veins are composed of sheared rock containing seams of quartz. The quartz seams generally show symmetrical banding and contain vugs. Many veins are associated with rhyolite dikes.

Ore. Lindgren considers the National veins to be low-grade silver veins with a small gold content, and the National gold shoot to be a unique occurrence. Stibnite is the most characteristic sulphide. Pyrite, chalcopyrite, arsenopyrite, zinc blende, and galena occur only in small grains. Cinnabar was noted in one vein. Calcite is rare and adularia was observed in quartz in one instance. The rich shoot in the National vein contained gold in the form of electrum.

Bibliography.

MR1908 I 490	MR1913 I 827	MR1918 I 241
MR1909 I 412	MR1914 I 685	MR1919 I 396
MR1910 I 517-8	MR1915 I 636	MR1920 I 324
MR1911 I 678-80	MR1916 I 481	MR1921 I 384
MR1912 I 797-8	MR1917 I 279	

USGS Paradise topographic map.

Cutler, H. C., "National, Nevada", M&SP 101 (1910) 606-7.
Hill 507 213.
Lindgren, W., "Geology of the National Mining District, Nevada", M&EW 35 (1911) 1175-6.

———"The Bonanza of National, Nevada", Wash. Acad. Sci. J 2 (1912), 107-8.

———"Geology and Mineral Deposits of the National Mining District, Nevada", USGS B 601 (1915); Abstract, Wash. Acad. Sci. J 5 (1915) 580-1.

Stuart NMR 125.

Weed MH 1156 Buckskin National Gold M. Co. 1236 National Ms. Co., National Nevada M. Co., National Treasure Ms. Co.

Winchell, A. N., "Geology of the National Mining District, Nevada", M&SP 105 (1912) 655-9

NEW GOLDFIELDS see REBEL CREEK

PARADISE VALLEY (Spring City, Mount Rose)
Silver, Gold, Placer Gold

Location. The Paradise Valley District is located on the E. slope of the Santa Rose Range on the edge of Paradise Valley. It is 8 m. N. W. of the town of Paradise Valley at the old camp of Spring City; and is 45 m. N. N. E. of Winnemucca which is on the S. P. R. R. and the W. P. R. R. The district was at one time known as the Mount Rose District.

History. The Paradise Valley District was discovered in 1868 and organized in 1873. The mines were most actively worked from 1879 to 1891, when operations ceased. A revival if interest in the district took place in 1907, and small productions of silver ore and of placer gold were made from 1909 to 1915.

Production. The early production was probably about $3,000,000, according to Lindgren.

Geology. According to R. S. Bolam, the country rock consists of calcareous slate cut by porphyry dikes, while to the E. is a great flow of rhyolite. The veins occur in two systems and contain ore only in their narrower portions. Hornsilver was present at the surface while below pyrite, ruby silver, and argentite were found. At depth sphalerite increased and precious metal values decreased.

Bibliography.

R1874 262, 263	MR1908 I 490	MR1912 I 798
R1875 189	MR1909 I 412	MR1913 I 827
SMN1877-8 67-8	MR1911 I 680	MR1914 I 685
MR1907 I 362		MR1915 I 636

USGS Paradise topographic map.

Davis 899. Hill 507 213-4.

Lindgren, W., "Geology and Mineral Deposits of the National Mining District", USGS B 601 (1915) 13-14, 16-17.

Stuart NMR 120. Thompson & West 451.

PUEBLO see WARM SPRINGS

RABBIT HOLE see SULPHUR

REBEL CREEK (New Goldfields)
Gold, Silver

Location. The Rebel Creek District is on Rebel Creek on the W. slope of the Santa Rosa Range. Winnemucca on the S. P. R. R. and the W. P. R. R. is 54 m. S.

History. Exploration work was conducted in the district in 1907 and in 1908 the Manitou M. Co. built a small mill. The Cottontail M. Co. of which J. J. Burke of Reno is Pres., has been active recently.

Geology. According to Hill, the country rocks are metamorphosed slates and granite and the ore deposits are veins.

Bibliography. MR1907 I 362 MR1910 I 518 MR1916 I 481
MR1908 I 490 MR1921 I 384
Hill 507 214.
Lindgren, W., "Geology and Mineral Deposits of the National Mining District, Nevada", USGS B 601 (1915) 12, 14, 17.
Weed MH 1180 Cottontail M. Co.

RED BUTTE

Copper, Antimony, (Mercury)

Location. The Red Butte District is at Red Butte on the W. flank of the Jackson Range near its S. end and just E. of the Black Rock Desert. Sulphur on the W. P. R. R. is 15 m. S.

History. The Red Butte District was discovered by A. D. Ramel in 1907. Antimony was found in the S. E. part of the district in 1908, and cinnabar was found in the S. W. part. Antimony ore was shipped in 1905, and copper ore in 1916.

Geology. The country rock of the Red Butte District is a gabbro which is cut by numerous dikes of aplite. Copper deposits occur in association with the aplitic dikes as veins and disseminations. At the Metallic property the ore consisted of cuprite, covellite, native copper, and chrysocolla associated with hematite, limonite, and a little barite, according to Ransome, and was irregularly distributed through an aplite dike; while at the Redeemer property a fissure zone in the gabbro carried streaks of chalcocite partly altered to chrysocolla, azurite, and malachite.

Bibliography. MR1916 I 481. Ransome 414 27-30

SHON

Silver, Gold

Location. The Shon District is situated in the Santa Rosa Range. Winnemucca on the S. P. R. R. is 28 m. S.

History. In 1878, the Eclipse was the principal mine and a number of prospectors were at work in the district. Prospecting was also active in the district in 1909. The Charleston Hill Mine was operated under lease in 1914, and again in 1917-1918, and the Charleston Hill National Ms., Inc., is now actively developing the property.

Geology. The veins worked in the early days are said to be in granite. The country rocks at the Charleston Hill Mine are of porphyritic and schistose varieties, and the veins are parallel and of contact and fissure types, according to a letter from J. P. Clough. The principal vein lies between a porphyritic hanging wall and a slate and schist footwall, and has a banded structure.

Mines. The Charleston Hill National Ms. has a capital stock of 1,000,000 shares of $1 par value. The home office is at Winnemucca. J. P. Clough is Pres.; M. L. Grantz, Sec.-Treas.; with R. C. Felter, additional Director. The property is opened by 5 tunnels. Machinery is now being installed and water developed, and the company intends to erect a mill in the near future.

Bibliography. SMN1877-8 65. USGS Paradise topographic map.
Stuart NMR 121. Weed MH 1163-4 Charleston Hill Gold M. Co.

SONOMA MOUNTAIN (Harmony)

Copper, Silver, Gold

Location. The Sonoma Mountain District formerly known as the Harmony

District occupies the N. end of the Sonoma Range in S. E. Humboldt Co. It adjoins the Golconda District on the W. and is 5 m. S. E. of Winnemucca which is on the S. P. R. R. and the W. P. R. R.

History. The district was discovered by M. Milleson and party in 1866. A little copper ore carrying silver and gold values was shipped from the district by the Wolverine Copper Co. in 1916 and 1917.

Geology. The ore occurs in veins.

Bibliography. SMN1866 54 MR1916 I 482 MR1917 I 379
Weed MH 1378 Wolverine C. Co.

SPRING CITY see **PARADISE VALLEY**

SULPHUR (Rabbit Hole)
Sulphur, Silver, (Potash, Petroleum)

Location. The Sulphur District is situated on the W. flank of the Kamma Mts. on the S. E. border of the Black Rock Desert. Sulphur station on the W. P. R. R. is 2 m. N. W. Prior to the construction of the W. P. R. R. the district was reached from the S. P. R. R. by a road via Rabbit Hole and was known as the Rabbit Hole District. The Rosebud District in Pershing Co. adjoins it on the S. E.

History. The sulphur deposits were known to the Indians who showed them to the white men. The first locations were made by McWorthy and Rover in 1875, and a little later others were made by Hale and Wright and sold to the Pacific Sulphur Co. According to Russell, the district was producing 6 tons of sulphur per day in 1882. The sulphur mines were acquired by the Nevada Sulphur Co. in 1900, and are now producing about 12 tons of sulphur per day. In 1917, I. C. Clark located alunite claims in the same district. Some years ago, a well which was drilled for water near Sulphur by the W. P. R. R. was said to have struck indications of oil, and in 1921 an oil prospecting well was sunk in the vicinity by private individuals. The Silver Camel mine has been worked for many years by leasers who are said to have produced $120,000 in silver, according to a letter from A. J. Crowley.

Geology. The N. part of the Kamma Mts. is composed of Tertiary rhyolite. according to Adams, bordered on the W. by water-laid tuffs belonging to the Truckee Miocene formation. The sulphur deposits occur as crystal masses on the walls of the irregular cavities and filling large chambers in the tuffs. A small amount of cinnabar, some gypsum, and a considerable amount of alunite are present. Clark states that the alunite contains from 7% to 10.5% of potash. Rich stringers of hornsilver occur in the Silver Camel Mine.

Mines. The Nevada Sulphur Co. is operating a mine and sulphur plant of which A. J. Crowley is Supt. The property contains some 300 sulphur deposits distributed over an area of about 1,200 acres. P. Webster is Pres. and Mgr. of the Silver Camel M. & Dev. Co.

Bibliography. SMN1875-6 64-6 MR1883-4 865-6 MR1916 II 410-1
MR1885 496 MR1918 II 406 Potash
USGE40th 2 (1877) 742 MR1886 644 MR1921 I 384

Adams, G. I., "The Rabbit Hole Sulphur Mines naer Humboldt House, Nevada", USGS B 225 (1904) 497-502.

Clark, I. C., "Recently Recognized Alunite Deposits at Sulphur, Humboldt County, Nevada", E&MJ 106 (1918) 159-163.

E&MJ, "Oil Wildcatting in Nevada", 109 (1920) 665.

——— "Three Sections Prospecting for Oil in Nevada", 110 (1920) 872.

——— 111 (1921) 25 Petroleum.
Knopf, A., "Some Cinnabar Deposits in Western Nevada", USGS B, 620D (1915) 65.
Russell, I. C., "Sulphur Deposits in Utah and Nevada", N. Y. Acad. Sci., T 1 (1882) 168-175; E&MJ 35 (1883) 31-2; Abstract AJS, III 25 (1883).
——— "Geological History of Lake Lahontan", USGS M 11 (1885) 54-5.
Stuart NMR 8.

VARYVILLE (Columbia)
Gold

Location. The Varyville District is situated at Varyville on Bartlett Creek in W. Humboldt Co. Winnemucca on the S. P. R. R. and the W. P. R. R. is 120 m. S. E.

History. The Varyville District was discovered by Vary in the early seventies and the town of Varyville named after him. The district was organized as the Columbia District and in 1875 had a number of mines and two mills. It declined rapidly and has long been inactive.

Geology. There were many small veins in the district, rich in gold and with but little silver.

Bibliography. R1874 263 R1875 189-190 SMN1873-4 53
USGS Disaster topographic map.
Hill 507 215.

VICKSBURG see WARM SPRINGS

VIRGIN VALLEY
Opal

Location. The Virgin Valley opal field is located on Virgin Creek in N. W. Humboldt Co., about 25 m. S. W. of Denio, Oregon.

History. Opal was discovered in Virgin Valley in 1908.

Production. About $200 worth of gems was produced in 1909, and a quantity was mined in 1912 and 1913. In 1919, a black opal weighing 16.95 Troy ozs. and valued by the owners at $250.000 was mined from the property of the Rainbow Ridge M. Co.

Geology. The opal deposits occur in Miocene beds consisting chiefly of ash and tuff and in part deposited in shallow lakes, according to Merriam. These beds occupy a synclinal basin of older formations consisting mainly of tuffs, ashes, and rhyolitic lavas, and are capped along the valley rims by dark-gray vesicular lava. The formation has been block faulted and tilted. Fossil mammals have been found in the upper beds, and plant remains and petrified wood are common in the middle beds.

Opal. The opal occurs in ash or fine tuff beds associated with petrified wood. The precious opal occurs chiefly as casts of portions of limbs and twigs of trees and also as coatings and fillings in cracks in ordinary petrified wood and in the country rock, according to Sterrett. The best gem opal is unexcelled in color, brilliance, and variety, but there is considerable brittle opal which checks and cracks after mining.

Bibliography. MR1908 II 831 MR1912 II 1049-1050 MR1919 II 177-8
MR1909 II 771 MR1913 II 677-9 MR1921 II 145
USGS Long Valley topographic map.
Kunz, G. F., "On the Occurrence of Opal in Northern Nevada and Idaho", Abstract, N. Y. Acad. Sci. Annals 21 (1912) 214-5.
Merriam, J. C., Science, NS 26 (1907) 380-2

WARM SPRINGS (Ashdown, Pueblo, Vicksburg)
Gold, Silver

Location. The Warm Springs District is located at Ashdown in N. W. Humboldt Co., on the Oregon boundary. Winnemucca on the S. P. R. R. and W. P. R. R. is 120 m. S. E. The district was formerly divided into the Pueblo District on the N. and the Vicksburg District on the S.

History. The district was located in 1863. A small mill built in 1864 was burned by the Indians. The Pine Forest Gold M. Co's. mill made the largest production in Humboldt Co. in 1905 and operated intermittently thereafter. The Ashdown Gold M. Co. had a new mill under construction in 1919.

Geology. According to Stretch, the country rocks are mica and clay slates intruded by a core of porphyry and granite which forms the central axis of the mountains, and flanked on the W. by basalt. The early ores were mainly silver ores in quartz gangue, sometimes with considerable copper and sometimes with galena. Gold has predominated in recent productions.

Bibliography.

B18856 128	MR1906 295	MR1916 I 483
B1867 332	MR1908 I 489	MR1917 I 280
SMN1866 45-6	MR1909 I 411	MR1918 I 242
MR1905 269	MR1915 I 637	MR1919 I 396
USGS Disaster topographic map.		MR1920 I 325

Hill 507 215. Thompson & West 453.
Weed MH 1137 Ashdown Gold M. Co.

WILLOW POINT
Copper, Silver

Location. The Willow Point District is situated 20 m. N. N. E. of Winnemucca which is on the S. P. R. R. and the W. P. R. R.

History. In 1912, the Little Humboldt property shipped oxidized copper ore containing silver.

Bibliography. MR1912 I 799. USGS Paradise topographic map.

WINNEMUCCA
Silver, Gold, Lead, Copper, Granite, Sandstone

Location. The Winnemucca District is on Winnemucca Mountain at the S. end of the Santa Rosa Range. Winnemucca on the S. P. R. R. and the W. P. R. R. is a few miles S. E. Winnemucca Mountain rises to an elevation of over 6,600 ft. and the Pride of the West Mine in its foothills lies at an elevation of 4,400 ft. while the Adamson Mine on its W. slope is at an altitude of 5,975 ft.

History. The Winnemucca District was discovered by young Winnemucca, an Indian, in 1863. A roasting plant and quartz mill were erected at Winnemucca by the Humboldt Reduction Works in 1872 and treated the ores of the district for a number of years.

Production. The Pride of the Mountain was the most important mine in the district and is credited with an early production of about $1,000,000. From 1907 to 1921, the district produced 5,771 tons of ore containing $72,664 in gold, 66,651 ozs. silver, 13,280 lbs. copper and 49,532 lbs. lead, valued in all at $123,190, according to Mineral Resources of the U. S. Geol. Survey.

Geology. The country rock at the Pride of the West Mine consists of Upper Triassic calcareous slates, in part metamorphosed to hornfels, with strata of gray limestone, according to Lindgren. The ore deposits are veins containing lead, silver and gold. On the E. side of the mountain, several deposits of oxidized copper ore occur along a small body of diorite intruded into the slate and are probably of contact-metamorphic origin. On the W. side of the mountain at the Adamson Mine a well-defined vein occurs in the slate. At one end this vein contains calcite, limonite, cinnabar, and traces of precious metals; while beyond a fault the other end contains quartz, barite, and pale native gold in a streak with clay and limonite.

Bibliography.

R1868 132	SMN1866 55	MR1909 I 414	MR1914 I 687
R1869 193	SMN1869-70 27-8	MR1910 I 519	MR1915 I 638
R1872 157	SMN1871-2 51-2	MR1911 I 681	MR1916 I 483
R1873 213-4	SMN1873-4 51-2	MR1912 I 799	MR1917 I 280
R1874 262	SMN1875-6 62	MR1913 I 828	MR1918 I 242
R1875 188-9	SMN1877-8 66	MR1913 II 1370 Gran-	MR1919 I 369
USGE40ths (1877) 737-9.		its, 1375 Sand-	MR1920 I 325
USGS Paradise topographic map.		stone	MR1921 I 384

Hill 507 215.

Lindgren, W., "Geology and Mineral Deposits of the National Mining District, Nevada", USGS B 601 (1915) 11-12, 13, 15-16.

Stuart NMR 120. Thompson & West 453.

Weed MH 1150 Bonanza M. Co. 1180 Craven Copper Co.
1221 Harmony Ms. Co. 1361 Too Close M. Co.
1377 Winnemucca Mountain M. Co.

LANDER COUNTY

AMADOR see REESE RIVER
AUSTIN see REESE RIVER
BANNOCK see BATTLE MOUNTAIN

BATTLE MOUNTAIN (Galena, Bannock, Telluride)

Silver, Gold, Lead, Placer Gold, Antimony, Arsenic

Location. The Battle Mountain District is located in the Battle Mt. Range principally in N. W. Lander Co. but extending into S. E. Humboldt Co. The N. section of the Lander Co. portion of the district lying about 5 m. W. S. W. of Battle Mt., and including Copper Basin, the old abandoned camp of Battle Mt. on Licking Creek, and Cottonwood Creek, is sometimes considered as the Battle Mountain District proper; while the S. section, lying about 15 m. S. W. of Battle Mt., and embracing Galena on Duck Creek, Copper Canyon, and the old abandoned camp of Bannock, is called the Galena or Bannock District. The elevation of Battle Mt. is 4513 ft., the old camp is at 6,132 ft., Galena at 5,900 ft.; and Antler and Sue Peaks in the Battle Mt. Range reach altitudes of nearly 8,500 ft.

History. The battle Mountain District was organized in 1866, having been discovered several years before. The Little Giant Mine was located in 1867 and became the most important early producer. By 1870, 32 mines, a mill and 2 smelters were in operation in the district. Between 1870 and 1880, the Battle Mt. M. Co., an English corporation, shipped over 40,000 tons of ore from Copper Canyon to Swansea, Wales. The antimony deposits on Cottonwood Creek were opened in 1871 and 50 tons shipped to San Francisco. The district was originally in Humboldt Co. but was ceded to Lander Co. in 1873. The mines of the district were idle from 1885 till 1897 when the Glasgow & Western M. Co. bought the Copper Canyon and Copper Basin Mines. There was a small gold rush to Bannock in 1909 and a gold discovery at Telluride, S. W. of Antler Peak, in 1910. In 1917, the holdings of the Glasgow & Western M. Co. at Copper Canyon and Copper Basin were acquired by the Copper Canyon M. Co., which has conducted extensive exploration work since. Large amounts of copper were shipped during the period of the Great War. In 1920, the Irish Rose Mine began shipping arsenic ore to the plant of the Toulon Arsenic Co. at Toulon.

Production. The Little Giant is credited with an early production of more than $1,000,000 in silver; the veins near Galena with one of between $4,000,000 and $6,000,000 in silver, lead and gold; and the Glasgow & Western M. Co. with several hundred thousand dollars worth of copper, according to Hill.

RECENT PRODUCTION OF THE BATTLE MOUNTAIN DISTRICT
(According to Mineral Resources of the United States, U. S. Geol. Survey)

Year	Ms.	Placers	Tons	Gold $	Silver Ozs.	Copper Lbs.	Lead Lbs.	Total $
1902			75		2,880			1,440
1903			625	5,180	121,095		1,120	56,108
1904			40	41	2,655		14,569	2,105
1905								
1906								
1907			1,452	1,362	8,645	287,300	22,210	65,706
1908	3		432	4,565	10,922		96,357	14,401
1909	4		292	12,855	9,238	354	6,930	18,003
1910	2	10	28	6,515	166			6,605
1911	4	10	114	5,861	2,613	551	22,467	8,326
1912	5	9	375	16,346	6,389	39,478	38,873	28,539
1913	15	12	546	30,676	13,910	116,574	28,134	58,384
1914	3	14	55	94,811	5,456		10,804	98,249
1915	6	17	135	162,444	2,532	1,920	20,060	165,008
1916	13	11	4,334	159,068	41,360	566,890	87,166	331,752
1917	20	33	18,162	71,627	19,055	2,431,247	98,978	759,570
1918	23	6	15,845	85,232	43,970	1,999,040	68,894	627,856
1919	9	8	3,544	51,017	15,203	254,860	50,145	118,106
1920	5	5	2,678	28,223	15,051	90,135	226,297	79,318
1921	3	12	947	75,998	3,237	103	12,834	79,825
Total 1902-21			49,679	$811,821	324,377	5,788,452	805,738	$2,463,583

Geology. According to Hill, the Paleozoic rocks of the district, beginning with the oldest, are shales, quartzite, conglomerates, and limestones. These sedimentary rocks have been intruded by small dikes and sheets of granite porphyry, monzonite, and quartz-diorite porphyry; and capped to the E. and W. by rhyolite and to the S., at Bannock, by augite-andesite.

Ore Deposits. The ore deposits occur for the most part along fissures or wide zones of fracturing, and as a rule are simple veins or replacement lodes though there is one lodelike contact-metamorphic deposit at Copper Canyon. There are four well-marked types of mineralization:—silver-lead lodes, copper deposits in the vicinity of intrusives, pyritic gold-quartz veins, and veins and replacement lodes carrying stibnite. The silver-lead lodes are fissure veins in the sandstone and quartzite and replacement lodes in the shales. The minerals present are galena, sphalerite, pyrite, and tetrahedrite. The oxide ore contains cerussite, anglesite and horn-silver, and the secondarily enriched ore polybasite, pyrargyrite, argentite and tetrahedrite. At Copper Canyon, the ore occurs in 3 fracture zones in altered quartzite intruded by a monzonite laccolith and dikes. The lodes are from 10 ft. to 30 ft. in width, with magnetite, pyrite, chalcopyrite, galena, and blende in the primary ore, chalcocite in the secondary, and cuprite and native copper in the oxide ore. At Copper Basin, limestones, shales, and quartzites are cut by numerous monzonite dikes; and disseminated primary mineralization with 0.3% copper has been enriched in the shale beds to ore. According to a letter from J. T. Reid, gold ore occurs replacing limestone on the property of the Buffalo Valley Ms. Co. in the N. W. part of the district.

Mines. Four properties have recently been active. The **Buckingham M. Co. of Nev.**, of which A. Johnson, 425 First Nat. Bank Bldg., San Francisco, is Pres., and C. W. Burge, Supt. at Battle Mt., owns 27 claims showing 2 veins developed by a shaft and tunnel. The **Copper Canyon M. Co.** was incorporated in Delaware in 1916 with a capital stock of 1,000,000 shares of $1 par value of which 900,000 have been issued. The stock is listed on the New York Curb. R. M. Atwater, 25 Broad St., New York City, is Pres.; L. E. Whicher, Treas.; F. S. Schmidt, Gen. Mgr.; with W. W. Cohen, S. H. March, and E. N. Skinner, Directors. The property consists of the Copper Canyon Group of 334 acres, the Copper Basin Group of 3,094 acres, and a camp site of 280 acres. Barling is Gen. Mgr. of the **Nicklas M. Co.** at Battle Mt., and James Dahl is Mgr. of the **Dahl Placer.**

Bibliography.

B1867 318-9, **412-3**	SMN1867-8 **42-3**	**MR1882 230, 438-9**
R1868 118-120	SMN1869-70 **29-36**	**MR1883-4** 643 Stibnite
R1869 190-1	SMN1871-2 45-50	**MR1885 40 Coal**
R1870 132-3	SMN1873-4 60-1	**MR1898 185** Copper
R1871 141, 213-7 Stibnite, 518	SMN1875-6 **77-82**	**MR1905 269**
	SMN1877-8 70-1	**MR1907 I 364**
R1872 158	USGE 40th 3, (1870) 317-9	MR1908 I 492
R1873 198-200		MR1909 I 415
R1874 240-1		
R1875 172, 189, 433		
MR1910 I 519-520	MR1914 I 688	MR1918 I 242
MR1911 I 682	MR1915 I 638	MR1919 I 396
MR1912 I 800	MR1916 I 483	MR1920 I 325, 58 Arsenic
MR1913 I 828	MR1917 I 280-1	MR1921 I 385

Bandman, C. J., "The Geology of the Battle Mountain Mining District, Nevada", M&EW 40 (1914) 933.

Cutler, H. C. "Telluride", M&SP 102 (1911) 845.

Hill 507 215, 216.

Hill 594 61-71.

Lawson, A. C., "Fanglomerate, a Detrital Rock at Battle Mountain, Nevada", Abstract GSA B 23 (1912) 72.

Martin, A. H., "The Bannock Mining District, Nev.,"' M&EW 32 (1910) 835.

Scott, W. A., "Operations at Battle Mountain, Nevada", M&EW 45, (1916) 327-8.

Stuart NMR 127, 129.
Thomas, C. S., "Bannock, Nevada", M&SP 99 (1909) 820-1.
Thompson & West 437-4.
Young, G. J., "A Cave Deposit", EG 10 (1915) 186-190.
Weed MH 1133 Antimony & Silver Ms. Co. 1156 Buckingham M. Co. of Nev. Buffalo Valley Ms Co.
1177 Copper Canyon M. Co. 1182 Dahl Placers.
1215 Gold Top M. Co. 1223 Hider Nevada M. Co.
1224 Homestake Cons. Placer Ms. Co. 1232 Joyce M. Co.
1266 Nevada-Calumet Copper Co. 1285 Nevada Silverfields Co.
1293 Nicklas M. Co. 1331 Silverfields M. Co., Ltd.

BIG CREEK
Antimony

Location. The Big Creek District is on Big Creek on the W. slope of the Toyabe Range. The elevation at the mouth of Big Creek Canyon is 6,400 ft. and Big Creek Peak rises above it to an altitude of 10,265 ft. Austin on the N. C. R. R. is 10 m. N. The Reese River District adjoins the Big Crek District on the N., the Birch Creek District adjoins it on the E., and the Kingston District on the S.

History. Antimony was known to be present in this district at least as early as 1870. What is now known as the Bray antimony mine was located as a silver mine in 1864, and its value for antimony was not recognized until 1891 when it was relocated by Joseph Bray. This property was worked at intervals up to 1898 and considerable 50% antimony ore was shipped from it. The Bray Mine lies S. of Big Creek. The Pine Mine, N. of Big Creek, was discovered in 1890 and a small amount of antimony ore shipped from it. The Big Creek mines were reopened in 1916 and shipped in 1917.

Geology. The country rock consists of shales and slates with occasional thin beds of quartzite and limestone and is believed to be of Ordovician age. At the Pine Mine the ore deposit is a silicified fault breccia cemented by white quartz intergrown with stibnite. At the Bray Mine there are two ore deposits one on the crest and one on the north flank of a zone of folding and faulting. That on the crest is composed of a network of quartz and stibnite stringers, while that on the flank consists of thin, tabular, irregular lenses of quartz and stibnite with a little tetrahedrite.

Bibliography.

B1867 402, 413, 431
SMN1866 97
SMN1869-70 44
MR1916 I 724
MR1917 I 657
Hill 594 114-123.
Spurr 208 93-7 Toyabe Range.
Stuart NMR 127.
Thompson & West 473.
Weed MH 1132 Antimony King Mine.

BIG SMOKY see BIRCH CREEK

BIRCH CREEK (Smoky Valley, Big Smoky)
Gold, Silver, (Molybdenum)

Location. The Birch Creek District is located on Birch Creek on the E. flank of the Toyabe Range W. of Smoky Valley. Austin on the N. C. R. R. is 12 m. N. The Reese River District adjoins the Birch Creek District on the N., the Big Creek antimony district adjoins it on the W., and the

Kingston District onthe S. W. Geneva Peak is the highest point in the district with an altitude of 10,994 ft.

History. The Birch Creek District was discovered in the early days and worked in the middle of the sixties. The Big Smoky Mine erected a 20-stamp mill at the mouth of Birch Creek which only ran for a few days as the ore proved too low grade to pay to work. Recently the Nevada Birch Creek M. Co. has been carrying on development work.

Geology. Granodiorite believed to be Cretaceous intrudes slates and shales with some limestone which are probably of Ordovician age, according to Hill. Quartz veins occur in the granodiorite and in the sedimentary rocks at and near granodiorite contacts. The common ore minerals are pyrite, galena, and sphalerite. Arsenical pryite occurs in the Smoky Valley contact vein and tetrahedrite in veins in granodiorite at one prospect. In the present prospects, gold values predominate over silver, but in the early days the reverse was the case and Raymond noted that the silver ores were silver-copper-glance and native silver. Some molybdenum prospects in the granodiorite N. of Birch Creek show quartz veins containing molybdenite flakes associated with chalcopyrite and a white micaceous mineral.

Mines. The Nevada Birch Creek M. Co. was incorporated in Nevada in 1919 with a capital stock of 1,500,000 shares. The office of the company is at Austin, and J. F. Bowler is Pres., and G. L. Belanger, Sec.-Treas. The company owns 8 claims on a contact vein which it has developed by 2 tunnels.

Bibliography. B1867 402, 413-4, 431. R1868 82-3. SMN1866 98
USGE 40th 3 (1870) 332-4, 347.
Hill 594 114-121, 125-7.
Spurr 208 93-7 Toyabe Range.
Weed MH 1264 Nevada Birch Creek M. Co.

BULLION (Lander, Campbell)

Silver, Gold, Lead, Copper, Arsenic

Location. The Bullion District is located on the E. slope of the Shoshone Range in N. E. Lander Co. Beowawe on the S. P. R. R. and the W. P. R. R. is 20 m. N. E. The district embraces Maysville, the Grey Eagle Mine and Mud Springs on the N.; Lander, 4 m. to the S. of Mud Springs; and Tenabo, 2 m. S. E. of Lander. The section of the district to the W. of Lander and Tenabo is sometimes called the Campbell District. Shoshone Peak which rises to an altitude of 9,760 ft. 2 m. W. of Maysville is the highest point in the Shoshone Range. The Hilltop District adjoins the Bullion District on the N. W.

History. Lander is the oldest camp in the Bullion District and was the milling center of the district in the seventies and eighties. The Lovie Mine was the principal property and worked its ore in a 5-stamp pan- amalgamation mill. The Grey Eagle Mine which also operated in the seventies and eighties was reopened for a short time in 1905. The Maysville mine worked its ore in a 4-pan mill about 1880, and has been operating recently. A rush to Tenabo took place in 1907 and Mud Springs was discovered the same year. In 1921, some ore from the Little Gem Mine was shipped to the plant of the Toulon Arsenic Co. at Toulon.

Production. The Lovie Mine is credited with a production of $300,000 in silver. From 1905 to 1921, the district produced 7,016 tons of ore containing $120,677 in gold, 538,545 ozs. silver, 365,221 lbs. copper, and 117,680 lbs. lead, valued in all at $624,546, according to Mineral Resources of the U. S. Geol. Survey.

Geology. The country rocks of the Bullion District are Carboniferous quartzites, shales, and limestones which have ben intruded by granodiorite and are capped in places by Tertiary andesite. Fissure veins occur in all these rocks. The Lovie vein is in siliceous shales and quartzite, according to Emmons, and its ore consists of quartz, iron oxides, hornsilver, lead and copper carbonates. The Maysville vein is in quartzite and its primary ore consists of quartz, pyrite, galena and gray copper. The Grey Eagle vein is in granodiorite which is strongly altered near the lode, and its sulphide ore consists of quartz, blende, galena, pyrite, and gray copper. The little Gem vein at Tenabo, is in andesite which is slightly altered near the lode and its sulphide ore is composed of quartz, arsenopyrite, chalcopyrite, galena, and blende with a little bornite and chalcocite.

Bibliography. MR1906 296 MR1913 I 828 MR1918 I 243
MR1907 I 364 MR1914 I 688 MR1919 I 396
MR1908 I 492 MR1915 I 638 MR1920 I 325
MR1909 I 415 MR1916 I 484 58 Arsenic
MR1912 I 800 MR1917 I 281 MR1921 I 385
Emmons 408 113-120 Hill 507 215 Stuart NMR 127.
Weed MH 1230 Jersey Valley Ms. Co.

BUNKER HILL see KINGSTON

CAMPBELL see BULLION

CORTEZ see EUREKA COUNTY

GALENA see BATTLE MOUNTAIN

GOLD BASIN

Gold, Silver

Location. The Gold Basin District is at Carroll in S. W. Lander Co. on the Churchill Co. border. The Eastgate District adjoins it on the S. W.

History. In 1912, the Gold Basin property produced gold-silver ore containing traces of copper and lead.

Bibliography. MR1912 I 800.

HILLTOP (Kimberly)

Gold, Silver, Copper, Lead

Location. The Hilltop District is at Hilltop on the N. E. slope of Shoshone Peak in N. E. Lander Co. Battle Mountain on the S. P. R. R. and the W. P. R. R. is 18 m. N. W. The camp lies at an elevation of 6,300 ft. and Shoshone Peak attains an altitude of 9,760 ft. The Bullion District adjoins the Hilltop District on the S. E., and the Lewis District on the N. W.

History. The district was discovered in 1906 and a boom occurred in 1908. The Hilltop Mlg. & Reduction Co. put a 10-stamp amalgamation mill into operation in 1912 which was changed later to a cyanide mill. The Kimberly Cons. Ms. Co. was the principal operator for a number of years, and its holdings were recently acquired by the Hilltop Nevada M. Co. which has just completed a 100-ton flotation mill.

Geology. The country rock of the Hilltop District is Carboniferous quartzite, accordng to Emmons, which includes fine siliceous shales with here and there some fine conglomerates, and is cut by altered dikes of granodiorite. The ore occurs in a zone of fractured quartzite cut by leached porphyry. Free gold is present in stringers of quartz and iron oxide; and the sulphide ore minerals are pyrite, galena, and a gray mineral said to contain bismuth.

Production. From 1909 to 1921, the Hilltop District produced 28,797 tons of ore containing $282,361 in gold, 214,415 ozs. silver, 349,488 lbs. copper and 480,057 lbs. lead, valued in all at $560,066, according to Mineral Resources of the U. S. Geol. Survey.

Mines. The Hilltop Nevada M. Co. was incorporated in Nevada in 1921 with a capital stock of 5,000,000 shares of $1 par value. The home office is at 1200 Liberty Bldg., Philadelphia. M. K. Harr is Pres., and the other Directors are J. L. Cox, W. H. Clark, L. L. Link, and W. T. Luttrell. At Hilltop, C. B. Harr is Mgr.; A. R. Kohlmetz, Gen. Supt.; C. Lakamp, Mine Supt., and G. Machan, Mill Supt. The property consists of 23 claims, a section of 640 acres, developed by tunnels, and equipped with a 100-ton flotation mill.

Bibliography.

MR1909 I 415	MR1913 I 829	MR1917 I 281
MR1910 I 520	MR1914 I 688	MR1918 I 243
MR1911 I 682	MR1915 I 638	MR1919 I 396
MR1912 I 800	MR1916 I 483	MR1920 I 325
		MR1921 I 385

Carpenter, J. A., "Kimberly, Nevada", M&SP 100 (1910) 482-3.
Emmons 408 120-1.
Hill 507 216.
Warren, S. P., Mill of the Hilltop Mlg. & Reduction Co., Colo. S. of M. Mag. 4 51; Abstract MI (1914) 357.
WeedMH 1149 Blue Dick M. Co.
1223 Hilltop Group, Hilltop Mlg. & Reduction Co.
1234-5 Kattenhorn Mine
1236 Kimberly Cons. Ms. Co., Kimberly Shipper M. Co.
1237 Kirk Mine.

JERSEY see PERSHING COUNTY

KIMBERLY see HILLTOP

KINGSTON (Santa Fe, Bunker Hill, Victorine, Summit)
Gold, Silver

Location. The Kingston District is at Kingston on the W. Flank of the Toyabe Range on the S. boundary of Lander Co. Austin on the N. C. R. R. is 24 m. N. The Kingston District, or Summit District as it was called in the early days, is sometimes divided into the Santa Fe District on the N. and the Bunker Hill or Victorine District on the S. The Big Creek and Birch Creek Districts adjoin the Kingston District on the N. and N. E. and the Washington District lies just to the S. W. in Nye County.

History. The Victorine Mine was discovered in 1852 and has been opened and shut down again repeatedly since that time. In 1875 there were 4 mills in Kingston Canyon, but operations were never very successful owing to the low grade of the ore. The Mother Lode Mine in Santa Fe Canyon was opened at an early date by the Centenary Co. which built a mill which was moved to the Newark District in 1867. The Kingston Co. has operated a 60-ton stamp mill intermittently since 1909.

Geology. The ore at the Victorine Mine occurs in quartz pockets in a bed of silicified limestone interbedded with shales and slates, according to Hill. These sedimentary beds are believed to be of Ordovician age. The surface ore consists of iron-stained quartz carrying limonite and copper carbonates which in depth changes to white and gray quartz containing pyrite galena, sphalerite, and tetrahedrite scattered irregularly through it. Argentite and silver chloride are said to have been found. Gold and silver values were about equal. The deposit at the Mother Lode Mine is of similar character except that more pyrite is present and the values are in gold.

Bibliography. B1867 402, 414, 431 | SMN1866 98 | MR1909 I 415
R1868 83-4 | SMN1871-2 65-6 | MR1915 I 638
R1874 239 | MR1920 I 325
USGE40th 334-6, 347.
Hill 594 114-121, 128-9. Spurr 208 93-7, Toyabe Range.
Thompson & West 472, 473, 476.

LANDER see BULLION

LEWIS (Pittsburg)

Silver, Gold

Location. The Lewis District is located in Lewis Canyon on the N. slope of Shoshone Peak in N. E. Lander Co. Battle Mountain on the S. P. R. R. and the W. P. R. R. is 14 m. N. The S. E. section of the district where the Pittsburg Mine is situated is sometimes known as the Pittsburg District. The Lewis District adjoins the Hilltop District on the N. W.

History. The Lewis District was discovered in the summer of 1847 by Jonathan Green and E. T. George. The Eagle Cons. Co. had a 10-stamp mill in operation in 1876; and the Starr Grove Mine was overhauling its mill in 1878. A narrow-gauge branch of the N. C. R. R. was built from Lewis Jct. to Lewis and afterwards extended to the Starr Grove Mine; but this line was dismantled many years ago. The Betty O'Neal mine was discovered in 1880. The gold deposits of the Pittsburg and Morning Star Mines were discovered shortly after the silver deposits of Lewis and were under development in 1882. These two mines were consolidated as the Cumberland Mines in 1904, but have not been operating recently. The Betty O'Neal Mine was shut down in 1882 as the result of a disastrous boiler explosion and again in 1918 on account of the death of the manager. It was reopened by N. H. Getchell in 1920, and a 100-ton flotation mill put into operation upon it in the fall of 1922.

Geology The country rocks of the Lewis District are quartizites, limestones, and shales of Carboniferous age which have been intruded by granodiorite porphyry and cut by quartz porphyry, according to Emmons. The Morning Star and Pittsburg veins are in granodiorite and quartzite, while the Starr Grove and Betty O'Neal mines are in limestone and quartzite. The gangue minerals are quartz with subordinate calcite, and barite. The ore minerals are pyrite, arsenopyrite, galena, blende, chalcopyrite, gray copper, free gold, and sometimes silver. The Betty O'Neal ore is a white sugary quartz tinted pink in places by manganese and carrying galena, gray copper, argentite, and stephanite.

Mines. The Betty O'Neal Mines have a capital stock of 1,000,000 shares of $5 par value of which 600,000 shares are held in reserve. The home office is at 68 Devonshire St., Boston. G. W. Sias is Pres.; N. H. Getchell, V. P.-Gen. Mgr., at Battle Mountain; and F. E. Nye, Sec.-Treas. The company owns 1260 acres of mineral land developed by tunnels and a 360-ft. shaft and equipped with hoist, compressors, and a 100-ton flotation mill.

Bibliography. R1875 189 | MR1909 I 415 | MR1917 I 281
MR1910 I 520 | MR1918 I 243
SMN1875-6 76 | MR1913 I 829 | MR1919 I 396

SMN1877-8 71 MR1914 I 688 MR1921 I 385
Bancroft 239. Emmons 408 122-6. Hill 507 215.
SLMR, "New Flotation Milling Plant for Betty O'Neal Mines Co.", 24 (1922) 9-10.
Stuart NMR 127.
Weed MH 1141 Battle Mountain Ms. & Dev. Co.
1144 Betty O'Neal Ms.

NEW PASS
Gold

Location. The New Pass District is located at New Pass in the New Pass Range on the boundary of Lander and Churchill Cos. Austin on the N. C. R. R. is 27 m. E.

History. It was discovered in 1865 and a small 5-stamp mill moved there from Austin in 1868. In recent years the Nevada Austin Mines Co. erected a 100--stamp cyanide mill at New Pass, but the operations of this company were unsuccessful and the mill was sold to W. C. Pitt of Lovelock.

Geology. Raymond states that the country rocks are limestone, porphyries, and gabbro and that the veins occur in the latter rock and are from 6 in. to 6 ft. in width. The gangue is quartz and the ore minerals are gold, argentiferous galena enclosing native gold, and auriferous pyrite, copper sulphide, azurite, and malachite.

Bibliography. B1867 420, 431 SMN1867-8 87-8 MR1918 I 243
R1868 76-7
Stuart NMR 127. Weed MH 1264 Nevada Austin Ms. Co.

PITTSBURG see LEWIS

RAVENSWOOD (Shoshone)
Copper, Lead, Silver, Gold

Location. The Ravenswood District is situated to the S. W. of Ravenswood Peak in the Shoshone Range in central Lander Co. Ravenswood camp is 7 m. W. of Silver Creek, a siding on the N. C. R. R.

History. The Ravenswood District was discovered in 1863 by prospectors from Austin and has been explored intermittently since that date, but the production has been insignificant.

Geology. The ore occurs in small lenslike quartz veins in Cambrian shales, quartzites, and limestones, according to Hill. The gangue consists of quartz and occasional barite; and the ore minerals are chalcopyrite, antimonial galena, and tetrahedrite, all of which are said to carry silver and a little gold.

Bibliography. B1867 413 SMN1869-70 44 MR1914 I 688.
Hill, J. M., "Notes on Some Mining Districts in Eastern Nevada", USGS B 648 (1915) 25, 28, 32, 36, 39, 42, 105-113.
Spurr 208 98-9 Reese River Range (Shoshone).

REESE RIVER (Austin, Amador, Yankee Blade)
Silver, Gold, Lead, Copper

Location. The Reese River District is situated E. of the Reese River in the Toyabe Range in S. Lander Co. Austin on the narrow-gauge N. C. R. R. is in the center of the district. The altitude of Austin is 6,800 ft.; Mt. Prometheus E. of the town rises to a height of 8,256 ft.; and Telegraph Peak 7 m. N. N. E. has an elevation of 9,500 ft. The principal mines are on Lander Hill, a spur of Mt. Prometheus. The N. and the N. E. sections of the district were formerly known as the Amador District and the

Yankee Blade District respectively. The Birch Creek District adjoins the Reese River District on the S. E., the Big Creek antimony district adjoins it on the S. W., and the Skookum District on the N. W.

History. The Reese River District was discovered by William M. Talcott on May 2, 1862. A rush ensued, the town of Austin was laid out, and on December 19, 1862, the new county of Lander was separated from Humboldt and Churchill Cos. to accommodate the Reese River miners. Austin was made the county seat on September 2, 1863. By 1867, there were 11 mills and more than 6,000 mining claims in the camp. The White Pine rush which took place that winter greatly reduced the population of Austin. During the period from 1872 to 1877, the Manhattan Silver M. Co. of Nev. acquired most of the properties on Lander Hill. This company continued its operations until 1887 when its mill, which had produced more than $19,000,000 from 100,000 tons of ore, was finally shut down. In 1880, the N. C. R.R. was completed to Austin. During recent years, nothing but small leasing operations have been carried on in the Reese River District.

Production. The early production of the Reese River District was about $50,-000,000. From 1902 to 1921, the district produced 16,126 tons of ore containing $52,451 in gold, 305,615 ozs., 17,851 lbs. copper, 166,307 lbs. lead, and 590 lbs. lead, valued in all at $265,688, according to Mineral Resources of the U. S. Geol. Survey.

RECENT PRODUCTION OF REESE RIVER DISTRICT LANDER COUNTY

(According to Mineral Resources of the United States, U. S. Geol. Survey)

Year	Mines	Tons	Gold $	Silver Ozs.	Copper Lbs.	Lead Lbs.	Zinc Lbs.	Total $
1902		50	550					550
1903		3,000	10,000					10,000
1904								
1905		205	1,605					15,087
1906	4	192	1,131	54,634	3,109	14,000	590	39,170
1907	2	586	667	27,788	2,640	528		19,535
1908	10	161	1,355	15,953	212	6,691		10,119
1909	3	618	3,272	9,127	1,562	19,837		9,074
1910	6	674	3,764	27,159	6,797	72,076		22,464
1911	7	3,043	21,801	101,592	1,972	29,067		77,199
1912	8	86	1,576	3,390	12	47		3,665
1913	8	3,952	4,123	18,256	348	922		15,238
1914	4	3,015	253	12,783	104	4,076		7,495
1915	4	63	264	4,745	96			2,687
1916	2	78	4	3,707	121			2,473
1917	7	101	155	3,455	284	10,476		3,980
1918	9	107	1,136	7,638	181	3,227		9,048
1919	4	55	53	4,013	35	43		4,557
1920	5	121	685	8,810	153	3,754		10,616
1921	3	18	57	2,575	225	1,563		.2,731
Total 1902-21	..	16,126	$52,451	305,615	17,851	166,307	590	$265,688

Geology. The principal country rock is quartz-monzonite which is probably of

Cretaceous age. It is intruded into quartzites and calcareous shales which are probably pre-Carboniferous. The summit of Mt. Prometheus and the E. slope of the Toyabe Range, to the E. and S. E. of Austin, are capped by Tertiary rhydolite. The quartz-monzonite is cut by a number of lamprophyric dikes which are more abundant on the W. slope of Lander Hill, to the W. of the rich high-grade area, than in the central part of Lander Hill, according to a letter from W. A. Marshall. E. of the town, the quartz-monzonite is also cut by a few aplitic dikes, according to Hill.

Ore Deposits. The orebodies of the district occur in lenticular shoots in fissure veins. These veins occur mostly in the quartz-monzonite, but they are also found in the schistose quartzites about 3½ m. N. of Austin. The walls of the veins are softened and bleached for a few inches, but this alteration is local in character and shows only at the surface, above the sulphide zone. While there are numerous small veins in the district, there are 5 principal veins on Lander Hill, and 3 principal veins on Central Hill to the S. E. of Lander Hill and across Pony Canyon. The main production of the camp has come from these larger veins, which are parallel, extend in a N. W.-S. E. direction for over ½ m., and dip to the N. E. They are cut by a series of N.-S. faults with a W. dip.

Ore. The ore occurs in shoots which vary in thickness from several inches up to 5 ft. and in length from 100 ft. to 1,200 ft. averaging from 12 in. to 14 in. of high-grade ore. The gangue at the surface consists of a somewhat rusty quartz strongly impregnated and blackened with pyrolusite and some psilomelane. In depth the gangue is white quartz and some dark quartz with rhodochrosite. The veins are sometimes banded, with quartz on the outside, quartz and rhodochrosite next, and quartz in the center. The ore minerals are chiefly associated with the quartz, and Hill says they were probably deposited in the following order: pyrite and chalcopyrite, arsenopyrite, galena, sphalerite, wurtzite, and tetrahedrite rich in silver. All the sulphides except the pyrite and possibly the primary sphalerite are argentiferous.

Mines. Three companies have been active recently. **The Austin Dakota Dev. Co.** has a capital stock of 1,500,000 shares of $1 par value, of which 1,200,000 shares have been issued. The home office is Valley City, N. D. F. White is Pres.; E. C. Cooper, V.-P.; and J. E. Buttree, Sec.-Treas., while C. F. Littrell is Supt. of Mines at Austin. The company owns 800 acres to the S. of Pony Canyon, developed by 4 shafts and equipped with hoist and compressor. The **Austin Manhattan Cons. Silver Ms.** is owned by Pittsburgh capitalists. W. A. Marshall is Supt. and Res. Agt. at Austin. This company owns 1,200 acres including the old Manhattan Mine on Lander Hill, which was the biggest producer of the early days. The property is developed by a 6,000-ft. tunnel, a number of vertical and inclined shafts, and miles of old workings. The **Nevada Equity Ms. Co.,** of which George D. Kilborn is Pres., and J. L. Madden, Res. Agt., owns ground adjoining the Austin Manhattan Cons. Silver Ms. on the W. and N. W., and covering the extension of the main Lander Hill veins. The next adjoining property to the S. W. is a small property owned by M. J. Escobar of Austin upon which a number of veins have been exposed by surface workings.

Bibliography. B1866 33, 84, 128-130, 245-7 B1867 395-403, 430-1 SMN1869-70 37-44 SMN1871-2 60-65 SMN1873-4 55-60 MR1911 I 682-3 MR1912 I 800-1 MR1913 I 829

R1868 77-80, 96-8	SMN1875-6 71-6	**MR1914** I 688-9
R1869 118, 128-140, 733-740	SMN1877-8 70	MR1915 I 638
R1870 111-4	MR1905 269	MR1916 I 484
R1871 141, 167-171, 181	MR1906 296	MR1917 I 281
R1873 193	517 Bismuth	**MR1918** I 243
R1875 171-2, 189,	MR1907 I 363	MR1919 I 396
446-451	MR1908 I 492	**MR1920** I 325
SMN1866 97, 99-101	MR1909 I 414	58 Arsenic
SMN1867-8 44-7	MR1910 I 519	MR1921 I 385

USGE 40th 2 (1877) 627-9, 3 (1870) 320-322, 347-393.

Bancroft 264-6.

Hill 507 216.

Hill 594 95-114.

Martin, A. H., "The Lander Mining District, Nevada", MS 61 (1919) 508-511.

Penrose, R. A. F., Jr., "Manganese; Its Uses, Ores, and Deposits", AR of the Geol. Sur. of Ark. for 1890, 1 (1891) 476-7.

Spurr 208 93-7.

Spurr, J. E., "Genetic Relations of Western Nevada Ores", T AIME 36. (1906) 400.

Stuart NMR 127, 128.

Taylor, N. B., "A study of Ores from Austin, Nevada", S of MQ 34 (1912) 32-9.

Thompson & West 473, 475-6.

Weed MH 1138 Austin Dakota Dev. Co.; Austin Goldfield M. Co.
1139 Austin Manhattan Cons. S. Ms.
Austin Nevada Cons. Ms. Co.
1179 Copper Prince M. Co.
1249 Maricopa Ms. Co. 1279 Nevada G. Ms. Co.

SANTA FE see KINGSTON

SHOSHONE see RAVENSWOOD

SKOOKUM

Silver, Gold

Location. The Skookum District lies at an elevation of 6,750 ft. in a low group of hills W. of Reese River in S. Lander Co. The C. P. R. R. skirts these hills just N. of Ledlie station. The Skookum District adjoins the Reese River District on the N. W.

History. The district was discovered in 1907 by an Indian who sold out to Lemaire Bris. of Battle Mountain. They commenced operations in 1908. when a small stampede took place. Some rich ore was shipped during the boom period and a little ore was shipped in 1914.

Geology. The veins of the Skookum District occur in dark fine-grained quartzites of Paleozoic age on the S. and S. W. sides of the hills, according to Hill. Small veins contain argentiferous tetrahedrite and a little pyrite intergrown with quartz.

Bibliography. MR1907 I 363-4 MR1914 I 688 USGE40th 2 (1877) 641-3.

Higgins, W. C., "Skookum, Nevada's New Chloride Camp", SLMR 10 (1908) 17-21,

Hill 507 216. Hill 594 92-5.

SMOKY VALLEY see BIRCH CREEK

SUMMIT see BATTLE MOUNTAIN

VICTORINE see KINGSTON

YANKEE BLADE see REESE RIVER

LINCOLN COUNTY

ATLANTA (Silver Park, Silver Springs)
Silver, Gold, (Radium)

Location. The Atlanta District, formerly known as the Silver Park or Silver Springs District, is located in a basin on the N. E. flank of the Fortification Range in N. E. Lincoln Co. Pioche on the U. P. R. R. is 47 m. S.

History. The district was discovered in 1869. In 1872, 2 mills were in operation on rich silver ore which soon gave out. In 1906 the Atlanta Group was acquired by the Atlanta Cons. Gold M. & M. Co. which was in turn taken over by the Atlanta-Home Gold M. Co. in 1915; and small shipments were made from 1913 to 1920.

Geology. The country rocks of the Atlanta District are quartzite and limestones which are probably of Cambrian age overlain by Tertiary rhyolite tuff upon which rest rhyolite flows. The ores occur in brecciated fault zones, cemented by quartz. The ore minerals are limonite, pyrolusite, hornsilver, ruby silver, minor amounts of copper and lead carbonates, and native gold. Gold preponderates over silver in value in the ore of the Atlanta Mine and finely crystalline carnotite occurs in joints in the gold ore and in the rhyolite tuff country rock, according to Hill.

Mines. The Atlanta-Home Gold M. Co. owns 11 patented claims equipped with hoist and compressor and developed by a 500-ft. shaft and drifts. F. R. McNamee, Pacific Electric Bldg., Los Angeles, is Pres.

Bibliography.

R1869 143	SMN1873-4 76	MR1920 I 326
SMN1869-70 97	SMN1875-6 172	MR1918 I 243
SMN1871-2 111	MR1914 I 689	MR1917 I 282

Abbott, J. W., "Present Conditions at Pioche, Nevada", M&SP 109 (1914) 486.

Davis 929.

Hewett, D. F., "Carnotite in Southern Nevada", E&MJP 115 (1913) 234.

Hill 648 33, 38, 39, 114-120.

Meyer, P. W., "Developments at Atlanta, Nevada", E&MJ 99 (1915) 541-2.

Stuart NMR 136.

Weed MH 1137 Atlanta-Home G. M. Co.

BRISTOL see JACK RABBIT

CALIENTE see CHIEF

CHIEF (Caliente)
Gold, Silver, Copper, Lead

Location. The Chief District is 8 m. N. N. W. of Caliente which is on the U. P. R. R.

History. The district was organized in 1870, and Raymond & Co. operated there in 1872. In 1909, gold ore was shipped. A 20-ton cyanide mill was in operation on the Gold Chief property in 1911 and a 50-ton mill in 1913 and 1914. Copper ore was produced in 1912, and lead ore and gold-silver ore in 1917.

Geology. The ore deposits of the Chief District are veins in Paleozoic sediments cut by basic dikes, according to Hill.

Bibliography.

R1872 186	MR1909 I 416	MR1913 I 830
SMN1871-2 98	MR1911 I 683	MR1914 I 690
SMN-875-6 92	MR1912 I 801	MR1917 I 282

USGS Pioche topographic map.

Davis 930. Hill 507 216-7. Thompson & West 484.

COMET

Silver, Lead, Gold, Copper, (Zinc, Tungsten)

Location. The Comet District is located on the S. W. flank of Comet Mt. on the W. side of the Highland Range. Pioche on the U. P. R. R. is 14 m. N. E., and the Pioche District adjoins the Comet District.

History. Comet was discovered in 1882. Shipments of silver-lead ore containing minor amounts of gold and copper were made from 1913 to 1920.

Geology. The ore deposits of the Comet District are silver-lead-copper-gold veins and replacements in Cambrian sediments cut by diorite porphyry. Zinc is said to occur in the ores of the Stella and Tungsten Comet Mines; and tungsten in those of the Silver Comet and Tungsten Comet Mines.

Mines. The Tungsten Comet M. Co. is the only company at present active. E. D. Smiley of Panaca is Pres.; and A. L. Scott is Res. Agt. at Pioche. The property is developed by a 500 ft. shaft and equipped with a hoist and compressor.

Bibliography. MR1913 I 530 MR1919 I 397 MR1920 I 326
MR1914 I 690 MR1921 I 386
USGS Pioche topographic map.
Anderson, J. C., "Ore Deposits of the Pioche District, Nevada", E&MJ 113 (1922) 281, 283.
Davis 943. Hill 507 217.
Weed MH 1227 Hybla M. Co. 1330 Silver Comet M. Co.
1345-6 Stella Ms. Co. 1362 Tungsten Comet M. Co.

DELAMAR see FERGUSON

EAGLE VALLEY (Fay, Stateline)

Gold, Silver, Lead

Location. The Eagle Valley District lies E. of Eagle Valley and N. of Fay in E. Lincoln Co. on the Utah border. Modena, Utah, on the U. P. R. R. is 21 m. S. E.

History. The Horseshoe Gold M. Co. treated its ore in a 90-ton cyanide plant at Fay in 1902, and in a 120-ton plant in 1908. The Newport Nevada operated a 30-ton amalgamation and cyanide plant in 1907, and the Iris a 30-ton arrastra in 1908. A mine in the Stateline section produced lead ore in 1908. The district continued to produce up to 1915.

Geology. The ore deposits of the Eagle Valley District are veins in Tertiary volcanic rocks, according to Hill. Gold values predominate over silver.

Bibliography. MR1905 270 MR1908 I 493 MR1913 I 830
MR1906 296 MR1909 I 416 MR1914 I 690
MR1907 I 366-7 MR1910 I 521 MR1915 I 639
MR1911 I 683
USGS Pioche topographic map.
Davis 945.
Gayford, E., "Details of Cyaniding", M&SP 85 (1920) 237-8.
Hill 507 217.

ELY see PIOCHE

FAY see EAGLE VALLEY

FERGUSON (Delamar)

Gold, Silver

Location. The Ferguson District is situated at Delamar on the W. slope of the Meadow Valley Range in central Lincoln Co. Caliente on the U. P. R. R. is 32 m. E. N. E.

History. The Ferguson District was discovered in 1892 and the principal claims purchased by Capt. John De Lamar in 1893. A barrel chlorination mill was installed in 1895 but soon discarded in favor of a cyanide plant which was gradually enlarged to a capacity of 300 tons. The Delamar Mine became the principal gold producer in Nevada, and held this position until 1900, shutting down in 1909. In 1920 and 1921 small shipments of silver ore were made from the Magnolia Mine.

Production. Miller estimates the production of the Delamar Mine from 1892 to 1908 at about $25,000,000. The production of the Ferguson District from 1902 to 1921, as recorded by Mineral Resources of the U. S. Geol. Survey, was $3,978,786, mostly in gold.

Geology. The principal country rock of the Ferguson District is Cambrian quartzite. This is cut at the mine by two dikes of granite porphyry which in turn are cut by a lamprophyric dike, according to Emmons. The main fracture lies within this latter dike, and the ore occurs in shoots of crushed quartzite adjoining this fracture at its intersections with the granite porphyry dikes. The quartzite in the ore has been altered to crystalline quartz and impregnated with pyrite, chalcopyrite, and an undetermined telluride.

Bibliography.

MR1903 183	MR1908 I 493	MR1912 I 801
MR1904 146	MR1909 I 416	MR1913 I 830
MR1905 116, 270	MR1910 I 520	MR1920 I 326
MR1906 122, 296	MR1911 I 683	MR1921 I 386

USGS Pioche topographic map.

Davis 320, 338, 341, 944-5.

Emmons, S. F., "The Delamar and Horn Silver Mines", T AIME 31 (1902) 658-675.

Hill 507 217.

Miller, G. W., "The De Lamar Mines, Lincoln County, Nevada", MS 58 (1908) 347-8.

FREIBERG (Worthington)

Location. The Freiberg District is situated in N. W. Lincoln Co. at the N. end of the Worthington Mts. on their E. flank. Pioche on the U. P. R. R. is 75 m. E.

History. In 1865, Didlake and Aikens were shown the district by an Indian and named it the Worthington District. In 1869, George Ernst and party found good ore and reorganized the district as the Freiberg District. In 1919 a little oxidized iron ore containing silver and lead was shipped from the Roadside property, and one lot of lead-silver ore was shipped in 1921. The Olympus M. & M. Co. of which J. Petrula, Frieberg via Sharp, is Mgr., is now operating.

Geology. The veins of the Freiberg District occur in rhyolite, according to Gilbert.

Bibliography.

SMN1866 64	SMN1871 113	MR1919 I 397
SMN1869-70 100-1	SMN1873-4 78	MR1921 I 386

USGSW 100th 3 (1875) 37.

Hill 507 217.

Spurr 208 76-7 Worthington Mountains.

Thompson & West 485.

GEYSER see PATTERSON

GROOM
Lead, Silver

Location. The Groom District is located in S. W. Lincoln Co. Indian Springs on the state highway and formerly on the L. V. & T. R. R. which was recently scrapped, is 60 m. S.

History. The district was organized in 1869 and worked for about 5 years, and abandoned. The Groom Mine was reopened recently and from 1915 to 1919 made a production of some $250,000.

Geology. The country rock of the Groom District consists of limestones and shales, and the ore is argentiferous galena.

Bibliography. SMN1869-70 101 SMN1871-2 97 SMN1873-4 66 MR1915 I 639 MR1916 I 484 MR1917 I 282 MR1918 I 243

Thompson & West 485.

Weed MH 1219 Groom Mine; Groom South End M. Co.

HIGHLAND see PIOCHE
HIKO see PAHRANAGAT

JACK RABBIT (Bristol)
Silver, Copper, Lead, Gold, Manganese

Location. The Jack Rabbit District is at Bristol in the Ely Range in N. E. Lincoln Co. Bristol is 16 m. N. N. W. of Pioche which is on the U. P. R. R., and the Jack Rabbit Mine, 2 m. to the N. E. of Bristol, is connected with Pioche by the narrow-gauge Pioche-Pacific R. R., while the Bristol Silver Mine is connected with the ore bins of tne narrow-gauge line by means of an aerial tramway 9,000 ft. long. The Jack Rabbit Mine is on the E. side of the range at an elevation of 6,446 ft., the summit of the ridge has an altitude of about 8,500 ft., and the Bristol mines on the W. flank are at elevations ranging from 7,200 ft. to 8,300 ft. The Silverhorn District adjoins the Jack Rabbit District on the W.

History. The Bristol District was organized in 1871 by Hardy, Hyatt, and Hall, and about 60 locations were made. The National Mine built a smelter. The Hillside Mine was discovered by John Blair in 1877, and the Hillside Co. took over the National Mine, well, and smelter. The Jack Rabbit District was organized in 1876 and the Day or Jack Rabbit Mine became an important producer. The narrow-gauge railroad from Pioche to Jack Rabbit was begun in 1891. The branch railroad connecting Pioche with Caliente on the transcontinental line was completed in 1907. A consolidation of the Bristol and Jack Rabbit properties was effected in 1911 by the Day-Bristol Cons. M. Co., and the aerial tramway from Bristol to Jack Rabbit was constructed in 1913. The mines were purchased by the Cons. Cal.-Nev. M. Co. in 1915, acquired by the Uvada Copper Co. in 1918, and taken over by the Bristol Silver Ms. Co. in 1919.

Production. The early production of the Jack Rabbit District is estimated at from $2,000,000 to $6,000,000. The recent production of the district up to 1911 and for 1914 and 1915 has been included in that of the Pioche District by Mineral Resources of the U. S. Geol. Survey.

RECENT PRODUCTION OF JACK RABBIT DISTRICT

(According to Mineral Resources of the United States, U. S. Geological Survey)

Year	Mines	Tons	Gold $	Silver Ozs.	Copper Lbs.	Lead Lbs.	Zinc Lbs.	Total $
1916	4	11,265	3,110	73,694	711,770	253,529		244,191
1917	4	6,709	1,466	45,059	381,315	146,368		155,282
1918	3	5,871	1,896	48,921	410,879	76,373	14,763	159,070
1919	5	8,288	1,745	75,242	106,582	100,738		110,070
1920	5	15,527	4,303	222,160	429,029	1,254,647		425,770
1921	5	14,156	5,200	189,558	335,922	1,732,548		316,057
Total 1916-21		61,871	$17,749	655,491	2,375,708	3,590,006		$1.411,709

Geology. The ore deposits of the Jack Rabbit District are replacements of Cambrian limestone, according to Hill. There are two sets of fissures in the district, a N. E. series with S. E. dip and a N. series with steep E. dip. As a rule, the replacement is most intense at the junctions of these two sets of more or less open fissures. The ores are almost entirely oxidized and include copper-silver ore, lead-silver ore, and manganese-silver ore. The silver is present as chloride and to a less extent as bromide and native silver, according to Goodale. The copper ores include carbonates, chrysocolla, red and black oxides, tetrahedrite and chalcocite. The lead ores are mostly carbonate with smaller amounts of anglesite and galena and containing some zinc carbonate and copper minerals. Manganese is present in the Jack Rabbit Mine chiefly as wad, psilomelane, and pyrolusite, according to Pardee and Jones, associated with more or less calcite, quartz, and limonite.

Mines. The companies which have recently been active in the Jack Rabbit District are the Black Metal Ms., Inc., and the Bristol S. Ms. Co. The Bristol Co. owns 550,000 of the 990,000 outstanding shares of the Black Metals M's., Inc. The Bristol S. Ms. Co. was incorporated in Nevada in 1919 with a capital stock of 1,000,000 shares of 10 cts. par value of which 800,000 shares have been issued. The home office is at 220 Felt Bldg., Salt Lake City. W. I. Snyder is Pres.; W. F. Snyder, V.-P.; E. H. Snyder, Sec.-Treas.; with G. W. Snyder and R. J. Evans, Directors. J. E. Tarbat is Supt. at Pioche. The company owns 324 acres developed by a 3,800 ft. tunnel and several miles of workings and equipped with hoist, 2 compressors, and a 9,000-ft. aerial tramway.

Bibliography. SMN1871-2 111-2
SMN1873-4 78
SMN1875-6 92
SMN1877-8 77
MR1901 174 Copper
MR1907 I 368
MR1908 I 493-4
MR1909 I 416-7
MR1910 I 521
MR1911 I 684
MR1912 I 801
MR1913 I 830
MR1914 I 690
MR1915 I 639
MR1916 I 484
MR1917 I 282
MR1918 I 243-4
MR1919 I 397
MR1920 I 326
MR1921 I 386

USGS Bristol Range topographic map.

Abbott, J. W., "Pioche, Nevada", M&SP 95 (1907) 178-9.

Anderson, J. C., "Ore Deposits of the Pioche District, Nevada", E&MJ 113 (1922) 279-285.

Goodale, S. L., "The Bristol Mines", M&M 30 (1910) 507-9.

Hill 507 216, 218.

Hill 618 24-6, 34, 39, 40, 41, 42, 124-137.

Pardee, J. T., & Jones, E. L., Jr., "Deposits of Manganese Ore in Nevada". USGS B 710F (1920) 221.

Spurr 208 46.

Stuart NMR 133.

Thompson & West 488-9.

Weed MH 1141 Bay State M. & Leasing Co. 1147 Black Met. Ms., Inc.
1152 Bristol Cons. M. & S. Co. 1152-3 Bristol Silver Ms. Co.
1170 Cons. Cal.-Nev. M. Co. 1182 Day Bristol Cons. Ms. Co.
1189 E. & F. M. Co. 1366 Uvada Copper Co.

PAHRANAGAT (Hiko)

Silver, Lead, Copper

Location. The Pahranagat District is situated on the E. flank of Mt. Irish 10 m. N. W. of Hiko. Caliente on the U. P. R. R. is 60 m. E.

History. The district was discovered by an Indian of the Pahranagat, or "squash eater", tribe who showed it to a party including Ely, Hatch, McClusky, Sanderson, Sayles, and Strutt in 1865. A thousand locations were made, expectations ran high, and Lincoln Co. was separated from Nye County on February 26, 1866, to accommodate the new district. The county seat was first located at Crystal Springs and then at Hiko. Raymond and Ely erected a 5-stamp mill, but little ore was found and the mill was moved to Pioche in 1869. In 1871, the county seat was moved to Pioche and the Pahranagat District was deserted until recently.

Geology. The ore deposits are veins in Paleozoic sediments containing silver-lead ore with a low copper content.

Mines. The Silver Prince Cons. M. Co. of which R. P. Jones of Caliente is V.-P., has recently been active.

Bibliography.

B1866 33	R1870 174	SMN1871-2 97
B1867 215, 339-340, 402, 426-9, 666-670	R1873 233	SMN1873-4 66
	SMN1866 64-6	SMN1875-6 90
R1868 112-5	SMN1867-8 80-5	MR1915 I 639
R1869 194-201	SMN1869-70 99	MR1920 I 326

Bancroft 271-2.

Davis 928-930.

Hill 507 217.

Spurr 208 152-3 Hiko Range.

Thompson & West 484-5.

Weed MH 1228 Irish Mountain S. Ms.
1334 Silver Prince Cons. M. Co.

PATTERSON (Geyser)

Silver, Gold

Location. The Patterson District is situated in Patterson Pass at the S. end of the Schell Creek Range in N. E. Lincoln Co. The postoffice is Geyser. Ely on the N. N. R. R. is 50 m. N., and Pioche on the U. P. R. R. an equal distance S. The elevation of the pass is 7,250 ft., while Patterson Peak, 6 m. to the N., is 10,000 ft in altitude.

History. The Patterson District was shown to R. G. Patterson by an Indian in January, 1869. Soon afterwards, 250 claims were located in the hills N. of the pass and a mill was erected. The rich ores soon gave out and the district was abandoned until recently. A little work was done in 1907, and in 1920, a shipment of siliceous silver ore was made by the Lake Valley M. Co. In 1921, this company shipped more ore and began the erection of a 50-ton cyanide plant.

Geology. The country rocks of the Patterson District are quartzites, shales and limestone of Cambrian age, cut by one narrow granite porphyry dike. acording to Hill. Small replacement veins of quartz and calcite occur in the limestone N. of the pass with occasional thin films of argentiferous lead carbonate and more rarely small pockets of silver-bearing copper carbonates. At one prospect copper carbonates occur in crushed shale, and at another the shale is altered to quartz, calcite, diopside, and fluorite with some sphalerite and pyrite. Sphalerite and pyrite also occur in joints in the quartzite at this property and a little molybdenite was observed in one joint.

Bibliography. R1869 178-9 | MR1905 I 270 | MR1914 I 690
SMN1869-70 93-4 | MR1907 I 367 | MR1920 I 326
SMN1873-4 78 | MR1921 I 386
Davis 930. Hill 507 218.
Hill 648 34, 40, 42, 120-4. Spurr 208 40.

PIOCHE (Ely, Highland)

(Silver, Lead, Gold, Copper, Zinc (Manganese)

Location. The Pioche District is situated at Pioche in the Pioche Range in N. E. Lincoln Co. The Pioche Range is a low E.-W. ridge, of which Mt. Ely, 7,150 ft., is the highest point; lying E. of the Highland Range in which Highland Peak attains an altitude of 9,395 ft.; and separating Duck Valley on the N. from Meadow Valley on the S. The town of Pioche is situated on the N. slope of the Pioche ridge at an elevation of 5,725 ft.; and is connected with Caliente which is on the main line of the U. P. R. R. by a branch line 32 m. in length. The district is sometimes divided into the Ely District on the E. and the Highland District on the W. The Comet District adjoins the Pioche District on the S. W.

History. Pahute Indians showed the Pioche District to William Hamblin, a Mormon missionary, in 1863; and the Meadow Valley District was organized in 1864. There was no activity of importance in the district until 1868 when F. L. A. Pioche sent C. E. Hoffman into the camp to purchase the claims which were afterwards worked by the Meadow Valley M. Co. Raymond and Ely brought their 5-stamp mill from Pahranagat to Bullionville, S. of Pioche, in 1869; and the district was reorganized as the Ely District that same year. There was a stampede to Pioche, which was made the county seat in 1871. In 1872 the boom reached its height and Pioche had a population of 6,000. The Meadow Valley M. Co. and the Raymond & Ely M. Co. were the two great rival mining companies of the early days. Bullionville became the milling center because of its water supply, and in 1873 the 21-m. narrow-gauge, Pioche & Bullionville R. R., was completed. The two big mining companies ceased their operations in 1876. There was a little activity in the nineties, but no important revival occurred until 1905 when the transcontinental line was completed through Caliente. In 1907, the Caliente and Pioche branch line went into operation, and since then the Pioche District has been an important producer. The Prince Cons. M. & S. Co. built a standard-gauge railway from Pioche to its mine in 1912; and in 1916 built a 200-ton flotation mill at Bullion to treat tailings, but this latter enterprise proved unsuccessful. The biggest producers in the camp within recent years have been the Prince Consolidated Mine and the adjoining Virginia Louise Mine which have shipped large tonnages of low-grade silver-lead ore containing 30

40% iron and manganese in excess of insoluble content making it a desirable flux for the Salt Lake smelters.

RECORDED PRODUCTION OF THE PIOCHE DISTRICT

Year	Ms.	Tons	Gold $	Silver Ozs.	Copper Lbs.	Lead Lbs.	Zinc Lbs.	Total $
1870								644,719*
1871								3,318,873*
1872								5,363,997*
1873								3,722,030*
1874								1,901,214*
1875								1,050,171*
1870								$16,001,004
1905	1	100		7,694				4,647**
1906	1	1,702	890	37,649	442,520	284,000		127,709**
1907	5	1,137	493	22,450	114,026	56,751	13,552	41,923**
1908	8	9,922	12,643	192,274	274,682	1,066,191		195,586***
1909	10	14,055	21,575	182,923	173,446	1,495,302		203,541***
1910	10	3,692	18,663	49,057	7,653	935,255		87,277***
1911	8	25,392	5,953	174,271	17,382	455,490		120,987***
1912	13	93,135	137,251	750,229	457,083	10,741,499	68,128	1,162,130***
1913	14	94,153	54,375	471,958	647,270	7,689,813	49,962	780,915***
1914	15	87,568	41,752	406,927	660,573	5,983,307	1,104,644	644,325***
1915	16	129,957	39,159	454,382	295,703	7,523,857	2,579,745	994,788***
1917	16	149,398	30,978	457,069	79,680	8,553,674	2,002,105	1,209,816**
1917	11	144,996	41,989	450,517	40,233	9,099,551	1,332,521	1,342,677**
1918	7	107,948	21,257	353,309	18,220	7,334,381	566,955	951,400**
1919	11	88,162	25,523	366,634	8,251	6,917,087	78,200	810,003**
1920	9	101,778	20,005	333,382	122,610	6,614,311		935,096**
1921	10	20,679	3,206	63,045	2,297	1,332,343		126,502**
Total 1905-1921		1,073,774	$475,712	4,773.770	3,361,629	76,082,812	7,795,812	$ 9,739,322

* Pack, F. J., School of Mines Quarterly, Vol. xxvii (1906) 285.

** Mineral Resources of the United States, U. S. Geol. Survey.

*** Includes Bristol and Jack Rabbit Districts, from Mineral Resources of the United States, U. S. Geol. Survey.

Geology. The Pioche Range is an anticlinal fold with an E.-W. axis in Cambrian sedimentary rocks. The oldest rock is the basal Cambrian quartzite which is over 1,500 ft. in thickness. Lying conformably above the quartzite is a series of alternating shales and limestones of Middle Cambrian age in which the shales preponderate for the first few hundred feet. In the Highland section, 10 m. W. of Pioche, there is an intrusion of granite into the limestone from which a rhyolite dike known as the Yuba dike extends E. to Pioche. Rhyolite dikes also occur near the Prince Consolidated Mine. Anderson has also noted diorite dikes in the neighborhood of Pioche. There are several series of mineralized fissures and of post-mineral faults. The ore deposits of the district may be divided into three groups: fissure veins in the quartzite and to a minor degree in the limestone and shale, contact veins along intrusive dikes, and bedded replacement deposits in the limestones and shales.

Fissure Veins. The Meadow Valley and Raymond & Ely Mines were located upon a fissure vein in the quartzite between the town of Pioche and the Yuba dike. The vein matter was siliceous and the ore minerals were mainly chloride of silver and carbonate of lead with a little gold and occasional sulphides. Fissures in the limestone and shale containing ore of a similar character occur at the Prince Consolidated Mine where they are of less importance than the bedded deposits.

Contact Veins. Silver-lead ore has been mined from contact veins along the Yuba dike, and zinc blende also exists in these veins.

Bedded Deposits. Anderson distinguishes three types of bedded deposits: thin, gold-silver-lead beds at limestone-shale junctions, silver-lead beds in limestone, and the large manganese-iron-silver-lead beds. Beds of the first type are now being mined by the Pioche Ms. Co.; while those of the second type have contributed the principal production of the Highland section. Beds of the third type are now the most important in the camp, being worked at the Prince Consolidated and Virginia Louise Mines on a large scale. The main mass of these ore deposits is made up of iron and manganese oxides through which are scattered calcite and quartz and the ore minerals, galena, lead carbonate and silver chloride. The big lead-zinc-silver-iron sulphide bed of the Combined Metals, Inc., is perhaps an unoxidized representative of this same type.

Mines. Four companies at present active. **The Combined Metals, Inc.,** was incorporated in Nevada in 1917 with a capital stock of 1,000,000 shares of 5 cts. par value of which 800,000 have been issued. E. H. Snyder, 516 Felt Bldg., Salt Lake City, is Pres., and Guido Cottina is Supt. at Pioche. This company has leased the Panaca Group of 6 claims belonging to the Amalgamated Pioche Ms. Corp. The property has 4,000 ft. of workings and is equipped with steam hoist and compressor. The **Pioche Ms. Co.** is operating under the superintendence of A. W. Thomas. The **Prince Cons. M. & S. Co.** was incorporated in Utah in 1907 and has a capital stock of 3,000,000 shares of 50 cts. par value of which 1,440,000 are outstanding. The stock is listed on the San Fancisco Stock Exchange and the New York Curb. A $300,000 bond issue was authorized in 1921. In 1915, 1916, and 1917, the stock paid dividends aggregating $574,924. The home office is at 1118 Newhouse Bldg., Salt Lake City. A. H. Godbe is Pres.; G. F. Wasson, V.-P.; M. C. Godbe, Sec.-Gen. Mgr.; with A. Thomas and F. C. Dern, Directors. The property consists of 386 acres, developed by a 550-ft. incline and 650-ft. shaft with several miles of workings, and equipped with 600 h. p. steam power, hoist, and compressor; 9-m. railway to Pioche; and 200-ton flotation plant. **The Virginia Louise M. Co.** was incorporated in 1912 and has a capital stock of 1,500,000 shares of $1 par value of which 1,112,000 have been issued. The stock is listed on the Los Angeles Stock Exchange. A bond issue of $60,000 was made in 1920. The home office is at 1203 Hibernian Bldg., Los Angeles. C. L. Horsey is Pres., and J. D. Thompson V.-P. L. G. Gillett is Supt. at Pioche and C. A. Thompson Sec.-Treas. The property consists of 3 claims developed by a 600-ft. shaft and 7,900 ft. of workings.

Bibliography. R1869 143 SMN1875-6 88-9, 92 MR1912 I 801
R1870 2, 164-8 SMN1877-8 73-7 MR1913 I 830
R1871 141, 223-249 MR1901 174 Copper MR1914 I 690-2

R1872 176-186
R1873 233-242
R1874 284-303
R1875 195-202, 268, 415
SMN1869-70 101-3
SMN1871-2 81-94, 97
SMN1873-4 65-6
MR1905 270
MR1906 296
MR1907 I 367
MR1908 I 493
MR1909 I 416-7
MR1910 I 521
MR1911 I 684
MR1915 I 639-40
MR1916 I 484-5
MR1917 I 282
MR1918 I 244
MR1919 I 397-8
MR1920 I 326-7
MR1921 I 386-7

USGSW100th 3 (1875) 243, 257-261.
USGS Pioche topographic map.

Abbott, J. W., "Pioche, Nevada", M&SP 95 (1907) 176-9.
——————"Present Conditions at Pioche, Nevada", E&MJ 109 (1914) 483-6.
Anderson, J. C., "Ore Deposits of the Pioche District, Nevada", E&MJ 113 (1922) 279-285.
Bancroft 271-3.
Bell, R. N., "The Pioche, Nevada, District", M&M 32 (1911) 243-4.
Davis 929-943.
Harder, E. C., "Manganese Deposits of the U. S.", USGS B 380 (1909) 153, 157.
Jessup, D. W., "Ore Deposits of the Prince Consolidated Mines", M&SP 106 (1913) 773-5.
——————"Mining the Prince Consolidated Ores", M&SP 106 (1913) 820-3.
M&EW, Staff Correspondence, "The Nevada-Utah Property at Pioche", 45 (1916) 1034.
M&SP, Editorial, "What Ails Pioche", 124 (1922) 248-9.
Maynard, G. W., "The Ore Deposits of Pioche, Nevada", E&MJ 51 (1891) 171-2.
Pack, F. J., "Geology of Pioche, Nevada, and Vicinity", SofMQ 27 (1906) 365-386.
Spurr 208 38-47 Schell Creek and Highland Ranges.
Stuart NMR 131-2.
Thompson & West 484-5.
Walcott, C. D., "Second Contribution to the Studies on the Cambrian Faunas of North America", USGS B30 (1886) 35, 36.
Weed MH 1130 Alps M. Co., Amalgamated Pioche Ms. & Sms. Corp.
1133 Arcane M. Co. 1166 Combined Metals, Inc.
1174 Cons. Nev.-Utah Corp. 1183 Demijohn Cons. M. Co.
1189 Eastern Prince G. & S. M. Co. 1221 Harney Group.
1224 Home Run C. Co. 1250 Mascot S. M. Co.
1253 Mendha-Nev. M. Co. 1286 Nev. S. Klondike M. Co.
1288..Nevada Volcano Ms. Co.
1304 Pioche Metals M. Co., Pioche Ms. Co.
1306 Prince Cons. M. & S. Co. 1328 Silgoled M. Co. of Nev.
1367 Victoria Nev. M. Co., Virginia Louise M. Co.
Wiltsie, E., "Notes on the Geology of the Half-Moon Mine, Pioche, Nevada", T AIME 21 (1893) 867-873.
Zalinski, E. R., "Ore Occurrence at Prince Consolidated", E&MJ 95 (1913) 809-812.

SILVERHORN

Silver

Location. The Silverhorn District is in N. E. Lincoln Co. adjoining the Jack Rabbit District on the W. Pioche on the U. P. R. R. is 23 m. S. E. The camp has an elevation of 6,200 ft.

History. The district was discovered by J. L. Whipple while searching for a stray horse in the fall of 1920. The original location was purchased by Crampton & Crampton of Santa Monica, Cal., and promoted as the Nevada Silver Horn M. Co. by Weir Bros. & Co. of New York. A boom took place early in 1921 which subsided later in the year.

Geology. The country rocks of the Silverhorn District are limestonees and shales of Upper Paleozoic age cut by an acid dike. The ore deposit at the Silver Horn and Silver Dale Mines is a siliceous replacement of fault breccia at a shale-limestone junction and is believed by Anderson to have a genetic relation with the neighboring acid dike. The ore consists mainly of quartz and the ore minerals are hornsilver and argentite. According to Marriage, nickel is also present.

Bibliography.

Anderson. J. C., "Ore Deposits of the Pioche District, Nevada", E&MJ 113 (1922) 281, 284.

Crampton, T. H. M., "The Silver Horn District, Near Pioche, Nevada", M&SP 122 (1921) 883-4.

Marriage, E. C. D., "A Method of Determining Nickel in Silverhorn Ores", E&MJ 112 (1921) 174.

Weed MH 1285 Nevada Silver Horn M. Co. 1329 Silver Carlisle M. Co.
1330 Silverdale Ex. M. Co., Silver Dale M. Co., Silver Dividend M. Co.
1333 Silver Horn M. & Dev. Co., Silver Peer M. Co.

SILVER PARK see ATLANTA

SILVER SPRINGS see ATLANTA

STATELINE see EAGLE VALLEY

TEM PIUTE

Silver

Location. The Tem Piute District is located at Tem Piute in the Timpahute Range in W. Lincoln Co.

History. It was discovered by Service and Plumb in 1868 and worked during the seventies but has been inactive since.

Geology. Silver chloride ore occurs in lodes in Paleozoic shale and limestone. Antimony, arsenic, copper, and lead are present in some of the ores.

Bibliography.

R1869 201 — SMN1867-8 85 — SMN1873-4 75-6
R1870 174 — SMN1869-70 99-100 — SMN1877-8 78-9
SMN1871-2 113-4

Davis 929. Thompson & West 486-7.

VIOLA

Copper, Silver

Loation. The Viola District is situated at Viola in the Mormon Range in S. E. Lincoln County.

History. A small shipment of copper-silver ore was made from the Viola property in 1917.

Bibliography. MR1917 I 282

WORTHINGTON see FFREIBERG

LYON COUNTY

COMO see PALMYRA

DAYTON see SILVER CITY

DEVILS GATE AND CHINATOWN see SILVER CITY

ELDORADO CANYON

Coal

Location. Eldorado Canyon lies on the boundary between Lyon and Ormsby Cos. The Eldorado Canyon District adjoins the Palmyra District in Lyon Co. on the E., and the Delaware District in Ormsby Co. on the W.

History. In the early sixties, lignite coal was discovered on Eldorado Canyon, and the Newcastle Co. marketed several hundred tons, but the deposit was small and the quality poor so the operation was abandoned. Another attempt to put these mines on a commercial basis was made in the seventies but given up after a few years.

Bibliography. B1867 313 | SMN1867-8 20, 88 | SMN1875-6 95
SMN1866 18 | SMN1873-4 62 | MR1885 40
USGS Carson topographic map.
Thompson & West 498.

GOLD CANYON see SILVER CITY

INDIAN SPRINGS see PALMYRA

MASON see YERINGTON

MOUND HOUSE

Gypsum

Location. The Mound House District lies on the E. flank of the Virginia Range W. of Moundhouse which is on the V. & T. R. R. and the S. P. R. R. The district is in Lyon Co. on the Washoe and Ormsby County borders, adjoining the Silver City District on the W.

Geology. On the E. slope of the Virginia Range in the neighborhood of Mound House, Triassic strata including limestones, gypsum, shales, and meta-andesites have been intruded and somewhat metamorphosed by granodiorite which is probably of Cretaceous age, acording to Jones. At the Pacific Portland Cement quarry, a thick lens of gypsum dipping at a high angle to the E. is interbedded with limestone. This bed thins to the N., is cut off by granodiorite on the S., and changes to anhydrite in depth. At the Regan quarry a thin bed of gypsum with a steep E. dip outcrops below limestone. The gypsite deposits to the E. have been formed from this integrated gypsum.

Mines. The gypsite was formerly treated at the Regan Mill just E. of Mound House on the S. P. R. R., but this mill has not been in operation for several years. The Pacific Portland Cement Co., Cons., is quarrying gypsum and transporting it by aerial tramway to its mill a short distance N. of Mound House on the V. & T. R. R. W. J. Walmsley is Supt. The home office of the Pacific Portland Cement Co., Cons., is in the Pacific Bldg., San Francisco. The company is now developing the large gypsum deposits in the Gerlach District in Washoe and Pershing Cos.

Bibliography. MR1902 907 | MR1910 II 727 | MR1916 II 260
MR1903 1039 | MR1912 II 648 | MR1917 II 89
MR1904 1044 | MR1913 II 367 | MR1918 II 290

MR1905 1108 MR1915 II 155 MR1919 II 105
USGS Carson topographic map.
Jones, J. C., "Nevada" in "Gypsum Deposits of the U. S.", USGS B 697 (1920) 150-153.
Louderback, G. D., "Gypsum Deposits in Nevada" in "Gypsum Deposits in the U. S.", USGS 8 223 (1904) 112-118.
Stone, R. W., "Gypsum Products; Their Preparation and Uses", USBM TP 155 (1917) 48, 56.
Stuart NMR 8, 141-2.

PALMYRA (Como, Indian Springs)

Gold, Silver

Location. The Palmyra District is located at Como in the Pine Nut Range in S. W. Lyon Co. Dayton on the S. P. R. R. is 12 m. N. W. The N. E. section of the district was formerly known as the Indian Springs District, and the district as a whole is sometimes called the Palmyra and Indian Springs District. The old Eldorado Canyon coal district adjoins it on the West.

History. The Palmyra District was discovered shortly after the Comstock Lode, and a mill was erected at Como in 1863. Como was a large town in the early days and the county seat of Lyon County, but the mines were unsuccessful and the site was completely abandoned in 1874. Since then, a number of attempts to work mines in the district have met with failure. The latest of these was that of the Como Cons. Ms. Co. which was incorporated in 1916 and made a considerable production in its 100-ton cyanide plant from 1918 to 1920.

Geology. Numerous quartz veins occur in hornbende-andesite at the Como Consolidated Mine. The ore consists mainly of quartz with small amounts of calcite and other minerals, according to Borzynski, and gold and silver in approximately equal proportions occur in a finely disseminated state. Varying amounts of copper and zinc are also said to be present.

Bibiiography. B1867 328 MR1906 297 MR1919 I 398-9
SMN1866 27 MR1911 I 684-5 MR1920 I 327
SMN1867-8 88 MR1918 I 245 MR1921 I 387
USGS Carson and Wabuska topographic maps.
Borzynski, F., "The Como Consolidated Mill", M&SP 120 (1920) 55-6.
Cutler, H. C., "Como, Nevada", M&SP 104 (1912) 539-540.
Weed MH 1166 Como Cons. Ms. Co. 1167 Como Group M. Co. 1301 Peer G. M. Co.
Thompson & West 490-500.

RAMSEY

Gold, Silver

Location. The Ramsey District is situated at Ramsey on the S. E. flank of the Flowery Range in N. Lyon Co. on the Storey Co. border. Clarks on the S. P. R. R. is 13 m. N. W. The Talapoosa District adjoins the Ramsey District on the E.

History. The Ramsey-Comstock orebody was discovered in 1906. The town of Ramsey was founded, and a 30-ton amalgamating mill was built. The mill operated till late in 1910, making a production of about $80,000, since which time the production has been small.

Geology. The ore at the Ramsey-Comstock Mine occurs in a silicified fracture zone, which may be an altered rhyolite dike, in quartz-bearing andesite porphyry of Tertiary age. The ore consists of quartz, pyrite, and free gold with minor silver values.

Mines. The Lahontan Ms. Co. took over the Ramsey-Comstock Mine in 1919. B. P. Howell is Pres.; M. Eddy, V.-P.; and P. Liddell, Box 414, Reno, is Sec.-Treas.

Bibliography. MR1907 I 370 MR1908 I 494 MR1909 I 418 MR1910 I 521 MR1911 I 684 MR1916 I 485 MR1917 I 283 MR1918 I 245

USGS Wabuska topographic map.

Hill 507 218.

Hill, J. M., "Notes on the Economic Geology of the Ramsey, Talapoosa, and White Horse Mining Districts in Lyon and Washoe Counties, Nevada", USGS B 470 (1921) 89-104, 106-7,

Stuart NMR 7, 142. Weed MH 1238 Lahontan Ms. Co.

SILVER CITY (Devils Gate and Chinatown, Gold Canyon, Dayton)

Gold, Silver

Location. The Silver City District is at Gold Canyon on the E. slope of the Virginia Range in W. Lyon Co. on the Storey Co. border. Gold Canyon extends from Devils Gate near Silver City on the Storey Co. boundary line in a S. E. direction to Dayton on the Carson River. Silver City is on the V. & T. R. R. and Dayton is on the S. P. R. R. The district is sometimes subdivided into the Devils Gate and Chinatown, or Silver City Lode District on the N., and the Gold Canyon, or Dayton Placer District on the S. The Comstock District in Storey Co. adjoins the Silver City District on the N., and the Mound House gypsum district adjoins it on the W.

History. Abner Blackburn, a Mormon immigrant, discovered placer gold at the mouth of Gold Canyon in July, 1849, and the little placer mining camp of Johntown was founded further up the canyon. In 1853, the Grosh Bros. found rich silver ore in the neighborhood of Gold Canyon, which perhaps came from one of the Silver City lodes and perhaps from the main Comstock Lode. Miners from Johntown working up Gold Canyon to its head discovered the Comstock Lode in 1859. With the coming of the Comstock rush, numerous lode claims were located in the Silver City District. While Silver City was established before Virginia City, the mines in its neighborhood were mainly low-grade, and its importance was chiefly due to the quartz mills erected in its neighborhood. A revival of interest took place in the seventies, and the camp has made an important production since. The Gold Canyon Dredging Co. erected a large gold dredge in 1920.

Production. The early production of the district is not definitely known. From 1920 to 1921, it produced 294,089 tons of ore containing $1,789,132 in gold, 737,219 ozs. silver, 590 lbs. of copper, and 2,698 lbs. lead, valued in all at $2,302,673, according to Mineral Resources of the U. S. Geol. Survey.

Geology. The ore deposits are veins in Tertiary andesite and rhyolite. The district is geologically an extension of the Comstock District, containing the extension of the S. E., or Silver City, Branch of the Comstock Lode, and the S. extension of the Brunswick Lode. The Gold Canyon placers were derived from the disintegration of the lodes of the Comstock and Silver City Districts.

Mines. Three companies are at present active. The property of the **Buckeye**

Cons. M. Co., consisting of 30 acres at the S. end of the Brunswick Lode said to have produced $1,000,000 in the early days, was purchased by H. P. Kervin and eastern asociates in 1922. The mine is developed by a 650-ft. incline and the new owners have installed electrically driven machinery. **The Gold Canyon Dredging Co.** is a subsidiary of the Metals Exploration Co., as is the United Comstock Ms. Co. in the neighboring Comstock District. B. Wells, 1213 Hobart Bldg., San Francisco, is Pres.; G. H. Hutton is Mgr.; while W. J. Harvey is Supt. at Dayton. The company was incorporated in Nevada in 1919 with a capital stock of 250,000 shares of $1 par value, all issued. It has leased 297 acres of patented placer ground which it is working with a 9 cu. ft. electrically driven 500 hp dredge of 5,000 cu. yds. daily capacity. **The Oest Cons. Ms. Co.** acquired the Oest Mine in 1921. L. M. Hall is Mgr. The mine is developed by a 200-ft. shaft and is credited with a production of $1,-000,000. The new company has installed transformers, a 100 hp electric hoist, compressor, and steel headframe, and is planning to sink the shaft to 500 ft.

Bibliography.

R1871 141	MR1906 297	MR1914 I 693
SMN1866 26	MR1907 I 359	MR1915 I 640
SMN1867-8 88	MR1908 I 494	MR1916 I 485
SMN1869-70 14	MR1909 I 417	MR1917 I 283
SMN1871-2 101-2	MR1910 I 521-2	MR1918 I 245
SMN1873-4 62-5	MR1911 I 694-5	MR1919 I 398-9
SMN1877-8 80	MR1912 I 802	MR1920 I 327
MR1905 271	MR1913 I 830-1	MR1921 I 387

USGS Carson topographic map.

Bancroft 96-100. Davis 315, 341

Hill 507 219. Stuart NMR 139.

Thompson & West 498, 502.

Young, G. J., "Cooperation Among Small Mines", E&MJ 106 (1918) 812.

——————"Gold Dredging Started at Dayton, Nev.", E&MJ 110 (1920) 640.

——————"Dredge Construction at Dayton, Nev.", E&MJ 112 (1921) 91-6.

——————"Transportation of Dredge Material at Dayton, Nev.", E&MJ 112 (1921) 374-5.

Weed 1155 Buckeye Cons. M. Co. 1168 Comstock Milwaukee M. Co. 1207 Gold Canyon Dredging Co. 1371 Well-Emma Ms. Corp.

TALAPOOSA

Silver, Gold, Copper

Location. The Talapoosa District is at Talapoosa in N. Lyon Co. Talapoosa is 14 m. S. of Fernley which is on the S. P. R. R. and 5½ m. N. W. of Hawes, a siding on the Tonopah Branch of the same railroad. The Talapoosa District adjoins the Ramsey on the E.

History. The Talapoosa District was discovered by prospectors from Virginia City as early as 1864. The principal property has been owned by the Talapoosa Co. since 1905 and has been worked in recent years by leasers.

Geology. The country rock is Tertiary rhyolite cut by andesite dikes and capped in places by basalt. According to Hill, the Justice vein is on the hanging wall of an altered andesite dike. This vein is from 8 ft. to 28 ft.

in width and is composed of bluish-gray quartz cut by stringers of white quartz and calcite. Pyrite and chalcopyrite, probably with some argentite and silver chloride, occur in the dark quartz.

Mines. The Talapoosa Co. has a capital stock of 1,500,000 shares of $1 par value of which 600,000 are in the treasury. The stock is listed on the New York Curb. W. S. Proskey, 2 Wall St., New York City, is Pres.; and W. H. Ferguson, Cons. Eng. The company owns 290 acres, developed by shafts and tunnels to a depth of 100 ft., and equipped with hoists.

Bibliography. MR1911 I 685 MR1917 I 283 MR1918 I 245
MR1913 I 831 MR1921 I 387
USGS Wabuska topographic map.
Hill 507 219.
Hill, J. M., "Notes on the Economic Geology of the Ramsey, Talapoosa, and White Horse Mining Districts in Lyon and Washoe Counties, Nev.", USGS B 470 (1911) 99-104, 108.
Weed MH 1348 Talapoosa Co.

YERINGTON (Mason)

Copper, Gold, Silver, Lead, Gypsum, Turquoise, (Granite)

Location. The Yerington District is situated in the neighborhood of Yerington and Mason which are on the N. C. B. R. R. in S. Lyon Co. The district extends from the Walker River Range at the Mineral Co. border on the E. across Lyon Co. to the base of the Pine Nut Range in Douglas Co. on the W.; but its principal mines are in the low Singatse Range just W. of Yerington and Mason. Yerington has an altitude of 4,405 ft. and Mason is 1 m. S. and 4,420 ft. in elevation. The N. C. B. R. R. continues S. from Mason to Mason Pass where it crosses the Singatse Range and turns N. to Ludwig which is 2 m. W. of Mason across the range at an elevation of 5,122 ft. The Buckskin District in Douglas Co. adjoins the Yerington District on the W., and the Mountain View District in Mineral Co. adjoins it on the E.

History. The Ludwig Mine was opened in 1865 and produced small quantities of high grade ore including natural bluestone which was used in the silver mills of the Comstock. A number of unsuccessful attempts were made to smelt the copper ores of the district; one at Ludwig in the early days, another at the Bluestone Mine in 1900, and a third at Yerington. In 1907, energetic companies with adequate capital purchased the principal properties. The N. C. B. R. R. was begun in 1909 and completed to Ludwig in 1911. The Mason Valley built an 1,800-ton smelter at Thompson which was blown in in 1912, ran till 1914, resumed operations in 1917, and again shut down in 1919. The Nevada- Douglas Cons. Copper Co. began work on a leaching plant in 1914 and had it in full operation in 1917 but shut it down as a failure the following year. The Bluestone M. & S. Co. built a 500-ton flotation plant in 1918, which was enlarged to 80 tons capacity and ran until 1921. The May Adams property in Mason Pass erected a 10-ton amalgamating mill in 1921.

Production./ The Yerington District has been the second largest copper producer in Nevada. Prior to 1905, the production was about 1,000,000 lbs. of copper. In recent years the production has been very erratic, exceeding 15,000,000 lbs. annually in 1912, 1913, 1917, and 1918, when the smelter was in full operation, and being low and irregular at other times, as shown in the following table.

RECENT PRODUCTION OF YERINGTON DISTRICT

(According to Mineral Resources of the United States, U. S. Geol. Survey)

Year	Mines	Tons	Gold $	Silver Ozs.	Copper Lbs.	Lead Lbs.	Total $
1905	2	1,382	2,921	1,510	294,320		49,747
1906	2	2,501			868,993		167,715
1907	2	1,074	10,302	274	290,787	12,960	70,527
1908	1	360			100,849		13,312
1909	2	88	578	8	None		582
1910—No Production.							
1911	5	120	2,179	27	2,727		2,534
1912	28	239,606	1,035	34	17,057,793		2,815,592
1913	25	206,558	727	295	15,106,475	4,595	2,343,611
1914	21	102,467	7,091	5,746	7,274,329		977,754
1915	12	4,736	354	1,132	700,874		123,581
1916	20	19,820	670	1,693	3,491,390		860,666
1917	33	322,945	1,236	1,573	19,396,536	2,046	5,297,964
1918	23	398,638	60	980	17,138,747		4,234,311
1919	12	85,924	281	4,069	3,076,647		577,094
1920	9	26,634	408	4,508	1,045,944		197,776
1921	3	96	2,181	1,346			3,527
Total 1905-21		1,412,994	$30,023	23,195	85,846,411	19,601	$17,735,294

Geology. The oldest rocks of the Yerington District are of Triassic age and comprise andesites, soda rhyolite-felsites, and limestones, with subordinate quartzite, shale, and gypsum, according to Knopf. These rocks have an aggregate thickness of 8,000 ft., of which 3,200 ft. are volcanic rocks. The Triassic rocks were intruded, probably in Cretaceous time, by granodiorite, which metamorphosed them strongly and converted large volumes of limestone into calcium silicate rocks made up of garnet. wollastonite, and allied silicates. After this metamorphism, dikes and bosses of quartz-monzonite were injected and subsequently faulting took place. Tertiary volcanic rocks 7,000 ft. in thickness rest unconformably upon the Mesozoic rocks. They consist of a lower division of quartz-latite, rhyoolite, and andesite breccia; a middle division of andesite flows, resting in places on the eroded edges of the rhyolite; and an upper division of conglomerate capped by basalt.

Ore Bodies. The orebodies are contact-metamorphic replacements of comparatively pure limestones, according to Knopf, and are related to faulting. The principal ore consists of pyrite and chalcopyrite in a gangue of pyroxene, garnet, and epidote. Orebodies may attain lengths up to 800 ft. and widths up to 100 ft.; but are rarely all ore, the ore commonly making up but a small fraction of the pyroxene-garnet-epidote rock.

Bluestone M. & S. Co. The Bluestone M. & S. Co. was incorporated in Maine in 1917 with a capital stock of 200,000 shares of $10 par value of which 100,000 are outstanding. The home office is at 42 Broadway, New York City; and the mine office is at Mason. D. M. Steindler is Pres.; J. Carson, V.-P.-Managing Director; with J. B. Waddell, H. M. Marler, and M. B. Davis, Directors, and H. M. Thompson is Sec.-Treas. The company owns 440 acres developed to a depth of 540 ft., having 2 m. of workings, and equipped with hoist, compressor, 800-ton flotation plant, and 2½-m. railway.

Mason Valley Ms. Co. The Mason Valley Ms. Co. was incorporated in Maine in 1907 and has a capital stock of 500,000 shares of $5 par value, all issued. The stock is listed on the Boston Stock Exchange and the New York Curb. The home office is at 14 Wall St., New York City, and the mine office at Thompson. A. J. McNab is Pres.; T. Schulze and E. O. Holter, V.-P.; H. E. Dodge, Sec.-Treas.; and H. E. Franklin, Asst. Sec.-Asst. Treas. The property consists of 170 acres of mineral land and a smelter and 1320-acre smelter site at Thompson; together with the Grey Eagle Mine in California. The Mason Valley Mine is developed to a depth of 550 ft. and connected with the N. C. B. R. R. by an aerial tram 6,250 ft. in length.

Nevada-Douglas Cons. C. Co. The Nevada-Douglas Cons. C. Co. was incorporated in Utah in 1915 with a capital stock of 1,000,000 shares of $5 par value of which 979,643 have been issued. Bonds to the amount of $215,100 are outstanding. The home office is that of A. J. Orem & Co., 79 Milk St., Boston, and the general office is at 222 S. W. Temple St., Salt Lake City. H. I. Moore is Pres.; J. G. Berryhill, V. P.; L. H. Curtis, V. P.; F. M. Orem, Sec.-Treas.; W. C. Orem, Gen. Mgr.; with J. J. Corum, A. J. Orem, J. B. Lord, H. Goodacre, F. J. Curtis, and J. G. Berryhill, Jr., Directors. G. E. Mainhart is Gen. Supt. at Ludwig. The property consists of 950 acres of mineral land, 2 mill sites, and miscellaneous lands making a total of 1070 acres including the Ludwig, Douglas, and Casting Copper Mines, opened by 8 m. of workings, and equipped with 4 hoists, compressor, and 17,000-ft. pipe line. The company also owns 88% of the capital stock of the N. C. B. R. R.

Walker River C. Co. The Walker River C. Co. was incorporated in Nevada in 1915 with a capital stock of 2,500,000 shares of $5 par value of which 675,000 shares are outstanding. The home office is at 402 Madison Ave., New York City. R. M. Atwater, Jr., is Pres.; W. M. Sanders, V. P. with J. H. Susmann, C. N. Edge, H. B. Lake, J. A. Martin, and W. J. Hand, Directors. G. F. Willis is Mgr. at Yerington. The property consists of 250 acres including millsite and a lease on the Empire-Nevada Copper M. & S. Co. It is developed by a 350-ft. shaft and equipped with a 50-ton leaching plant.

Gypsum. A deposit of gypsum 4,000 ft. in length and with a maximum thickness of 200 ft. outcrops in the Triassic limestone 150 ft. W. of the gossan at the Ludwig Mine. The gypsum is very pure, snow white, and coarsely crystalline. It changes to anhydrite in depth. A similar deposit occurs 5 m. S. E. of Yerington, according to Jones. The Standard Gypsum Co. has recently purchased the Ludwig deposit and installed a two-kettle plaster plant which is now in operation.

Turquoise. Turquoise has been mined by Otto Taubert from two properties on the W. side of the Singatse Range, according to Sterrett. The turquoise occurs as seams and nodules in altered porphyritic rock.

BIBLIOGRAPHY OF YERINGTON DISTRICT

General

B 1867 317-8
MR 1883-4 545 Cobalt
MR 190 159
MR1910 II 885-6
Turquoise
MR 1911 I 685
MR 1914 I 693-4
MR 1915 I 640-2
MR 1916 I 485-6

MR 1905 271
MR 1906 297, 401
MR 1907 I 611, 370
MR 1908 I 213, 494
MR 1909 I 408, 418; II 788 Turquoise
MR 1910 I 510
MR 1912 I 802-3
MR1912 II 648 Gypsum
MR 1913 I 831
MR 1913 II 367 Gypsum; 1369, Granite
MR 1917 I 283
MR 1918 I 245-6
MR 1919 I 399-400
MR 1920 I 328
MR 1921 I 387

Carpenter, J. A., "The Yerington Copper District, Nevada", M&SP 101 (1910) 4-9.

Durand, C. S., The Yerington District, Lyon County, Nevada", M&M 31, (1910) 24-5.

Hill 507 219.

Scott, W. A., "Nevada-Douglas Mines and Mill", M&EW 45 (1916) 277-8.

Stuart NMR 140.

Tyssowski, J., "Conditions in the Yerington Copper District", E&MJ 89. (1910) 764-6.

WeedMH 1148-1381 Describes 21 Yerington companies.

Zehring, W. S., "The Nevada-Douglas Copper Properties, Nevada", MW 30 (1909) 736-738.

Geology

Goodale, F. A., "Yerington-Buckskin Copper District, Nevada", Colo S. of M. Mag., 1 (1911) 3-8.

Hodges, A. D., Jr., "Note on an Occurrence of Nickel and Cobalt in Nevada", T AIME 13 (1885) 657; "Carter's Copper Mine, Near Mason Valley, Esmeralda County".

Jennings, E. P., "Genesis of the Copper Deposits of Yerington, Nevada", E&MJ 83 (1907) 1143-4; Canadian M. J., 28 (1907) 365-6.

———"Secondary Copper Ores of the Ludwig Mine, Yerington, Nevada", Canadian M. Inst., J. 11 (1908) 463-6.

———"The Location of Ore Bodies and the Occurrence of Shoots in Metalliferous Deposits", EG 4 (1909) 255-7.

———"The Copper Deposits of Yerington, Nevada," AMC 12th Ann. Sess. R. of P (1909) 423-7.

Jones, J. C., "The Origin of the Anhydrite at the Ludwig Mine, Lyon County, Nevada", Discussion, EG 7 (1912) 400-2.

———"Nevada" in "Gypsum Deposits of the U. S.", USGS B 697 (1920) 153-5.

Knopf, A., "Geology and Ore Deposits of the Yerington District, Nevada", USGS PP 114 (1918) 68; Abstract E&MJ 107 (1919) 627; Abstract M&SP 118 (1919 455-6; Discussion, W. E. Gaby, "The Yerington District", M&SP 118 (1919) 615-6; Review by W. H. Emmons, E&MJ 107 (1919) 1079-1080.

Ransome, F. L., "The Yerington Copper District, Nevada", USGS B 380 (1909) 99-119; Abstract M&SP 100 (1910) 54-6.

Read, T. T., "The Nevada-Douglas Mines, Lyon County, Nevada", M&SP 105 (1912) 205-7.

Rogers, A. F., "A New Synthesis and New Occurrences of Covellite," SofM. Q 32 (1911) 298-304.

———"The Occurrence and Origin of Gypsum and Anhydrite at the Ludwig Mine, Lyon Co., Nev.", EG 7 (1912) 185-9.

Rogers, A. F., "Gypsum and Anhydrite from the Ludwig Mine, Lyon County, Nevada", Abstract, GSA B 24 (1913) 94.

Smith, Dwight T., "The Geology of the Upper Region of the Main Walker River, Nevada", U. of Cal. Dept. of Geol. B 4 (1905); Abstract

in M&SP 90 (1905) 183-4.
Spurr 208 117 "Smith Valley Range".

Metallurgy

M&SP, "Copper Leaching at the Nevada-Douglas Property", 107 (1913) 127.
——— "Leaching at Yerington," 111 (1915) 94.
——— "Leaching Copper Ores," 111 (1915) 485.
Orem, A. J., "Nevada-Douglas Con. Copper Company's Process of Ore Leaching", M&EW 43 (1915) 601-3.
Perry, R. W., "Leaching of Oxidized Copper Ores with Ferric Chloride," M&SP 118 (1919) 669-74.
Westby, C., "Sulphatizing with Weak Sulphurous Gases" E&MJ 104 (1917) 119-122.

Metallurgy

Aikens, W. "Hydro-Electric Power at Mason Valley, Nevada", M&EW 38 (1913) 1083-6.
Arentz, S. S., "A Standard Ore Chute", E&MJ 92 (1911) 1216.
USGS Wellington topographic map.

MINERAL COUNTY

ACME see FITTING

AURORA (Esmeralda)
Gold, Silver, (Chalcedony)

Location. The Aurora District is located at Aurora in W. Mineral Co. Aurora is 37 m. by road S. W. of Thorne, a station on the S. P. R. R. The town lies 3 m. E. of the California boundary, at an elevation of 7,415 ft. The district was formerly known as the Esmeralda District.

History. E. R. Hicks, who was prospecting with J. M. Corey and J. M. Braley, discovered the Aurora District while hunting for game on August 26, 1860. A spectacular rush ensued, and the camp at Esmeralda near the discovery proving unsatisfactory, the new town of Aurora was started on the flat to the N. By act of the first territorial legislature on November 25, 1861, Esmeralda Co., named after the Esmeralda District, was made one of the nine original countries and Aurora was named as the county seat. The district remained part of Esmeralda County until February 10, 1911, when Mineral Co. was separated from Esmeralda Co. and the Aurora District became part of the new country. In 1864, Aurora had a population of 10,000, but by 1869, the early bonanzas had become exhausted, and a decline began, although the district continued to be productive up to 1882. The Aurora Cons. Ms. Co., incorporated in 1912, was operated under the control of the Goldfield Cons. Ms. Co. from 1914 to 1918, when it was shut down and its 40-stamp mill moved to Goldfield.

Production. The production records for the early days are not compleete. $27,000,000 in bullion was shipped through Wells, Fargo & Co. up to 1869; and Wasson gives figures showing that $2,365,969 in bullion was shipped from 1861 to 1869 without insurance; so the output for 1861 to 1869 may be placed at about $30,000,000. The recent production came almost entirely from thee Aurora Consolidated Mine. From 1910 to 1920, the district produced $1,882,861 in gold and 128,808 ozs. silver, valued in all at $1,974,290, according to Mineral Resources of the U. S. Geol. Survey.

Geology. The rocks of the Aurora District are volcanic rocks, probably of Tertiary age, which have poured out on a basement the only portion of which exposed is a porphyritic granite. The oldest flows are biotite-quartz-latites with associated andesites at least 900 ft. thick, according to Hill. Above these come rhyolite flows up to 1000 ft. in thickness; while above the latite and rhyolite lies a basalt with a thickness up to 600 ft. All of these flows dip gently to the N. N. W. and rest upon uneven surfaces indicating erosion intervals between the extrusions.

Ore Deposits. The ore deposits are quartz veins filling fissures in the biotite-quartz-latite and associated andesites. They are not simple clean-cut veins, but send off numerous small interlacing branches. The country rock is altered for considerable distances from them.

Ore. The principal vein filling is a fine-grained white quartz deposited in bands of different-sized grains, and small druses lined with minute clear quartz crystals are present in all the veins. The rich ore is marked by irregular streaks of what looks like dark quartz, but is in reality made up of quartz, adularia, argentiferous tetrahedrite, pyrite, chalcopyrite, and a soft bluish-gray mineral supposed to be a combination of gold and possibly silver with selenium. Free gold is present in the richest ore.

Chalcedony. Beautiful rose and lilac chalcedony occurs resting upon a base of almost white chalcedony in an army gdaloidal rock, according to Kunz.

Bibliography.

B1866 28, 33, 107-118, 125-7	SMN1875-6 37	MR1910 I 522
B1867 335-6	SMN1877-8 24	MR1911 I 686
R1871 141	MR1904 955, Chalcedony	MR1912 I 803
SMN1866 30-36	MR1905 267	MR1913 I 832
SMN1867-8 92-95	MR1906 294	MR1914 I 695
SMN1869-70 108-9	MR1907 I 349	MR1915 I 642
SMN1871-2 31-2	MR1908 I 479	MR1916 I 486
SMN1873-4 19	MR1909 I 401-2	MR1917 I 284
		MR1918 I 247

USGS Hawthorne topographic map.

Bancroft 259.
Clark, H. G., "Aurora, Nevada", S. of M. Q. 3 (1882) 133-6.
Davis, 848-855. Hill507 205. Hill594.
M&EW, "Aurora Cons. Mines Passes to Goldfield Cons. Co." 41 (1914) 62-3
M&SP, "The New Aurora Mill", 109 (1914) 57-9.
WeedMH, 1138. Spurr 208. Stuart, NMR 47.
Thompson & West 414-416.
Wasson, J., "Bodie and Esmeralda", M&SP, San Francisco, (1878).

BASALT see BUENA VISTA

BELL (Simon, Omoc, Cedar Mountain)
Silver, Lead, Zinc, Gold

Location. The Bell District is located at the N. end of the Cedar Mt. Range in E. Mineral Co. on the Nye Co. border. The silver-lead-zinc camp of Simon is 22 m N. E. of Mina which is on the S. P. R. R.; and the former gold camp of Omco lies 4 m. to the N. of Simon. Simon has an elevation of 6,700 ft. and Omco one of 6,000 ft, while the highest point in the range, Little Pilot Peak, rises to an altitude of 8,046 ft. The district is sometimes subdivided into the Omco District on the N. and the Simon District on the S. The Goldyke District in Nye County adjoins the Bell District on the N. E. and the Athens District adjoins it on the E.

History. The Simon Mine was discovered in 1879 and a little oxidized lead ore was shipped from the gossan, but its importance was not recognized till

40 yrs. later. The Olympic Mine was discovered by J. P. Nelson in 1915 and promoted by F. J. Siebert. The Olympic Ms. Co. was incorporated in 1916, developed the property, built a 70-ton cyanide mill, and gave the name Omco to the camp. The mill burned down in 1919 and was rebuilt in 1920. The ore became exhausted in 1921 and the mine shut down after having made a production of over $700,000. P. A. Simon obtained control of the Simon Mine in 1916 and began prospecting work on it. By 1919, he had developed a large body of ore and the Simon Silver-Lead Ms. Co. was incorporated to work the property. In 1921, this company put a 150-ton preferential flotation mill in operation and obtained control of the Kirk-Simon Smelting Co. at Harbor City, Cal. In 1922, the adjoining Simon Contact and Simon Sterling properties were taken over by the Simon Silver-Lead Ms. Co.

Geology. According to Knopf, a great transverse fault crosses Cedar Mt. at Simon. To the N. of this fault, the range consists of Tertiary rocks which are mainly rhyolites and andesites; while to the S. it is composed of Triassic limestones with interbedded lavas and tuffs. The Triassic formations are cut by granodiorite intrusions and allied dike rocks which were intruded at the end of the Jurassic or early in the Cretaceous period. Lake bed deposits comprising sandstones, shales and limestones belonging to the Miocene Esmeralda formation lie on both flanks of Cedar Mt.

Silver-Lead-Zinc Orebodies. At the Simon Mine, an alaskite dike has been intruded along a reverse-fault contact between Triassic limestone and quartz-keratophyre which is probably also of Triassic age. This dike dips 70 degrees N. E. and the quartz-keratophyre forms the hanging wall at the surface, but both walls are in limestone at a depth of 230 ft. The orebodies are replacements of the limestone near this dike. Argentiferous galena and blende are inclosed in a fine-grained aggregate of quartz with subordinate amounts of pyrite and arsenopyrite. The gangue consists of calcite and limestone. Cerussite and plumbojarosite occur in the gossan; and minor amounts of calamine, smithsonite, and adamite, have been noted by Knopf on the 230-ft. level.

Gold Veins. A few gold veins occur in the Tertiary volcanic rocks at the N. end of Cedar Mt. The most important of these is the Olympic vein which is mainly in rhyolite. It has been cut by a number of faults and its extension has been lost or eroded. The vein filling consists of quartz and more or less silicified rhyolite. The quartz is white and sugary and shows evidence of having been derived from the replacement of calcite. The gold is invisible; but, since the ore contains equal parts of gold and silver by weight, is believed to be in the form of electrum.

Mines. The **Simon Silver-Lead Ms. Co.** is incorporated in Nevada and has a capital stock of 3,000,000 shares of $1 par value. The stock is listed on the San Francisco and Los Angeles Stock Exchanges and the New York Curb. The principal office is at Mina. P. A. Simon is Pres.; C. D. Kaeding, V. P.; B. F. Baker, Sec.-Treas.; with C. C. Boak and J. W. Mason, Directoros. J. F. Thorn is Gen. Mgr. The company is at present engaged in sinking a 600-ft. 3-compartment shaft and driving levels from it; and in enlarging its flotation mill to a capacity of 250 tons.

Bibliography.
MR1916 I 486 MR1918 I 247-8 MR1920 I 328
MR1917 I 284 MR1919 I 400 MR1921 I 388
USGS Tonopah topographic map.
Knopf, A., "Ore Deposits of Cedar Mountain, Mineral County, Nevada", USGS B 725 H (1921) 361-382.

Merriam, J. C., "Tertiary Vertebrate Fauna from the Cedar Mountain Region of Western Nevada", B Cal. U. Dept. Geol. 9 (1915), 161-198.

Siebert, F. J., "Nevada's Latest Gold District", M&SP 114 (1917) 449-450.

Stevens, G. R., "Geology of the Cedar Range", M&SP 114 (1917) 130.

Tiernan, A. K., "The Cedar Range Gold District of Western Nevada", SLMR 19 (1917) 23-5.

Weed MH 1177-1349 Describes 29 Simon companies.

BLACK MOUNTAIN see SILVER STAR

BOVARD see RAND

BUCKLEY see WALKER LAKE

BUENA VISTA (Oneota, Basalt, Mt. Montgomery)

Silver, Lead, Gold, Copper, Mercury, (Tungston)

Location. The Buena Vista District is at Oneota at the N. end of the White Mountain Range on the border of Mineral and Esmeralda Cos. near the California bountary line. It lies immediately to the S. E. of Buena Vista and Mt. Montgomery on the narrow-gage Nevada and California Branch of the S. P. R. R., a few miles S. of Basalt siding and a few miles E. of Queen, all on the same railroad.

History. A lode was discovered in this district in 1862 and a district organized which was abandoned shortly afterward. In 1870, Wetherell was shown rich float by an Indian and located the Indian Queen Mine, and the Oneota District was organized. A 4-stamp mill was erected at this mine in 1873, and a notable production of high grade ore was made for a number of years. A revival took place in 1905 and the Indian Queen Mine began to produce again in 1907. In 1915, the Tip-Top Mine was equipped with a cyanide plant and the I. X. L. property with an amalgamation mill. The Red Rose Mercury Mine erected 2 D retorts in 1917 and produced up to 1919.

Geology. According to Whitehill, the Indian Queen vein lies between granite and slate and its ore is very base. Mines Handbook states that the Tip-Top Mine has 2 veins between andesite and rhyolite carrying values in gold and silver. The Red Rose mercury mine is 5 m. from Mt. Montgomery station at an altitude of 9,100 ft. According to Ransome, the ore occurs in a remnant of a rhyolitic flow resting on andesite. The ore consists largely of a soft powdery mixture of cinnabar with altered, white, pulverulent rhyolite. Three miles N. E. of Queen, contact-metamorphic deposits containing scheelite occur in sediments consisting of quartzite, slaty rocks and hornfels with some interbedded marble, near their contact with intrusive diorite, according to Hess & Larsen.

Tip Top Mine. The Louisiana Cons. M. Co., described under the Tybo District in Nye Co., owns the Tip-Top Mine. J. Siegbert is Pres., at 900 Broadway; and L. A. Dessar, Sec. at 111 Broadway, New York City. The Tip-Top property is developed by tunnels and has 1000 ft. of workings. It is equipped with a 10-stamp cyanide mill of 50 tons capacity, and is said to have produced $130,000 up to 1918.

Bibliography. SMN1871-2 38-40
SMN1873-4 19-20
SMN1875-6 36-7
MR1905 268
MR1906 295
MR1907 I 349
MR1908 I 479
MR1909 I 402
MR1911 I 686
MR1912 I 792
MR1919 I 821, 832
MR1914 I 695
MR1915 I 631
MR1916 I 476
MR1917 I 284
MR1918 I 248
MR1919 I 400
MR1920 I 328

Davis 857.

Hess, F. L., & Larsen, E. S., "Contact-Metamorphic Tungsten Deposits of the U. S.", USGS B 725D 277.
Hill 507 206.
Spurr 208 206-212 White Mountain Range.
Thompson & West 417.
Weed MH 1242 Louisiana Cons. M. Co.

CANDELARIA (Columbus)

Silver, Gold, Lead, Copper, Nickel, Variscite, Turquoise

Location. The Candelaria District is in S .E. Mineral Co. near the Esmeralda Co. border. Candelaria is on the narrow-gage Nevada & California Branch of the S. P. R. R. at an altitude of 5,665 ft.

History. The Candelaria District, formerly known as the Columbus District, was discovered and organized by Mexicans in 1864. Its largest mine, the Northern Belle was located in 1865, but allowed to lapse and relocated in 1870. The town of Columbus was situated on the edge of the Columbus Salt Marsh, in Esmeralda Co., 5 m. S. E. of the silver mines, and in 1876 when these became of importance, the new town of Candelaria was founded near the mines. The first 20-stamp mill was erected at Belleville, 8 m. W. of the mines in 1873; and the second 20-stamp mill at the same place in 1876. The Northern Belle Mine began paying dividends in 1875 and produced a million dollars annually for 10 years. Litigation begun in 1883 led to the consolidation of the Northern Belle and the Holmes Mines as the Argentum in 1884. The Mount Diablo Mine began paying dividends in 1883, but by 1892, the output of the camp had sunk into insignificance. A revival occurred in 1918, when the Candelaria Ms. Co. secured control of the Argentum, Mount Diablo, Lucky Hill and other properties.

Production. The district produced about $20,000,000 in the early days, chiefly in silver. Of this sum, the Northern Belle produced about $15,000,000 and the Mount Diablo $4,000,000. Within recent years the U. S. Geol. Survey has kept statistics of the production of the camp, which show that from 1903 to 1920 inclusive, it produced 6,475.13 oozs. gold, 1,021,867 ozs. silver, 50,129 lbs. of copper, and 653,982 lbs. of lead, having a total value of $977,868. In 1882, 10 tons of high grade nickel ore were shipped to Swansea from near Columbus. In 1908, variscite and turquoise were discovered in the Candelaria Mts. and considerable amounts were shipped for gem purposes before the closing of the mines a few years later.

Geology. The sedimentary formations of the Candelaria District have been called the Columbus chert and Candeleria shale in an unpublished report by J. A. Burgess.* The Columbus chert is the oldest formation. It is a thin-bedded white or black flinty chert with a few small interbedded strata of limestone. The Candelaria shale rests unconformably upon it. The lowest member of this formation is a graywacke, followed in turn by bedded shale and massive shale. The Candelaria Mts. are part of the N. limb of a great anticline, and the sedimentary rocks dip deeply to the N. They have been intruded by a basic rock now altered to serpentine, by granodiorite, and by andesite. The veins were formed after these intrusions had taken place. Vein formation was followed by faulting, erosion, and the eruption of the Tertiary rhyolite and basalt capping the older rocks.

* Burgess, J. A., "Report on the Geology of the Northern Belle Mine, Candelaria Nevada", (1922), an unpublished report obtained through the courtesy of the Candelaria Ms. Co.

Orebodies. The principal orebodies in the Candelaria District occur in the Candelaria shale near its lower border and in the vicinity of the large intrusive masses of igneous rock. The vein system is a zone of fissuring 100 ft. to 150 ft. in width with an E.-W. strike and roughly parallel to the bedding of the shale. Within this zone are more or less parallel and overlapping veins, cross veins, and irregular fractures. The principal orebodies are from 10 ft. to 20 ft. in width.

Ore. The ore is a highly oxidized manganiferous and ferruginous silver ore. The hot solutions which formed the lodes caused replacement by tourmaline, sericite, and dolomite, according to Knopf, and deposited manganiferous ferrodolomite containing pyrite, sphalerite, and jamesonite. This primary mineralization was of low grade and was enriched by oxidation. The manganiferous ferrodolomite was changed to manganese oxide and limonite, the pyrite to limonite, the sphalerite to calamine and possibly also to smithsonite, and the jamesonite probably to bindheimite. The form in which the silver occurs has not been determined.

Variscite and Turquoise. Variscite was discovered in the foothills of the Candelaria Mts. by L. A. Dees and Edward Murphy in 1908. The Los Angeles Gem Co. purchased their claims and worked them for several years. The variscite occurs in rhyolite, limestone, and sandy shale, and in some instances is associated with small trachyte dikes, according to Sterrett, and the deposits are small, irregular, filled fissures, replacements, and segregations. The claims are in two localities, 2 m. apart; and turquoise has been found upon one of these groups.

Mines. The only mines of present importance in the Candelaira District are thost of the Candelaria Mines Co. This company holds the Mount Diablo and Argentum groups under lease and bond, and owns outright the Lucky Hill group and other miscellaneous claims. It is incorporated in Nevada with a capital stock of 3,000,000 shares of $1 par value, of which 2,300,000 have been issued. The stock is listed on the Los Angeles Stock Exchange, and the Boston and New York Curbs. The principal office is at 648 Mills Building, San Francisco, and the mine office at Candelaria. C. D. Kaeding is Pres.-Treas.; S. Rossiter, V. P.; V. Wimberly, Sec.-Asst. Treas.; J. C. Peebles, Asst. Sec.-Asst. Treas.; with F. M. Manson and O. W. Jones, additional Directors. A 400-ton cyanide mill was erected in 1922 and began operating late that year.

Bibliography.

B1866 126	SMN1877-8 25-6	MR1912 I 804
B1867 335, 337-8, 419	MR1882 404 nickel	MR1913 I 832
R1868 115-6	MR1905 267	MR1915 I 642
R1872 174-5	MR1906 294	MR1916 I 487
R1874 283	MR1907 I 349	MR1917 I 285
SMN1866 41-2	MR1909 I 406	MR1918 I 248
SMN1867-8 96	II 877-8 turquoiose.	MR1919 I 400-1
SMN1869-70 108	796-801 variscite.	MR1921 I 388
SMN1871-2 34-40	MR1910 I 522,	MR1920 I 328
SMN1873-4 20-21	II 886-7 turquoise,	
SMN1875-6 34-5	890 variscite.	

USGS Hawthorne topographic map.

Bancroft 259-260.

Knopf, A., "The Candelaria Silver District, Nevada", USGS B 735A, (1922) 1-22.

Robie, E. H., "The Rejuvenation of Candelaira", E&MJ 115 (1923) 61-2.

Spurr 208 113-4.
Thompson & West, 413-4.
Weed MH 1161-2 Candelaria Ms. Co.

CAT CREEK see WALKER LAKE

CEDAR MOUNTAIN see BELL

COLUMBUS see CANDELARIA

DOUBLE SPRINGS MARSH

Soda, Sodium Sulphate

Location. Double Springs Marsh is in N. Mineral Co., 7 m. E. of Schurz which is on the S. P. R. R.

History. The Occidental Alkali Co. was operating a plant on this marsh in 1898, and had produced a considerable amount of high grade soda.

Geology. The deposit comprises 800 acres, 500 of which are covered with a layer of salts varying from 2 in. to 14 in. in thickness and averaging 6 in., according to Knapp. Below this is a soda-bearing clay which renews the soda incrustation after it has been stripped. The average composition of the incrustation is as follows:

Sodium Carbonate	20%
Sodium Bicarbonate	25%
Sodium Sulphate	15%
Sodium Chloride	10%
Water	15%
Sand and Insoluble Matter	15%
	100%

Bibliography. Knapp, S. A., "Occurrence and Recovery of Sodium Carbonate in the Great Basin Region", MI 7 (1898) 627, 631-4.
———"Occurrence and Treatment of the Carbonate of Soda Deposits of the Great Basin Region", M&SP 77 (1898) 448.

EAST WALKER (Mount Grant)

Gold, Silver

Location. The East Walker District lies E. of the East Walker River on the W. slope of the Walker River Range. The Walke Lake District adjoins the East Walker District on the E. Mount Grant, 11,303 ft. in elevation, is located in the S. part of the district.

History. The East Walker District produced gold ore in 1912. The 2-stamp mill on the Mammoth property has operated intermittently since 1914.

Bibliography. MR1912 I 802 MR1916 I 487 MR1917 I 285
MR1914 I 695 MR1921 I 388
USGS Hawthorne topographic map.

ESMERALDA see AURORA

FITTING (Acme)

Gold, Silver, Lead, Copper

Location. The Fitting District is located at the S. E. end of the Gillis Range, N. of Acme, a siding on the S. P. R. R. The altitude of Acme is about 4,450 ft. and the highest point in this section of the Gillis Range is 7,342 ft.

Production. The Montreal Mine, 2 m. N. of Acme, is said to have produced $1,500,000 in gold; and the Silver King, 2 m. N. W. of Acme, $500,000 in silver and lead, according to a letter from H. J. Fick.

Geology. The country rocks are shales, cherts, and limestones perhaps of Triassic age, intruded by Cretaceous quartz monzonite, cut by rhyolite dikes which are probably Tertiary, and capped in places by Tertiary andesite At the Acme copper mine, 6 m. N. of Acme, the sedimentary rocks have been cut by quartz-monzonite and the limestone replaced by copper ores in the form of stringers and small lenses following the bedding planes, according to an unpublished report by F. C. Lincoln. The ore minerals are chrysocolla, malachite and azurite carrying a little silver probably as cerargyrite; and the gangue minerals are quartz, calcite, limonite, hematite, and a yellow basic sulphate of iron. At the Silver Chief Mine, 12 m. N. of Luning, silver ore is said to occur at a rhyolite-granite contact.

Mines. The Nevada Ore & C. Co. is the only company operating in the district at the present time. It is incorporated for 2,000,000 shares of $1 par value. J. L. Robinson is Pres.; A. W. Pingrey, V. P.; J. A. Champion, Sec.-Treas.; and H. J. Fick, Supt.-Mgr. at Hawthorne. The company owns four groups;—the Gold Basin Group adjoining the old Montreal Mine, the Greenman Group near the old Silver King; the Paymaster Group 7 m. N. of Acme; and the 1917 Group, 2½ m. S. W. of Acme in the Hawthorne District. These properties are developed by shafts 400 ft. and 560 ft. in depth, and about 2 m. of workings. They are equipped with 2 gasoline hoists and a 50-ton cyanide plant which is located at Kinkead siding, W. of Acme, on the S. P. R. R.

Bibliography. MR1908 I 484 MR1909 I 406
USGS Hawthorne topographic map.
Weed MH 1282 Nevada Ore & Copper Co. 1329 Silver Chief M. Co.

GARFIELD

Silver, Gold, Copper, Lead, (Marble)

Location. The Garfield District lies N. of Garfield Flat in central Mineral Co. The Garfield Mine is 6 m. S. of Acme, a siding on the S. P. R. R. The Garfield District adjoins the Hawthorne District on the E., and is sometimes considered a part of it.

History. The Garfield Mine was a big producer of rich silver ore from 1880 to 1887. Interest in the district revived a few years ago, small productions were made, and the West End Cons. M. Co. of Tonopah purchased the Mabel Mine which adjoins the old Garfield Mine on the W.

Production. The Garfield Mine is said to have produced more than $6,000,000.

Bibliography. MR1905 268 MR1908 I 484 MR1920 I 328
1055 Marble MR1921 I 388
USGS Hawthorne topographic map.
Hill 594 176.
Weed MH 1140 Balea Gold M. Co. 1149 Blue Light Copper Co.
1343 Sparta M. Co. 1372 West End Cons. M. Co.

GOLD RANGE see SILVER STAR

GRANITE see MOUNTAIN VIEW

HAWTHORNE (Lucky Boy, Pamlico)

Silver, Gold, Lead, Copper, Barite, (Gypsum, Tungsten)

Location. The Hawthorne District is located S. of the town of Hawthorne in central Mineral Co. The W. section of the district, where the Lucky Boy mine is situated is sometimes known as the Lucky Boy District; and the E. section in the neighborhood of the Pamlico Mine as the Pamlico District. Hawthorne is 7 m. S. S. W. of Thorne, a station on the S. P. R. R. It has an elevation of 4,325 ft. and the Lucky Boy Mine 7 m. to the

S. S. W. in the foothills of the Wassuk Range is 6,225 ft. high, while the Lucky Boy Pass crosses the range at an altitude of 10,516 ft. The La Panta and Pamlico Mines are 10 m. E. S. E. of Hawthorne in a lower range of the mountains. The Garfield District adjoins the Hawthorne District on the E. and the Walker Lake District adjoins it on the N. W.

History. The Pamlico and La Panta Mines were the most important early mines in the district, and have been worked intermittently by leasers of recent years. The Pamlico Mine is equipped with a 20-stamp mill, which is now being operated by leasers on the Gold Bug property. The Lucky Boy Mine was discovered by men repairing the stage road over Lucky Boy Pass in 1906, and has been worked mainly by leasers who have made large productions, and are still actively at work. The placers in the neighborhood of the Pamlico Mine were worked by drift mining from 1915 to 1917. Barytes was mined near Kinkead siding on the S. P. R. R. and shipped in 1917, 1918 and 1919. The Mandich M. Co. erected a 5-stamp amalgamation mill in 1921.

Production. The Pamlico Mine is credited with an early production of $500,-000; the La Panta with $200,000; the Good Hope Mine with $150,000; the War Eagle and the New York Central with $100,000 each; and the Evening Star, now part of the Gold Bug, with $30,000. The bulk of the recent production of the district has come from the Lucky Boy Mine. The production of the Hawthorne District from 1907 to 1921 as recorded by Mineral Resources of the U. S. Geol. Survey amounted to 11,201 tons containing $55,400 gold, 746,265 ozs. silver, 194,434 lbs. copper and 1,124,449 lbs. lead, valued in all at $572,220.

Geology—Lucky Boy. At the Lucky Boy Mine, granodrite is intrusive into limestone, according to Hill, and the ore deposit is a complex contact vein. This vein strikes E. and dips steeply to the S.; and the ore occurs as lenses and shoots with a steep W. pitch, which seem to be closely related to branch veins. The richest ore carries 2,000 to 3,000 ozs. of silver per ton and occurs in the ore shoots as veinlets of argentiferous tetrahedrite in a gangue of quartz and barite. Argentite, native silver, and copper carbonates stains have also been noted in this ore. The low grade ore consists of crushed and silicified wall rock carrying disseminated galena, tetrahedrite, sphalerite, and pyrite.

Geology—Pamlico. The principal country rock of the Pamlico section is limestone. This has been tilted by granite which intrudes it in the N. part of the district, is cut by porphyry dikes in the S. and W. portions, and is capped in places by rhyolite and basalt flows, according to a letter from F. R. Dodge. The Pamlico vein is in rhyolite and porphyry. The La Panta ore deposit is in dolomitic limestone near the intrusive granite. It consists of a series of irregular replacement orebodies occupying a fracture zone which appears to have been produced by the intrusion of a diorite dike along a fault. The ore of the La Panta and Pamlico Mines consists of red- and yellow stained sugary quartz, according to Hill, which carries rather coarse free gold as nuggets and wires, and some argentiferous galena; and the orebodies are small and irregularly distributed. The Gold Bug ore deposit is a contact vein which follows the limestone-granite contact for part of its course and then enters the granite. It contains shoots of copper ore and pockets of high-grade gold ore.

Mines. Four companies are at present active. F. R. Dodge of Hawthorne is Pres. of the **Gold Bug Cons. Ms. Co.** which is being operated by leasers. **The Greek Hills Trimetal Co.** is actively developing its property. T. M.

Conolly of Hawthorne is Pres.- Gen. Mgr. **The Lucky Boy Cons. Ms. Co.** is controlled by the Knight Investment Co. of Utah. The company is incorporated in Utah and has issued 2,681,763 shares. It has paid dividends to the amount of $135,000 and in 1919 levied an assessment of 1 ct. per share. O. R. Knight is Pres.; J. H. Miller of Hawthorne, V. P.-Gen. Mgr.; and R. E. Allen, Sec.-Treas. The company owns the Lucky Boy Mine which is developed by a 1000-ft shaft and 6,200-ft. tunnel and equipped with compressor, electric motors, machine drills, and an old 10-stamp mill. **The Nevada Ore & C. Co.** which owns the 1917 Group in this district is described under the Fitting District where it owns 3 groups of claims.

Gypsum and Tungsten. A large body of gypsum interbedded with white limestone outcrops 3 m. W. of Hawthorne, according to Jones. Scheelite occurs in sediments metamorphosed by granodiorite for a distance of 4 m. along the E. slope of the Wassuk Range, 7 m. S. of Hawthorne according to Hess & Larsen.

Bibliography.

MR1882 230	MR1914 I 696	MR1912 II 648 Gypsum
MR1907 I 353-4	MR1915 I 642	MR1913 II 367 Gypsum
MR1908 I 482-3	MR1916 I 487	MR1915 II 176 Barite
MR1909 I 405	MR1917 I 285	MR1916 II 250 Barite
MR1910 I 522-3	MR1918 I 248	MR1917 II 288 Barite
MR1911 I 686	MR1919 I 401	MR1918 II 955 Barite
MR1912 I 804	MR1920 I 228	MR1919 II 340 Barite
MR1913 I 832	MR1921 I 388	MR1920 II 194 Barite

USGS Hawthorne topographic map.

Hess, F. L. & Larsen, E. S., "Contact-Metamorphic Tungsten Deposits of the U. S.", USGS B 725D (1921) 280-1.

Hill 507 208.

Hill 594 151-7.

Jones, J. C., "Nevada" in "Gypsum Deposits of the U. S.", USGS B 697, (1920) 155.

Stuart NMR 46, 61-3.

Weed MH 1243 Lucky Boy Cons. Ms. Co. 1282 Nevada Ore & Co. Co.

Wroth, J. S., "Geology of the Lucky Boy Mine, Nevada", M&SP 97, (1908) 251.

LUCKY BOY see HAWTHORNE

LUNING see SANTA FE

MARIETTA see SILVER STAR

MINA see SILVER STAR

MOUNT GRANT see EAST WALKER

MT. MONTGOMERY see BUENA VISTA

MOUNTAIN VIEW (Granite, Reservation)

Copper, Gold, Silver, Lead

Location. The Mountain View District is situated in the neighborhood of the abandoned camps of Mountain View and Granite in the Walker River Range in N. W. Mineral Co. on the Lyon Co. border. It is a few miles W. of Reservation station and of Schurz, both on the S. P. R. R. The elevation of Schurz is 4,128 ft., while Black Mt. in the S. part of the Mountain View District rises to an altitude of 8,136 ft. The Yerington District in Lyon Co. adjoins the Mountain View District on the W.

History. In 1908, 55 lessees were active at Granite and the first production was made. $4500 in gold and silver was produced the following year. Lead-silver ore was shipped from the Flynn Mine in 1912. The Yerington Mt. Copper Co. was under development in 1912 and made productions of copper ore containing a little silver and gold up to 1916, when it shut down.

Geology. The main mass of the Walker River Range is composed of granodiorite which is probably of Cretaceous age, according to Hill. This rock is cut by dikes of granodiorite porphyry, aplite, and augite-andesite; and is capped on the W. side of the range by flows of Tertiary lavas. The vein on the property of the Yerington Mt. C. Co. is in granodiorite which has been cut by aplite dikes, and its outcrop has been partially covered by augite-andesite flows. The ore consists of crushed granodiorite cemented by quartz in which occur pyrite, copper carbonates, and chalcocite. The Mountain View vein occurs in a fracture along a dark dike of granodiorite, and to the E. on the Big Twenty ground it splits into 3 fracture zones. Gold occurs in this vein in small irregular pockets. The lead-silver ore at the Flynn Mine comes from a vein in granodiorite.

Mines. Three companies have been active recently. O. Anderson, Schurz, is Mgr. of the **Anderson Copper M. Co. The Pack Saddle Gold Ms. Co.** was incorporated in Nevada in 1916 with a capital stock of 1,000,000 shares of $1 par value of which 665,000 are outstanding. The home office is in Carson City. E. H. Walker is Pres.; H. R. Mighels, V. P.; E. W. Miller, Sec.-Treas.; W. E. Casson, Mgr.; with F. M. Baker, Directors. The company owns 7 claims developed by a 224-ft. tunnel. The Reservation Hill M. & M. Co. was incorporated in Nevada in 1907 with a capital stock of 1,000,000 shares of $1 par value of which 788,000 have been issued. The home office is in Carson City. R. A. Trimble is Pres. and W. E. Casson is Mgr. The company owns 2 claims developed by a short tunnel and shallow shaft.

Bibliography. MR1907 I 355 MR1913 I 832 MR1914 I 696 MR1916 I 488
USGS Hawthorne topographic map.
Hill 507 207. Hill 594 129-133.
Ransome, F. L., "The Yerington Copper District, Nev.", USGS B 300 (1909) 118-9.
Spurr 208 115-7 Walker River Range.
Weed MH 1286 Nevada S. M. & Dev. Co. 1300 Pack Saddle G. Ms. Co.
1315 Reservation Hill M. & M. Co. 1316 Review S. M. Co.
1368 V7 M. Co. 1381 Yerington Mt. C. Co.

OMCO see BELL

ONEOTA see BUENA VISTA

PAMLICO see HAWTHORNE

PILOT MOUNTAINS (Sodaville)

Mercury, Silver, Gold, Lead, Turquoise, (Tungsten)

Location. The Pilot Mountains or Sodaville District is situated in the S. part of the Pilot Mts. E. of Sodaville in S. E. Mineral Co. near the Esmeralda Co. boundary. Sodaville is on the S. P. R. R., and has an elevation of about 4,550 ft. while Pilot Peak, the highest point in the Pilot Mts. rises to a height of 9,207 ft.

Geology. The country rocks of the Pilot Mountains are limestone, graywacke, slate, and chert which are probably of Paleozoic age, according to Knopf. These sedimentary rocks are intruded by granodiorite and capped in places by Tertiary volcanics. Gold-silver veins occur in granodiorote porphyry at the Aureola Mine 5 m. S. E. of Sodaville, according to Weed. Argentiferous galena occurs with tungsten in the E. part of the district as noted below.

Mercury. Cinnabar deposits occur in the Pilot Mts. in a belt 2 m. in length having a N. E. trend and an elevation of 11,300 ft. These deposits are 12 m. by road E. S. E. of Mina, according to Knopf, and were rediscovered in 1913 by Pepper & Keough, the original discoverer being unknown. On Cinnabar Mt. where the discovery was made, the deposits occur in fracture zones in limestone. Cinnabar is intergrown with calcite or dolomite in veinlets in the limestone or occurs as replacements of the adjoining wall rock, and in one instance it is associated with stibnite. To the N. cinnabar is found in a gangue of barite and with brecciated chert as country rock; while still farther N., the country rock is graywacke and the cinnabar is disseminated through a siliceous gangue.

Turquoise. The Montezuma mine of the German American Turquoise Co. is in the foothills of the Pilot Mts. at an elevation of 5900 ft., according to Sterrett, and is 20 m. by road E. of Sodaville. The turquoise occurs as seams, veinlets, and nodules up to 1 inch or more in thickness in an altered porphyritic rock which is probably trachyte. The turquoise varies from hard gem material of fine color to soft material of poor color, and the bulk of the output was low grade.

Tungsten. The tungsten deposits are located near Graham Spring on the E. slope of the Pilot Mts. about 23 m. by road from Mina, according to Hess & Larsen. The orebodies are at or near the contact of granodiorite and limestone. The ore occurs in 3 forms, contact-metamorphic deposits, a quartz-calcite-scheelite contact vein, and bunches of quartz, calcite, galena, and scheelite, rich in silver. The contact-metamorphic ore is composed mainly of garnet with some quartz, calcite, and diopside.

Bibliography. MR1905 268 MR1909 II 785-6 Turquoise MR1920 I 432-3
MR1908 I 484 Mercury
USGS Hawthorne and Tonopah topographic maps.
Hess, F. L., & Larsen, E. S., "Contract-Metamorphic Tungsten Deposits of the U. S.", USGS B 725D (1921) 278-280.
Hill 507 209-210.
Knopf, A., "Some Cinnabar Deposits in Western Nevada", USGS B 620D (1915) 59-62.
Spurr 208 103-5 Pilot Mts.
Weed MH 1138 Aureola Ms. Co.

PINE GROVE (Wilson, Rockland)
Gold, Silver

Location. The Pine Grove District is situated at Pine Grove on the E. flank of the Smith Valley Range in W. Mineral County. It is 20 m. S. of Yerington, which is on the N. C. B. R. R. Pine Grove camp is at an elevation of 6,700 ft. in a canyon to the N. of Pine Grove Summit which reaches an altitude of 8,695 ft. The camp of Rockland is 3 m. S. E. of Pine Grove to the E. of Pine Grove Summit and at an elevation of 7,500 ft.

History. The Pine Grove District was discovered by William Wilson in 1866. He located the Wilson Mine; and the Wheeler Mine was located shortly afterwards; while the Rockland Mine was discovered in 1868. In 1868 there were a 5-stamp mill, a 10-stamp mill, and an arrastra at Pine Grove, and about 1870 a 10-stamp mill, was erected at Rockland but burned down a few months afterwards. In 1882, the Wheeler Mine was operating a 15-stamp mill and the Wilson Mine a 10-stamp mill. A 5-stamp mill was constructed for the Rockland Mine in 1902 and a 15-ton cyanide plant added later; and in 1907 a 60-ton dry-crushing and leaching plant was erected. The new mill proved unsuccessful, and the property was taken over by the Pittsburg-Dolores M. Co. in 1912. This company reconstructed the milling plant and operated from 1915 to 1918 when it shut down.

Production. The production of the Wilson Mine up to 1893 is estimated at about $5,000,000; and that of the Wheeler Mine at $3,000,000. Since that time these properties have only been worked intermittently. The Rockland Mine is credited with a production of $700,000 up to 1921.

Geology. The principal country rock of the Pine Grove District is quartz-monzonite which is probably of Cretaceous age. This has been intruded by dikes of granite porphyry and capped by Tertiary rhyolite and basalt. The orebodies in the Wilson and Wheeler Mines lie in a zone of intense crushing and alteration in the quartz-monzonite immediately S. of a well-marked fault, according to Hill. This zone contains considerable disseminated pyrite and a large number of interlacing quartz stringers carrying pyrite, as well as lenses of quartz and pyrite up to 2 ft. or 3 ft. thick and 10 ft. to 150 ft. in length. The stringers and lenses constitute the ore. A minor amount of chalcopyrite occurs with the pyrite. The principal valuable constituent of the ore is gold with a minor amount of silver. At the Rockland Mine, according to data furnished by E. J. Schrader, the vein is entirely in grandorite at its N. end, then follows a rhyolite dike as a hanging wall, passes into the dike, and at the S. end has the dike as footwall. The vein is a filled fissure with some replacement of the walls by ore. Gold values predominate but there is little free gold. The silver minerals are argentite, stephanite and pyrarggrite.

Bibliography.

B1867 336-7	SMN1875-6 37-8	MR1912 I 804
R1868 116-7	MR1905 268	MR1913 I 832-3
R1871 141	MR1906 294	MR1914 I 696-7
SMN1866 40	MR1907 I 359	MR1915 I 643
SMN1867-8 89-91	MR1908 I 484	MR1916 I 488
SMN1869-70 109	MR1909 I 406	MR1917 I 285
SMN1871-2 32-4	MR1910 I 523	MR1918 I 249
SMN1873-4 24	MR1911 I 686	MR1919 I 401

USGS Wellington topographic map.

Davis 857, 952. Hill 507 209. Hill 594 133-140.

C&ME, "Cyaniding by Continuous Decantation at Two Nevada Silver Mills", 14 (1916) 435-7.

Schrader, J. F., "Pittsburg-Dolores M. Co.'s Mill at Rockland, Nevada", E&MJ 99 (1915) 653-4.

———"The Pittsburg-Dolores Mill", E&MJ 104 (1917) 155-6.

StuartNMR 46. Thompson & West 41. WeedMH 1320 Rockland M. Co.

Hill 594 133-140. M&SP 117 (1918) 538.

Olson, C. R., "Specific Gravity Apparatus", E&MJ 102 (1916) 305.

RAND (BOVARD)

Silver, Gold, Copper, Lead, (Potash, Turquoise)

Location. The Rand or Bovard District is situated just W. of Bovard on the N. E. slope of the Gabbs Valley Range in N. E. Mineral Co. Bovard is 17 m. N. E. of Nolan station, (Rand postoffice), which is on the S. P.

R. R. The altitude of Bovard is 4,900 ft. and the highest peak in the district is 6,624 ft.

History. The district was discovered by prospectors from Rawhide in 1907, and the Nevada Rand and Gold Pen Mines were located in 1908. High-grade silver-gold ores were treated at the Rawhide mill when that was in operation; and more recently have been shipped to the smelters. Schrader discovered alunite in the ore deposits in 1911. J. H. Malloy is reported to have discovered turquoise in the district in 1915. The Copper Mountain Mine was under option to the Jumbo Copper Mountain M. Co., a subsidiary of the Jumbo Ex. M. Co. from 1917 to 1919, when it reverted to the original owners. The Gold Pen Ms. Co. purchased the Gold Pen Mine in 1919, and built a 20-ton amalgamating mill which was put in operation in 1920; but got into financial difficulties and the property was bought by Ferretto Bros. in 1921.

Production. The Copper Mountain Mine is credited with a production of more than $125,000; the Gold Pen Mine is said to have produced $200,000; the Nevada Rand Mine has made a production of $50,000; and the Lone Star $25,000.

Geology. The country rocks are Tertiary volcanics 2,000 ft. in thickness resting upon highly disturbed Triassic limestone which has been intruded by Cretaceous granodiorite. At the Gold Pen Mine in the S. E. section of the district, the ore deposit is a brecciated quartz vein in rhyolite. The rhyolite is much altered, and the vein is separated from it by sheets of fairly pure alunite. The ore consists of native gold alloyed with silver, cerargyrite, and argentite in a gangue of quartz, kaolin and rhyolite. At the Nevada Rand Mine in the N. W. part of the camp, the ore deposit consists of replacement veins in latite which has been highly propylitized in their neighborhood. The ore minerals are cerargyrite, argentite, native gold alloyed with silver, and some pyrite in a gangue of quartz, adularia, sericite, calcite, kaolin, iron and manganese oxides. At the Copper Mt. Mine, a limestone-monzonite contact carries carbonates on the surface and chalcopyrite and chalcocite in depth.

Alunite. Besides the occurrence of alunite at the Gold Pen Mine, Schrader noted its presence at the Valley View prospect 1 m. distant and another prospect in the vicinity. At the Valley View, alunite occurs as a vein in Triassic limestone. The alunite is high in soda and is too low in potash to be of probable value for that substance.

Mines. Several properties in the Rand District are carrying on development work and shipping ore extracted in the course of this work, among which is the Nevada Rand. The Nevada Rand Ms. Co. was incorporated in 1916 and has a capital stock of 1,500,000 shares of 10 cts. par value of which 548,000 remain in the treasury. The home office is in the Fordonia Bldg., Reno. W. V. Rudderow is Pres.-Treas.; S. W. Wilcox, V. P.; and C. F. Watkins, Sec., all Directors; while A. E. Painter is Res. Agt. at Reno. The property consists of 7 claims developed by a 450-ft. shaft and 2500 ft. of workings, and equipped with hoist and assay office.

Bibliography

MR1908 I 485	MR1916 I 488	MR1921 I 388	
MR1910 I 522	MR1917 I 285	MR1912 II 899	Potash
MR1914 I 697	MR1918 I 249	MR1916 II 113	"
MR1915 I 643	MR1919 I 401	MR1915 II 856	Turquoise
	MR1920 I 328-9		

USGS Hawthorne topographic map.

Hill 507 **206.**

Schrader, F. C., "Alunite in Patagonia, Arizona, and Bovard, Nevada", EG 8 (1913) 752-767.

——"Alunite at Bovard, Nevada", USGS B 540 (1914) 351-6.

Spurr 107-9 Gabbs Valley and Gabbs Valley Range.

WeedMH 1179 Copper Mt. Mine. 1208 Golden Pen Cons. M. Co. 1213 Gold Pen Ms. Co. 1214 Gold Pen-Rand Ms. Co. 1284 Nevada Rand Ms. Co. 1308-9 Queen Regent Merger Ms. Co.

RAWHIDE (Regent)

Gold, Silver, Placer Gold, Copper, Lead

Location. The Rawhide or Regent District is situated in the vicinity of Rawhide and Regent in N. E. Mineral Co. on the Churchill Co. border. Rawhide has an elevation of 5152 ft. and is 25 m. E. of Schurz, which is on the S. P. R. R. The district adjoins the Eagleville District in Churchill Co. on the S.

History. Locations were made in the Rawhide District in 1906, and a rush which took place in 1908 brought the population temporarily to 4,000. The district was opened up by the leasing system. George Graham Rice promoted the Rawhide Queen M. Co., the Rawhide Coalition M. Co., and the Black Eagle M. & M. Co. In 1909, there were two small and peculiar mills in the district—one employing a Tadmor mill for crushing and the other three Knight cannon-ball mills. In 1911, a 10-stamp amalgamating and concentrating mill and 35-ton cyanide plant were put in operation at the Rawhide Queen. In 1912, the Nevada New Mines Co. was incorporated and took over the George Graham Rice promotions which were the most important mines in the camp. This company worked mainly through lessees, making an attempt to operate on its own account in 1916, but returning to the leasing system the following year. Lessees are active in the camp at the present time. A considerable placer production has come from dry washing.

Production. From 1908 to 1920, the camp produced 68,933 tons of ore containing $993,888 in gold, 696,673 ozs. silver, 24,895 lbs. copper and 1,472 lbs. lead, valued in all at $1,418,680, according to Mineral Resources of the U. S. Geol. Survey.

Geology. The lodes of the Rawhide District occur in Tertiary rhyolite. There are also later flows of andesite and rhyolite containing briotite, according to Lindgren, and still later of basalt, according to Wolcott. The ore occurs in quartz veins and in lodes of kaolinized rhyolite penetrated by stringers of quartz. The ore minerals are native gold alloyed with silver, argentite, and ruby silver.

Bibliography. MR1907 I 356 MR1912 I 804 MR1917 I 205
MR1908 I 484-5 MR1913 I 833 MR1918 I 249
MR1909 I 406 MR1914 I 697 MR1919 I 401
MR1910 I 523 MR1915 I 643 MR1920 I 329
MR1911 I 687 MR1916 I 488 MR1921 I 388

USGS Carson Sink topographic map.

Del Mar, A., "Rawhide, Nevada", E&MJ 85 (1908) 853-4.

Gehrmann, C. A., "The Gold Camp of Rawhide, Esmeralda County, Nevada", MS 57 (1908) 305-6.

Hill 507 209.

M&SP Editorial, "Rawhide, Nevada", 96 (1908) 238.

——Excerpt from letter, "Notes on Rawhide, Nevada", 96 (1908) 424-5.

——"Rawhide Queen Mill", 103 (1911) 49.

Rogers, A. F., "Orthoclase-bearing Veins from Rawhide, Nevada, and

Weehawken, New Jersey", EG 6 (1911) 790-8.
Shanklin, E. S., Discussion, "Rawhide, Nevada", M&SP 96 (1908) 557-8.
Stuart NMR 63-8.
Weed MH 1160 Cal.-Nev. Mlg. Co. 1281 Nev. New Ms. Co.
1326 Seminole-Regent M. Co.
Wolcott, G. E., "Mining and Milling at Rawhide, Nev.", E&MJ 87 (1909) 345-8.
Whytock, P. R., "The Rawhide District, Nevada," M&EW 31 (1909) 266.
——"Rawhide, Its Past, Present, and Future",M&EW 33 (1910) 89-90.

REGENT see RAWHIDE

RESERVATION see MOUNTAIN VIEW

RHODES MARSH (Virginia Marsh)

Borax, Salt

Location. Rhodes, or Virginia, Marsh lies just E. of Rhodes siding on the S. P. R. R. in S. E. Mineral Co. It has an altitude of slightly under 4,400 ft. and occupies 3.2 sq. m. in the lowest part of the S. end of Soda Spring Valley which has an area of 232 sq. m. The marsh adjoins the Gold Range District on the S. E.

History. Up to 1862, the Comstock Lode obtained the supply of salt for its silver mills from San Francisco at a cost of $150 per ton. In that year salt was transported to Virginia City from Rhodes Marsh by camels which had been imported for the purpose, and the price of salt was reduced one-half. Sand Springs Marsh was discovered the following year and captured the Comstock market. In 1869, Rhodes Marsh was supplying salt to the mills of Columbus and Belmont. About 1874, Rhodes and Wasson were marketing 1 ton concentrated borax a day, and in 1882, the Nevada Salt & Borax Co. was producing about 1 ton refined borax a day besides table and mill salt.

Geology. Rhodes Marsh is ordinarily dry and encrusted with salt at the surface. Pure salt occurs in the central part of the playa surrounded by a zone containing borax, ulexite, sodium sulphate, and sodium carbonate. The potash content of the salines is too low to be of commercial interest.

Bibliography. B1867 310,311 SMN1869-70 108 SMN1875-6 38
R1872 175 SMN1871-2 36 MR1882 544 Salt; 569-570
SMN1873-4 24

USGS Hawthorne topographic map.

Free, E. E., "The Topographic Features of the Desert Basins ot the U. S. with Reference to the Possible Occurrence of Potash", U. S. Dept. Agr. B 54 (1914) 32.
Hanks, H. G., "Report on the Borax Deposits of California and Nevada", Cal. S. M. Bur. 3rd AR (1883) II 48-53, 55.
Russell, I. C., "Geological History of Lake Lahontan", USGS M 11 (1885) 85.
Thompson & West 419.
Whitfield, J. E., "Analyses of Natural Borates and Borosilicates", USGS B 55 (1889) 58-9 Analysis of ulexite; Also in B419 (1910) 301.
Young, G. J., "Potash Salts and Other Salines in the Great Basin Region" U. S. Dept. Agr. B 61 (1914) 1, 33, 38, 39, 42-3.

ROCKLAND see PINE GROVE

SANTA FE (Luning)

Copper, Silver, Gold, Lead, Antimony, Marble, (Granite, Barite)

Location. The Santa Fe, or Luning, District lies on the W. flank of the mountain range to the E. of Luning in E. Mineral Co. Luning is on the S. P. R. R., and the mines are a few miles to the E. and N. E. of the town. The elevation of Luning is about 4,500 ft.; and that of Volcano Peak at the S. end of the district about 6,500 ft., while the highest point in the range to the E. of the mines is 8,309 ft.

History. The Santa Fe silver mine was discovered in 1879, and a number of other silver, silver-lead, and silver-copper claims were located in the district. These silver properties were exploited up to 1893. Work upon the copper-lead deposits began about 1906 and was encouraged by the blowing in of the Thompson smelter in 1912. In that same year the Nevada Marble Co. operated a quarry in the district, according to Hill, and antimony ore was said to have been mined on Volcano Peak. During the World War the camp was an important copper producer.

Production. From 1906 to 1921 the district produced 88,019 tons of or containing 8,849,597 lbs. copper, $123,146 in gold, 233,058 ozs. silver, and 253,019 lbs. of lead, valued in all at $2,406,829, according to Mineral Resources of the U. S. Geol. Survey.

Geology. The country rock consists of white and gray crystalline limestones probably of Triassic age, according to Hill, which have been intruded by granitoid rocks ranging from quartz-monzonite to quartz-diorite that are probably Cretaceous. Tertiary volcanics overlie these rocks to the N.; and to the S. of Volcano Peak, a fault has let them down and exposed overlying conglomerates and sandstones.

Ore Deposits. The commonest type of ore deposit in the Santa Fe District is the typical contact-metamorphic deposit carrying ores of copper and lead. Replacement bodies of copper and lead ores also occur, and are at a distance from known intrusives. Veins are less common. The Santa Fe vein is in altered quartz-monzonite; the Nogal vein of the Luning Consolidated is a lead-silver-gold vein in the same country rock; while fissure veins in granite and quartzite are said to occur on the Luning-Idaho property. Stibnite and antimony oxide are said to be present in stringers on Volcano Peak.

Mines. Three properties have recently been active. **The Luning Cons. S. Ms. Co.** was incorporated in Nevada in 1920 with a capital stock of 1,500,000 shares of 10 cts. par value of which 750,000 shares remain in the treasury. The home office is at 265 Russ Bldg., San Francisco. L. A. Browne is Pres.; H. Zadig, V. P.; with A. S. Wollberg, J. N. Rogers, and A. E. Lowe, Directors. A. E. Lowe is Supt. at Luning. The property consists of 23 claims, developed by 3 shafts the deepest of which is 125 ft., having 1 m. of workings, and equipped with steam power, hoist, and aerial tram. **The Luning-Idaho M. Co.** was incorporated in 1910 with a capital stock of 1,500,000 shares of $1 par value of which 1,116,533 have been issued. The home office is in Reno. R. B. Todd is Pres.; R. G. Withers, V. P.; G. M. Todd, Sec-Treas.; with M. W. O'Boyle and G. B. Brest, Directors. The property consists of 13 claims developed to a depth of 200 ft. by 3000 ft. of workings and equipped with hoist and compresor. **The Mineral Buttes M. Co.** was incorporated in Nevada in 1921 with a capital stock of 1,500,000 shares of 10 cts. par value. The home office is in the Fordonia Bldg., Reno, and C. L. Richards is Pres. H. A. Davison is Supt. at Luning. The company owns 3 claims developed by a 150-ft. shaft and 600 ft. of workings and equipped with gasoline hoist and compressor.

Bibliography.

MR1882 230	MR1913 I 833	MR1916 I 488
MR1905 1055 Marble	MR1913 II 1369	MR1917 I 285-6
MR1908 I 484	1374 Marble	MR1918 I 249
MR1909 I 406-7	Granite	MR1919 I 401
MR1910 I 523	MR1914 I 697-8	MR1920 I 329
MR1911 I 687	MR1915 I 643	MR1921 I 388
MR1912 I 804	MR1915 II 175 Barite	

USGS Hawthorne topographic map.

Hill507 209.

McCormick, E., "The Copper Deposits of Southwestern Nevada," M&SP 81 (190)) 401.

———— "The Santa Fe Mining District, Nevada," M&M 21 (1901) 407.

Stuart NMR 68-9.

Weed, W. H., "The Copper Mines of the U. S. in 1905," USGS B 285 (1906) 116.

WeedMH 1159 Calavad C. Co. 1160 Cal. Crude Oil Co.
1170 Congres Copper Co. 1194 Esmeralda C. & S. Co.
1227 Ideal C. Co. 1229 Iron Gate Mine, Iroquois, C. Co.
1244 Luning Cons. S. Mg. Co., Luning-Idaho M. Co.
1286Nevada Standard C. Co. 1303 Pilot C. Co.
1305 Polaris G. & S. M. Co. 1310 Redmond Cons. M. Co.
1327 Shiper C. M. Co. 1328 Silvada Ms. Co.
1369 Wall Street C. Co. 1371 Wedge C. Co.

SILVER STAR (GOLD RANGE, MINA, BLACK MOUNTAIN, MARIETTA)

Silver, Lead, Copper, Gold, Tungsten, (Variscite, Turquoise)

Location. The Silver Star District is situated in the Excelsior Mountains. It is frequently subdivided into the Gold Range, or Mina District, on the N. E. in the vicinity of Mina; and the Black Mountin, or Marietta District, on the S. W. in the vicinity of Marietta. Mina is on the Hazen Branch of the S. P. R. R. and is the town from which the narrow-gauge S. P. line runs into California by way of Belleville. Marietta is 10 m. from Belleville and 26 m. from Mina. Mina has an elevation of 4552 ft. and the Mina or Gold Range section lies to the S. W. of it Marietta is immediately N. of Teels Marsh at an elevation of 4900 ft., and the Black Mountain or Marietta section lies to the N. of it. The highest peak in the Excelsior Range within the Silver Star District has an elevation of 8,766 ft.

History. The Endowment Mine in the Black Mountain section was found shortly after the discovery of the Aurora District and produced rich silver-lead ore for a number of years. Ore was discovered in the neighborhood of Silver Star camp in the Gold Range section by Pepper, Grassie, and Robb in 1893. Variscite and turquoise were located at the E. end of the Excelsior Mts. in 1910.

Production. The Endowment Mine is credited with a production of $1,500,000; and the mines near Silver Star from 1893 to 1901 are estimated to have produced $300,000. From 1902 to 1921, the district produced 13,874 tons of ore containing $97,672 in gold, 121,891 ozs. silver, 1,124,631 lbs. copper, and 596,500 lbs. lead, valued in all at $475,959, according to Mineral Resources of the U. S. Geol. Survey.

Geology. The Excelsior Mts. are composed mainly of Triassic sediments ranging from fine-grained argillites to coarse-grained conglomerates. These bedded rocks have been intruded by Cretaceous quartz-monzonite and younger augite-andesite and capped by Tertiary andesite, rhyolite, and basalt. Near Silver Star in the Gold Range section numerous E-W. veins occur in the conglomerates of the Triassic sediments. Their filling consists of white sugary quartz, adularia, and a little calcite or siderite, according to the Hill, and gold and silver are their only metallic constituents. The E. W. veins in the Black Mountain section occur in black quartzite and white quartzite conglomerate not far from a quartz-monzonite intrusion. They contain base metal minerals including galena, sphalerite, pyrite, and chalcopyrite and more rarely tetrahedrite and argentite; and at the surface they were rich in silver.

Mines. Two companies are at present active in the Silver Star District. **The Black Mountain S. M. Co.,** was incorporated in 1919 with a capital stock of 1,500,000 shares of 10 cts. par value of which 1,046,000 shares are outstanding. L. W. Whiting is Pres.-Mgr.; A. E. Sparks, V. P.; and W. W. Whiting, Sec.-Treas. The office is at Mina. The property consists of 7 claims aggregating 110 acres. **The Marietta Group** is under the management of J. L. Giroux of Marietta.

Tungsten. According to Mines Handbook, the property now belonging to the Silver Dyke & Tungsten Ms. Inc., 12 m. S.W. of Mina, was located for gold and silver in 1915. It became a tungsten producer under the title Silver Dike Tungsten Mine and continued to produce tungsten ore under different operators up to 1918. The orebody is said to occur in a fracture zone in a shale-limestone formation intruded by diorite.

Bibliography, SMN1877-8 25
MR1905 268
MR1906 294
MR1907 I 354
MR1908 I 484,486
MR1909 I 407
MR1910 I 522,523
MR1911 I 686,687
MR1912 I 804-5
MR1913 I 833
MR1914 I 696,697-8
MR1915 I 643
MR1916 I 488-9
MR1917 I 286,944
MR1918 I 249,974
MR1919 I 401
MR1920 I 329
MR1910 II 887
Turquoise; 894
Variscite

USGS Hawthorne topographic map.

Hill507 209. Hill594 171-18-. StuartNMR 46,47.

WeedMH 1143 Berundy Tungsten Ms. Co. 1148 Black Mountain S. M. Co.
1187 Douglas M. Co. 1228 Irondike M. Co.
1249 Marieta Group. 1264 Nevada Black Hawk S. M. Co.
1276 Nevada C. Co. 1287 Nevada Tri-Metallic Cons. M. Co.
1327 Shepherd S. M. Co. 1329 Silver Center M. Co.
1311 Silver Dyke & Tungsten Ms. Co. Silver Gulch M. Co.
1346 Sultana S. Ms. Co. 1374 Western S. Ms. Co.

SIMON see BELL

LEADVILLE see PILOT MOUNTAINS

SULPHIDE

Gold (Tungsten)

Location. The sulphide District is 18 m. S.E. of Hawthorne and 3 1-2 m. E. of Whisky Spring. It adjoins the Whisky Flat District on the N.

History. A test lot of ore from the Tango claim was treated at the Pamlico

Mill in 1914, and gold ore was shipped from the same property in 1915.

Geology. According to a letter from F. R. Dodge, scheelite occurs with the gold.

Biblography MR1914 I 698 MR1915 I 643
USGS Hawthorne topographic map.

TEELS MARSH
Borax, Salt

Location. Teels Marsh lies S. of Marietta in S. Mineral Co. Its elevation is a little less than 4900 ft., and its area is about 8 sq. mi. The Marsh adjoins the Black Mountain District on the S.E.

History. Teels Marsh was first worked as a salt deposit, salt being shipped from it to the silver mills at Aurora as early at 1867. F. M. Smith discovered borax in Teels Marsh in 1872, and he and his brother erected a borax plant in 1873. This borax plant was the largest in the state and kept up its production for many years.

Geology. Teels Marsh is a typical playa deposit.

Bibliography. B1867 311 Salt. SMN1869-70 108 Salt MR1882 567,570
R1875 458. SMN1875-6 38 MR1911 II 858
SMN1866 32 Salt SMN1873-4 24-5 MR1889-1890 503
SMN1867-8 95 "
USGS Hawthorne topographic map.

Free, E. E., "The Topographic Features of the Desert Basins of the U. S. with Reference to the Possible Occurrence of Potash," U. S. Dept. Agr. B 54 (1914) 32.

Hanks, H. G., "Report on the Borax Deposits of California and Nevada," Cal. S. M. Bur. 3rd AR (1883) II 46, 55.

StuartNMR 8.

Young, G. J., "Potash Salts and Other Salines in the Great Basin Region," U. S. Dept. Agr. B 61 (1914) 1,33,38,39,42.

VIRGINIA MARSH see RHODES MARSH
WALKER LAKE (AT CAT CREEK, BUCKLEY)
Gold, Silver, Copper

Location. The Walker Lake District is situated W. of Walker Lake on the E. slope of the Walker River or Wassuk Range. It extends from Reese River Canyon on the N. to Cat Creek on the S. The East Walker District adjoins it on the W., and the Hawthorne District on the S.E. The old Buckley District on the opposite side of the lake was sometimes considered part of the Walker Lake District.

History. The Walker Lake District was discovered in 1866. The Cat Creek gold-copper deposit was discovered in the early eighties. A revival of interest occurred in recent years, development work was in progress in 1907 and 1908, and silver ore was shipped in 1909. In 1911, the Walker Lake Mine operated a 5-stamp amalgamation and concentrating mill on gold-silver ore. Gold ore was produced in the district in 1914 and 1916. In 1917, the Northern Light property produced a little copper ore. This mine was operated by the Mason Valley Ms. Co., under leasee in 1919 and 1920 and produced 2800 tons of 4.8 per cent. copper ore, while in 1921, 4000 tons of ore were mined and stored. Gold bullion was produced from the Hematite claim in 1921.

Geology. The country rock at Cat Creek is granodiorite cut by dikes of basic granodiorte and aplite. The ore occurs in a brecciated fault zone, according to Hill, and the material of this zone is somewhat siliceous and contains disseminated pyrite and chalcopyrite with low gold values. Chalcocite, bornite, and native copper are present as secondary minerals.

Bibliography. SMN1866 39-41 MR1909 I 405 MR1917 I 286
MR1907 I 353-4 MR1911 I 687-8 MR1919 I 402
MR1908 I 482-3 MR1914 I 698 MR1921 I 389
MR1916 I 489
USGS Hawthorne topographic map.
Hill594 151, 156.
WeedMH 1215 Nevada Cons. Ms. & Selling Co.
1294 Northern Light C. Co.

WASHINGTON

Copper, Gold, Silver.
Coal, (Petroleum)

Location. The Washington District is located on the East Fork of the Walker River in the vicinity of Washington and Wichman. It is about 30 m. S. of Yerington which is on the N. C. B. R. R.

History. The Washington District was organized in 1861, and a discovery of silver chloride and sulphide associated with lead, antimony, and copper sulphide was made in 1867. A few tons of copper ore containing a little gold and silver were shipped from the district in 1915. In 1919, the Nevada Coal & Oil Co. exploited the lignite coal deposits of the district, and in 1920 drilled for oil there without success. Some lignite has been mined and used locally, but its high ash content precludes its use for most purposes.

Bibliography. B1867 337,431. SMN1867-8 91 MR1915 I 643
SMN1869-70 109
SMN1871-2 43
USGS Hawthorne topographic map.
E&MJ "Oil Wildcatting in Nevada," 109 (1920) 665.

WHISKY FLAT

Copper, Silver, Gold

Location. The Whisky Flat District is located at the S. end of Whisky Flat on the N. slope of the Excelsior Mts. It is 16 m. S.E. of Hawthorne which in turn is 7 m. S.W. of Thorne, which is on the S. P. R. R. The Whisky Flat District adjoins the Black Mountain or Marietta District on the W., and the Sulphide District on the S.

History. The district was first worked in 1882, when copper ores rich in silver and carrying a little gold were smelted in a 400-lb. Mexican furnace. The Excelsior Mountain C. Co. was active from 1907 to 1914.

Geology. A narrow zone of rich silver ore associated with chalcocite occurs at a granite-limestone contact ,and away from the contact in the limestone is a broad zone of oxide and sulphide copper ores. The ore minerals are copper carbonate, chalcocite, pyrite and chalcopyrite. Garnet is present as a gangue mineral.

Bibliography. USGS Hawthorne topographic map.
Hill594 157. WeedMH 1199 Excelsior Mountain C. Co.

WILSON see PINE GROVE

NYE COUNTY

ANTELOPE SPRINGS
Silver, Gold

Location. The Antelope Springs District lies at the S.E. end of the Cactus Range. Goldfield on the T. & G. R. R. is 30 m. W. N. W. The highest mountain is Antelope Peak with an elevation of 7,600 ft. The Wellington District adjoins the Antelope Springs District on the S. W. and the Trapmanns District adjoins it on the S. E.

History. The district was discovered by the Bailey Bros. in 1903. Jordan and Reilly found high grade ore in 1911, a small boom ensued, and intermittent shipments were made from 1912 to 1917.

Geology. The Cactus Range is composed mainly of Tertiary eruptives, according to Schrader, which rest unconformably upon Paleozoic sedimentary rocks which have been intruded by Cretaceous granite and diorite porphry. The country rock of the Antelope Springs District is Tertiary rhyolite. The ore occurs in quartz veins filling fissures and in the altered rhyolite adjoining these veins. A little adularia is present with the quartz in the veins, and the rhyolite has been altered by silicification, sericitization, kaolinization, and alunitization. The ore minerals are hornsilver, argentite, and native gold, much of which contains silver. Limonite, hematite, and manganese oxide are common.

Bibliography. MR1912 I 805 MR1913 I 834 MR1916 I 489 MR1917 I 287
USGS Kawich topographic map.
Ball 308 89-95.
Schrader, F. C., "Notes on the Antelope District, Nevada," USGS B 530 J (1912) 1-14, and USGS B 530 (1913) 87-98; Abstract, M&SP 105 (1912) 170; M&EW 37 (1912) 201-2.

ARROWHEAD
Silver, Gold

Location. The Arrowhead District is situated at Arrowhead at the N. end of the Reveille Range. Tonopah on the T. & G. R. R. is 65 m. W. The Reveille District adjoins the Arrowhead District on the S.

Geology. The Reveille Range is composed of Paleozoic quartzite and limestone which have been folded into a monocline on its W. flank, according to Ball, and are capped by flows of Tertiary rhyolite and basalt on the E. The country rocks of the Arrowhead Mine are andesite and rhyolite. The orebody is a replacement vein from 3 ft. to 8 ft. in width carrying pyrargyrite and argentite as the principal ore minerals. A minor amount of gold is present.

Mines. Two companies are at present active in the district. **The Arrowhead Ex. M. Co.** was incorporated in Nevada in 1919 with a capital stock of 1,500,000 shares of 10 cts. par value. The home office is at 130 Main St., Tonopah, and J. B. Kendall is Pres. M. J. McVeigh is Gen. Mgr. at Arrowhead. The company owns 100 acres adjoining the Arrowhead Mine and developed by a 150-ft. shaft. **The Arrowhead M. Co.** was incorporated in Nevada in 1919 with a capital stock of 1,500,000 shares at 10 cts. par value of which 750,000 shares are treasury stock. R. L. Jones is Pres. at Tonopah; and P. Fox is Supt. at Arrowhead. The company owns 4 claims including thee Arrowhead Mine which is developed by a 345-ft. shaft and equipped with a gasoline hoist.

Bibliography. MR1920 I 329 MR1921 I 390

Ball307 114-7 Reveille Range. Spurr208 161-3 Reveille Range.
WeedMH 1135 Arrowhead Annex M. Co.
1136 Arrowhead Bonanza M. Co., Arrowhead Cons. M. Co., Arrowhead Esperanza Mc. Co., Arrowhead Ex. M. Co., Arrowhead Inspiration Ms. Co., Arrowhead M. Co.
1137 Arrowhead Silver King Mg. Co., Arrowhead Silver Signal Ms., Arrowhead Syndicate Ms. Co., Arrowhead Wonder Ms. Co.
1141 Sunny Divide M. Co. 1371 West Arrowhead Ms. Co.

ASH MEADOWS

Fuller's Earth, (Potash)

Location. Ash Meadows is in S. Nye Co. near the California boundary.

History. In 1917, clays in this district were located for their potash content. The Standard Oil Co. mined clay for use as Fuller's earth in this locality in 1918 and 1919.

Bibliography. MR1919 II 423-4 Potash MR1918 II 145 Fuller's Earth

ATHENS

Gold, Silver ..

Location. The Athens District is situated at Athens in N. W. Nye Co., on the Mineral Co. border. Athens is about 30 m. by road N. E. of Mina which is on the S. P. R. R.; and is 4 m. N. of Simon silver-lead camp and 4 m. E. of the former gold camp of Omco. * The Fairplay District adjoins it on the N.; and the Bell District in Mineral Co. adjoins it on the W.

History. The Warrior Gold Mining Co. developed a mine in this district and produced some $20,000 in gold bullion. The Warrior Mine has recently been acquired by the Aladdin Divide M. Co. which is continuing the exploration of the property. The Lucky Boy Divide M. Co. has secured the adjoining ground and proposes to undertake active development work upon it. H. McNamara of Tonopah is Pres.

Geology. The country rocks at the Warrior Mine consist of early Tertiary rhyolite, andesite, and dacite, together with later lake beds. There are said to be two veins, one in rhyolite, and the other at a rhyolite-dacite contact. These veins are cut by a flat fault.

Bibliography. USGS Tonopah topographic map.
WeedMH 1128 Aladdin D. M. C. 1243 Lucky Boy D. M. Co.
1369 Warrior G. M. Co.

ATWOOD see FAIRPLAY
BARCELONA see BELMONT
BARE MOUNTAIN see FLUORINE

BELLEHELEN

Silver, Gold

Location. The Bellehelen District is situated at Bellehelen at the N. end of the Kawich Range in central Nye County. Bellehelen is 50 m. E. of Tonopah which is on the T. & G. R. R.

Geology. The Kawich Range is composed of Tertiary eruptive rocks. The country rock of the district is rhyolite in which a number of parallel veins occur. Silver values predominate on some properties and gold values on others.

Mines. The Bellehelen Merger Ms. Co., Tonopah, is now building a 50-ton

* The date for this descriptfon was kindly furnished by H. McNamara.

cyanide plant at Bellehelen. M. M. Green is Pres.; W. F. Gray, V. P.-Gen. Mgr.; and E. W. McClave, Sec.-Treas.

Bibliography.

MR1905 272	MR1912 I 805	MR1917 I 287
MR1906 298	MR1913 I 834	MR1918 I 250
MR1907 I 372	MR1914 I 699	MR1919 I 402
MR1908 I 496	MR1915 I 644	MR1920 I 229
MR1909 I 419	MR1916 I 489	MR1921 I 389

Hill 507 219-220. Spurr208 181 Kawich Range. StuartNMR 91-2.
WeedMH 1141 Bellehelen Merger Ms. 1165 Clifford Silver Ms. Co.
1190 Elgin-Bellehelen D. M. Co.

BELMONT (PHILADELPHIA, SILVER BEND, SPANISH BELT, BARCELONA)

Silver, Gold, Turquoise, (Mercury, Molybdenum)

Location. The Belmont or Philadelphia District is located in the neighborhood of Belmont on the E. flank of the Toquima Range. Tonopah on the T. & G. R. R., is 50 m. S. W. Silver Bend was an early name for the district. The N. W. section where the Barcelona Mine is situated is sometimes considered as a separate district and called the Spanish Belt or Barcelona District. The Manhattan District adjoins the Belmont District on the S. W., and the Round Mountain District adjoins it on the N. W.

History. The Belmont District was discovered by C. L. Straight and others in 1865. By 1867, it was a prosperous camp and was made the county seat of Nye Co. The mines were active up to 1885, and the county seat remained at Belmont until the Tonopah boom. In 1909, turquoise was discovered in the district by Mrs. E. S. Weber. The Monitor-Belmont M. Co. acquired the old mines at Belmont in 1914, and erected a 10-stamp 100-ton flotation plan in 1915 which shut down in 1917. The Nevada Wonder M. Co. took a lease on the mine in 1918, but relinquished it after making a considerable expenditure. The Cons. Spanish Belt Silver M. Co., acquired the old Barcelona Mine and put a 10-stamp 50-ton flotation and concentration plant in operation there in 1921, shutting down in 1922. Belmont tailings were treated in a 30-ton cyanide plant in 1921.

Production. The Belmont District is credited with a production of more than $15,000,000 from 1865 to 1885.

Geology. The country rocks of the Belmont District are Ordovician shales and limestones cut by intrusive granite. At the Monitor-Belmont, quartz veins and lenses occur in the sedimentary rocks near a granite contact. According to Emmons, the ore consists principally of stetefeldtite, an argentiferous antimony ore, with which are combined lead, silver, copper, and iron. The metallic minerals are scattered through the quartz in bunches and diseminated particles and rarely occur in bands. The Barcelona ore consists of quartz banded with sulphides of iron, zinc, copper, and lead. Becker noted the occurrence of cinnabar at the Barcelona Mine. At the Belmont Big Four Mine, a quartz vein in granite is said to carry 2 per cent molybdenite besides silver and copper values.

Bibliography.

B1866 129	R1873 230-3	SMN1871-2 102-6
B1867 402, 404, 420-3, 431	R1874 279-280	SMN1873-4 67-70
R1868 109-111	R1875 415	SMN1875-6 102-4
R1870 128-130	SMN1866 62	SMN1877-8 89-90

R1871 141-182 SMN1867-8 70-3 MR1920 I 329
R1872 172 SMN1869-70 90 MR1921 I 389,390
USGE40th 3 (1870) 393-405.
Becker, G. F., "Geology of the Quicksilver Deposits of the Pacific Slope," USGS M 13 (1888) 385.
Hill507 220.
Hughes, W. W., "The Belmont Camp, Nevada," E&MJ 103 (1917) 1007-8.
Spurr, J. E., "Quartz-Muscovite Rock from Belmont, Nevada." AJS IV 10 (1900) 351-8.
Spurr208 90-93 Toquima Range.
Stuart NMR 84-5.
Thompson & West 519-520.
WeedMH 1142 Belmont Big Four M. Co. 1175 Cons. Spanish Belt S. M. Co. 1260 Monitor-Belmont M. Co. 1342 Spanish Belt Ex. S. M. Co. 1374 West Spanish Belt S. M. Co.

BERLIN see UNION

BIG DUNE see LEE

BLACK SPRING

Diatomaceous Earth.

Location. Diatomaceous earth occurs near Black Spring N. W. of the Cloverdale District in N. W. Nye Co.

Analysis. Material from the mine of the Nature Products Co., was aassayed by the Industrial Research Co. with the following results:

Loss on Ignition	10.91%
Silica	84.39
Iron Oxide and Alumina	3.07
Soda	1.54
Potash	Trace
Lime	None
Magnesia	None
Carbonic Acid	None
	99.90%

Mines. The **Nature Products Co.** owns over 200 acres of diatomaceous earth lands in this district. J. M. Fenwick is Pres.; E. N. Richardson, V. P.; and G. A. Foster, Sec.-Treas. The company is mining the earth and using it as a base for Superdent Tooth Power and Super Dental Cream.

BLAKES CAMP see GOLDEN ARROW

BRUNER (PHONOLITE)

Gold, Silver

Location. The Bruner District is located at Bruner and Phonolite in N. W. Nye Co. on the Lander Co. and Churchill Co. border. Luning on the S. P. R. R. is 50 m. S. W. The Lodi District adjoins the Bruner District on the S.

History. The Paymaster Mine made shipments of gold ore from 1912 to 1915. In 1915, the Kansas City-Nevada Cons. Ms. Co. consolidated the Paymaster Mine with the Big Henry, Silent Friend, and Duluth properties. A 50-ton cyanide mill was constructed by this company in 1919 and remodeled in 1921. Ore from the Broken Hills Silver Corp. Mine is hauled 12 m. to this mill and treated under a contract made in 1921.

Geology. The ore deposit at the Paymaster Mine is a fissure vein in rhyolite and andesite carrying values in gold and silver.

Mines. The Kansas City-Nevada Cons. Ms. Co. was incorporated in Nevada in 1915 with a capital stock of 6,000,000 shares of $1 par value, all issued. The home office is at 510 Commerce Bldg., Kansas City. W. P. Neff is Pres. and H. Crawford is Sec. W. H. Kinnon is Engr. at Bruner. The company owns 25 patented claims aggregating 487 acres developed by a 300-ft. shaft and 1-2 m. workings and equipped with a 50-ton cyanide mill.

Bibliography. MR1912 I 805 MR1914 I 699 MR1915 I 644 MR1920 I 329 WeedMH 1234 Kansas City-Nevada Cons. Ms. Co.

BULLFROG (PIONEER)
Gold, Silver, Copper, Lead

Location. The Bullfrog District is located in the Bullfrog Hills in S. Nye Co. on the California border. It is W. and N. W. of Beatty on the T. & G. R. R. and extends from Bullfrog and Rhyolite on the S. to Pioneer on the N. Beatty has an elevation of 3309 ft. and the highest point in the Bullfrog Hills reaches one of 6,235 ft. The N. section of the Bullfrog District is sometimes considered as a separate district and called the Pioneer District.

History. Bullfrog was discovered by Frank ("Shorty") Harris in 1904 and a rush of prospectors into the new camp ensued. In 1906, the L. V. & T. R. R. reached the district; and in 1907 the B. G. R. R. and the T. & T. R. R. ,while the L. V. & T. R. R. was extended to Goldfield. The Montgomery-Shoshone Mine was the most important in the district. This property was equipped with a 200-ton amalgamating, concentrating, and cyaniding mill which began operations in 1907, and shut down in 1910. In 1911, the Mayflower Mine was operating a 15-stamp mill, and in 1912 the Tramp Mine was also running a mill. The Pioneer 10-stamp mill began operations late in 1913, and the Sunset 10-stamp mill was erected in 1914. The Pioneer mill shut down in 1917, and activities in the camp were at low ebb until 1921 when the mines in the Pioneer section resumed active development work.

Production. The production of the Bullfrog District from 1905 to 1921 was 286,664 tons of ore containing 111,805.16 ozs. gold; 868,749 ozs. silver, 5,294 lbs. copper, and 11,897 lbs. lead, valued in all at $2,792,930, according to Mineral Resources of the U. S. Geol. Survey.

Geology. Tertiary volcanic rocks having an aggregate thickness of at least 6000 ft. cover most of the district, according to Ransome, Emmons, and Garrey, and contain all the important orebodies. Rhyolite is the predominant rock. Sixteen successive rhyolite formations have been recognized; intercalated between them are 5 flows of plagioclae basalt, 1 of quartz-latite, and some stratified tuffs; while capping the whole series is a flow of quartz-bearing basalt. The series of volcanic rocks is divided into eastward tipping blocks by faults. The ore deposits are mostly nearly vertical mineralized faults or fault zones in rhyolite. The common type of lode is a fissure zone with veinlets parallel to the sides linked by irregular cross stringers. The primary ore consists of auriferous pyrite in a gangue of quartz and calcite. In the oxide ore, native gold, cerargyrite, and limonite occur. A little chalcocite and galena have been observed. Alunite is present but not in sufficient amount to be of commercial interest.

Mines. The Cons. Mayflower Ms. Co. is incorporated in Nevada with a capital stock of 2,500,000 shares of 10 cts. par value of which 1,500,000 are outstanding. The stock is listed on the New York Curb and the San Francisco Stock Exchange. The main office is at Pioneer. J. S. Gibbons is Pres.; W. J. Tobin, V. P.-Sec.-Treas.; with C. W. Taylor, M. H. Olmert, and W. O. Lamping, Directors. The property is at Pioneer and consists of 11 patented claims, developed by 4 shafts the deepest of which is 530 ft., and equipped with hoist, compressor, pipe line, and 15-stamp mill. **The Reorg. Pioneer Ms. Co.** was incorporated in Nevada in 1918 with a capital stock of 5,000,000 shares of 10 cts. par value of which 2,250,000 are outsanding. The home office is at Goldfield. W. J. Tobin is Pres.; J. K. Turner, V. P.; B. Gill, Sec.-Treas.; with E. A. Blakesley, additional Director. The property consists of 12 claims including old workings from which $500,000 is said to have been extracted, developed by an 800-ft. shaft, and equipped with engine, hoist, compressor, and 10-stamp mill.

Bibliography.

MR1904 198	MR1910 I 524-5	MR1915 II 112	Potash
MR1905 116,272	MR1911 I 688	MR1916 I 491	
MR1906 298	MR1912 I 805-6	MR1917 I 287	
MR1907 I 372,394-8	MR1913 I 834	MR1918 I 250	
MR1908 I 496-7	MR1914 I 699	MR1919 I 402-3	
MR1909 I 419-420	MR1915 I 644	MR1921 I 389	

USGS Furnace Creek and Bullfrog Special topographic maps.

Ayres, E. R., "The Bullfrog Cyanide Mill," E&MJ 83 (1907) 376-8.

Davis 967.

E&MJ, "From Goldfield to Rhyolite," 82 (1906) 815-6.

Emmons, W. H., "Normal Faulting in the Bullfrog District," Science, N. S. 25 (1907) 221-2.

Hill507 220.

Martin, A. H., Montgomery-Shoshone Mines," M&SP 100 (1910) 289-290.

Ransome, F. L., "Preliminary account of Goldfield, Bulfrog and other mining districts in Southern Nevada, USGS B 303 (1907) 40-62.

Ransome, F. L., Emmons, W. H., & Garrey, G. H., "Geology and Ore Deposits of the Bullfrog District, Nevada," USGS B 407 (1910).

Rice, C. T., "The Bullfrog Mining District, Nevada," E&MJ 82 (1906) 534-6.

Spurr, J. E., "Genetic Relations of Western Nevada Ores," T AIME 36 (1906) 382-3.

StuartNMR 89-91.

Taft, H. H., "The Bullfrog Mining District," E&MJ 81 (1906) 322; also in T AIME 37 (1907) 184-6.

Tallman, C., "The Bullfrog District," P AMC 12th Ann. Sess. (1909) 428-437.

Van Saun, P. E., "Cyaniding at the Montgomery-Shoshone Mill," M&M 28 (1908) 385.

——— "Cyaniding at the Montgomery-Shoshone Mill," E&MJ 89 (1910) 217-9.

WeedMH 1174 Cons. Mayflower Ms. Co.
1261 Montgomery-Shoshone Cons. M. Co.
1314 Reorg. Pioneer Ms. Co. 1347 Sunset M. & Dev. Co.
1252 Mayflower Bullfrog Cons. M. Co.
1304 Pioneer Ex. Ms. Co. 1327 Shoshone Polaris M. Co.

Young, G. J., "Cooperation Among Small Mines," E&MJ 106 (1918) 813.

BUTTERFIELD MARSH (RAILROAD VALLEY MARSH)

Salt (Soda, Potash)

Location. Butterfield Marsh is situated in N. Railroad Valley in N. E. Nye

Co. It is about 110 m. E. N. E. of Tonopah which is on the T. & G. R. R. and 80 m. S. W. of Ely which is on the N. N. R. R.

History. Salt was mined from Butterfield Marsh in the early days for use in the silver mills of Tybo. The Railroad Valley Co. drilled the marsh in 1912 and 1913, putting down 7 wells aggregating more than 10,000 ft., in an unsuccessful attempt to find potash salts in commercial amounts.

Geology. Railroad Valley is a typical desert basin extending for 100 m. in a N.-S. direction and having a width of from 10 to 20 m. Its drainage area is about 6,000 sq. m. and the flat central portion has an area of 200 sq. m Butterfield Marsh is in the lowest portion of the valley and has an area of 40 sq. m. The valley was formerly occupied by a lake whose level was from 50 to 300 ft. above that of the playa. Butterfield Marsh is commonly covered with a thin crust of salt and toward its N. end are irregular salt pans where the salt incrustation is thicker, and from which the production has come. Potash occurs in these efflorescences but drilling by the Railroad Valley Co. failed to disclose any potash salts in depth although soda-bearing beds consisting chiefly of gaylussite, hydrous sodium-calcium carbonate, were encountered.

Bibliography. SMN1869-70 92 MR1912 II 882-3, 891 MR1913 II 85-6

Free, E. E., "The Search for Potash in the Desert Basin Region," USGS B 530 (1913) 305.

———"The Topographic Features of the Desert Basins of the U. S. with Reference to the Possible Occurrence of Potash," U. S. Dept. Agr. B 54 (1914) 35-6.

Gale, H. S., "The Search for Potash in the Desert Basin Region," USGS B 530 (1913) 305.

Hance, J. H., "Potash in Western Saline Deposits," USGS B 540 (1914) 457-462.

Phalen, W. C., "Salt Resources of the U. S.," USGS B 669 (1919) 144-5.

Sheldon, G. L., "Railroad Valley Potash Fields," M&SP 105 (1912) 502-3.

Thompson & West 364.

Young, G. J., "Potash Salts and Other Salines in the Great Basin Region," U. S. Dept. Agr. B 61 (1914) 8, 13, 32, 39, 56-9, 64, 83-5.

CACTUS SPRINGS

Silver, Gold, (Turquoise)

Location. The Cactus Springs District is at Cactus Springs at the N. W. end of the Cactus Range. Goldfield on the T. & G. R. R. and the T. & T. R. R. is 24 m. W.

History. Turquoise was discovered on Cactus Peak by William Petry in 1901. The Cactus Nevada Silver Mine was discovered in 1904 and has shipped a little silver ore.

Geology. The ore occurs as fissure fillings and replacements in silicified tertiary rhyolite, according to Ball. Quartz veins in the altered rhyolite contain a small amount of pyrite and chalcopyrite, and these ore minerals are still more sparingly disseminated in the adjoining country rock. The outcrops are heavily stained with limonite, and free gold is said to be present.

Bibliography. MR1901 760-1 Turquoise MR1916 II 491 Potash
MR1909 I 420 MR1920 I 329

Ball285 86. Ball 388 46, 89-96. Hill 507 220-1.

WeedMH 1159Cactus Cons. Silver Ms. Co. Cactus Leona Silver Corp. Cactus Nevada Silver Ms. Co.

CARRARA
Marble

Location. The Carrara marble deposit is located on the S. W. flank of Bare Mt. in S. Nye Co. It is about 3 m. N. E. of the town of Carrara which is on the T. & T. R. R. The Carrara District adjoins the Bare Mt. District on the S. W.

Geology. The marble occurs in clearly defined beds which dip into the mountain at about 60 degrees, according to a report by R. D. Salisbury. The outcrop has a length of more than 2 m., and the thickness is estimated at 2,500 ft. While the proportion of this marble which is of commercial quality has not as yet been determined; the quantity is known to be very great. There is about 60 ft. of pure white marble, much of which is of statuary grade; besides several hundred feet of commercial grade white marble. There are also large amounts of black marble, including some of very high grade; of gray marble; and of blue marble; together with smaller quantities of marbles of less usual colors.

Quarries. The American Carrara Marble Co. owns and operates the Carrara quarries. P. V. Perkins is Pres. with office at 303 Chapman Bldg., Los Angeles; and the sales office is at 803 Central Bldg., Los Angeles. The marble is shipped to Los Angeles and to a less extent to San Francisco.

Bibliography. MR1913 II 1374 USGS Furnace Creek topographic Map.

CLIFFORD
Silver, Gold

Location. The Clifford District occupies a small hill on the W. flank of the Kawich Range in central Nye Co. Clifford is on the state highway from Tonopah to Ely about 35 m. E. of Tonopah which is on the T. & G. R. R.

Geology. Clifford Hill is composed mainly of thin-bedded rhyolitic sandstone and pyroclastic rocks, rhyolite tuff, and breccia, while andesite outcrops at the extreme W., according to Ferguson. There are numerous shallow workings in which the ore consists of heavily iron-stained tuff cut by small quartz veins that contain small irregular masses of limonite. Cerargyrite, native silver, and jarosite occur in the rich ore, with rare specks of a silver sulphide mineral and of pyrite; and a small amount of pale yellow gold may be obtained by panning. Sulphide ore has been mined in rhyolitic agglomerate at its contact with the andesite which appears to have been faulted against it. The agglomerate has been highly pyritized. The rich ore contains less pyrite and is cut by quartz veinlets in the vugs of which have been deposited stephanite, pyrargyrite, and proustite, with pyrite and more rarely marcasite; while the silver sulphide minerals also occur in minute streaks between the quartz and the wall rock.

Bibliography. MR1905 272

Ferguson, H. G., "The Golden Arrow, Clifford, and Elendale District, Nye County, Nev.," USGS B 640F (1916) 113-5, 121-2.

CLOVERDALE (GOLDEN)
Silver, Gold, Lead, Copper, Placer Gold

Location. The Cloverdale District is situated in the neighborhood of Cloverdale Ranch and S. of Golden in N. W. Nye Co. The district is 38 m. by road N. of Millers which is on the T. & G. R. R. The Black Spring diatomaceous earth district adjoins it on the N. W.

History. The Cloverdale District was being explored and developed in 1905.

Production began in 1906 and continued up to 1919. The Orizaba Mine is the principal property. A little placer gold has been mined 4 m. E of Cloverdale Ranch.

Geology. At the Orizaba Mine, according to Mines Handbook, veins from 2 ft. to 7 ft. wide occur at a limestone-granite contact. Gold-silver-copper and lead ores occur in these veins.

Bibliography. MR1905 272, MR1907 I 373, MR1908 I 497, MR1909 I 420, MR1910 I 525, MR1911 I 688, MR1913 I 834, MR1914 I 699, MR1915 I 645, MR1916 I 491, MR1917 I 287, MR1918 I 250, MR1919 I 403

USGS Tonopah topographic map.

Hill507 221. WeedMH Orizaba M. & M. Co.

CROW SPRINGS see ESMERALDA COUNTY

CURRANT

Gold, Lead, Copper, (Oil Shale, Petroleum)

Location. The Currant District lies E. of Currant in N. E. Nye Co.

History. Gold ore containing a little lead and copper was produced from the Sheperd property in 1914; and lead ore with some silver and copper from the Sunrise mine in 1916.

Geology. According to a letter from E. S. Giles, there is an extensive deposit of oil shale 10 m. E. of Currant, and the limestone above this contains globules of oil.

Bibliography. MR1915 I 699 MR1916 I 495

DANVILLE

Silver, Gold

Location. The Danville District is at Danville in the Monitor Range in N. Nye Co.

History. The Danville District was discovered by P. W. Mansfield in 1866, and reorganized in 1870. By 1881, it was practically abandoned.

Geology. Argentiferous quartz veins containing a little gold occur in limestone, according to Thompson and West.

Bibliography. B1867 403-6, 423-4 Thompson & West, 516-7

EDEN

Silver, Gold

Location. The Eden District is located at Eden on the E. slope of the Kawich Range in central Nye Co. It is 55 m. by read E. of Tonopah which is on the T. & G. R. R. The elevation of Eden is 6,700 ft., while Kawich Peak to the W. rises to a height of 9,500 ft.

History. The Eden District was discovered by John Adams in 1905. Recently, the Eden Creek M. & Mlg. Co. has been carrying on exploration work in the district under the management of Mark Bradshaw.

Geology. The country rock of the Eden District is Tertiary rhyolite. The ore deposits consist of 3 mineralized zones, according to Ball. These mineralized zones vary from quartz veins to silicified rhyolite containing numerous quartz stringers. Gold occurs in quartz of a flinty variety, and silver as ruby silver, native silver, and hornsilver. A small amount of pyrite is present in the quartz and the altered rhyolite.

Bibliography. USGS Kawich topographic map.

Ball285 66 Ball308 110-1 Hill507 221

ELLENDALE
Gold, Silver, Copper

Location. The Ellendale District is situated in W. Nye Co., a few miles E. of Tonopah and a short distance S. of the state highway from Tonopah to Ely.

Geology. Most of the workings are in rhyolite near its contact with andesite porphyry, according to Ferguson. The mineralization consists of numerous irregular stringers in the rhyolite composed of iron-stained quartz and of the silicification and slighter sericitization of the adjacent rock. Brown specks in the rhyolite resulting from the alteration of pyrite contain jarosite, and minute veinlets of this mineral occur in the rhyolite of the mineralized zone.

Bibliography. MR1910 I 525 MR1914 I 699 MR1917 I 287
MR1911 I 689 MR1915 I 635 MR1918 I 250
USGS Tonopah topographic map.
Ferguson, H. G., "The Golden Arrow, Clifford, and Ellendale Districts, Nye County, Nev., USGS B 640F (1916) 113-5, 122-3.
Stuart NMR92.

ELLSWORTH see LODI

FAIRPLAY (GOLDYKE, ATWOOD)
Gold, Silver, Copper, Lead, (Tungsten)

Location. The Fairplay District is located at the S. end of the Paradise Range in the neighborhood of Goldyke and Atwood in N. W. Nye Co. on the Mineral Co. border. Goldyke is 35 m. by road N. E. of Luning which is on the S. P. R. R. The Athens District adjoins the Fairplay District on the S.

History. The Fairplay District was discovered in 1903 and made small annual productions up to 1911. Two properties produced in 1920.

Geology. In the Paradise Range 60 m. N. of Tonopah, according to Hess, a quartz veing cutting granite contains hubnerite and scheelite accompanied by gold, silver, copper, fluorite, biotite, and pyrite.

Bibliography. MR1905 268 MR1908 I 497, 724 Tunsgten MR1910 I 525
MR1906 298 MR1909 I MR1911 I 689
MR1907 I 372-3 MR1920 I 329
USGS Tonopah topographic map.
Hill507 221-2 StuartNMR 92-3

FLUORINE (BARE MOUNTAIN)
Fluorite, Gold, Mercury, Kaolin, (Silver, Opal)

Location. The Fluorine District occupies the N. part of Bare Mt. in S. Nye Co. just E. of the town of Beatty on the T. & T. It adjoins the Bullfrog District on the E. and the Carrara District on the N. E. The elevation of Beatty is 3,309 ft. and the highest peak of Bare Mt. is 6,235 ft. above sea-level.

History. Gold ore was discovered on the E. slope of Bare Mt. in 1905; and the camp of Telluride sprang up. The Bull Moose property produced gold ore from 1913 to 1915. Cinnabar was discovered by J. B. Kiernan and A. A. Turner on the E. flank of Bare Mt. in 1908, and in 1912 a 10-ton Scott mercury furnace was erected there by the Telluride Cons. Quicksilver M. Co. and made small annual productions from 1912 to 1916. In 1913, opal containing cinnabar from this deposit was cut for gem purposes. A few carloads of kaolin were shipped from three properties to a California pottery plant in 1917 and 1918. Fluorspar claims in the Fluorine District were located in 1918 by J. Irving Crowell, and sold to the Spar Products Corp. which gave the Continental Fluorspar Co. a 20-year lease. The first fluorite production of Nevada was made in 1919 when 700 tons were shipped by the Continental Co. In 1920 shipments amounting to 632 tons were made.

Geology. The main mass of Bare Mt. is composed of Paleozoic sedimentary rocks which have been intruded by pegmatites and monzonite porphyry, while at the N. end the hills are composed of flows of Tertiary rhyolite and basalt. Ball gives the following sedimentary section, from the top down:

Schist	200 feet
Quartzite with some schist and limestone	1000 "
Limestone with schist beds near top	3000-4000 feet
Quartzite and schist	100-200 "

These rocks form a faulted monocline striking N. W. and dipping N. E. with minor cross folds. They have been much disturbed by folding and faulting. According to Knopf, the limestone is a fine-grained gray dolomite of Silurian age.

Precious Metals. Small irregular quartz veins are common in the dolomite and are chiefly gold-bearing, although Ball notes some which contain silver ore. The minerals present are: chalcopyrite, malachite, azurite, galena, pyrite, limonite, hematite, fluorite, and gypsum.

Mercury Ore and Opal. There are two groups of mercury deposits, one on the E. slope of Bare Mt. and the other 3 m. N. E. on the N. end of Yucca Mt. The country rock on Bare Mt. is dolomite, while on Yucca Mt. it is rhyolite and rhyolite tuff. The deposits in the dolomite consist of mases of opal or of cryptocrystalline silica carying cinnabar; while those in the rhyolite consist of masses of opal and alunite carrying cinnabar, according to Knopf. Coarsely crystalline calcite and barite were noted with cinnabar in shattered dolomite in one instance. The mercury ore occurs in irregular, erratic shoots in the siliceous masses.

Kaolin The Shepard kaolin deposit lies in the Tertiary volcanic area 9 m. E. of Beatty, according to Buwalda, and is probably an alteration product of quartz porphyry or tuff. It was opened as a cinnabar prospect and contains a little of that mineral. A carload of the clay has been shipped, but it seems to be lacking in plasticity for pottery use. The Kiernan and the Bond & Marks properties are 8 m. S E. of Beatty, high up on the flank of Bare Mt. in the area of Paleozoic sedimentary rocks. Clays from these deposits were shipped to California and found to contain too much hematite for use in pottery-making.

Fluorite. There are two groups of fluorite deposits, both in dolomite. One group lies 5 m. E. of Beatty, and the other 4 m. S. of the first. The fluorite is said to occur in veins from 8 ft. to over 150 ft. in width and up to 1-2 m. in length. The material averages 83% calcium fluoride with less than 5% silica as mined.

Mines. The **Continental Fluorspar Co.,** Beatty, of which J. I. Crowell is Pres., is at present active. This company owns 23 claims in 2 groups which are developed to a depth of 300 ft. A 150-ton fluorite mill has recently been completed at Beatty driven by semi-Diesel engines and embracing Blake and gyratory crushers, 2 36-in. emery grinding mills, 3 17-ft. Plat-O concentrating tables, and 3 rotary driers.

Bibliography.

MR1911 I 909 Mercury	MR1915 I 272 Mercury, 645, II 851 Opal
MR1912 I 945 "	MR1920 I 72 "
MR1913 I 208-9 " 834, II 680 Opal	MR1921 I 42 "
MR1914 I 328 " 699	

USGS Furnace Creek topographic map.

Ball285 71-2. Ball308 153-7.

Buwalda, J. P., "Nevada" in "High-Grade Clays of the Eastern U. S. with Notes on Some Western Clays," USGS B 708 (1922) 124-6.

Hill507 219.

Knopf, A., "Some Cinnabar Deposits in Western Nev.," USGS B 620D (1915) 62-7.

"Flourspar Company operates at Beatty," M&SP 120 (1920) 426.

GOLD CRATER

Gold, Silver

Location. Gold Crater is on Pahute Mesa, 27 m. from Goldfield which is on the T. & G. R. R. and the T. & T. R. R. The Gold Crater District adjoins the Wellington District on the S. W.

History. The district was discovered in May, 1904, and a small rush to the camp took place in the fall. The Gold Prince M. & Leasing Co. was incorporated in 1914, took a lease and bond on the property of the Gold Crater Cons. M. Co., and in 1916 operated a 25-ton amalgamation and concentration mill.

Geology. Gold Crater is an inlier of Tertiary andesite which protrudes as rounded hills through the later basalt flow of Pahute Mesa, according to Ball. This andesite has been silicified and kaolinized. The ore appears to have been formed by the impregnation of the andesite along faults, brecciated zones, and joints with silica, pyrite, and copper sulphide; and the later oxidation of the sulphides to lminonite and a little chrysocolla with free gold

Bibliography. MR1916 I 491.

USGS Kawich topographic map.

Ball285 69-70. Ball308 131-7, 139-40. Hill507 221.

WeedMH 1208 Gold Crater Cons. M. Co. 1214 Gold Prince M. & Leasing Co.

GOLDEN see CLOVERDALE

GOLDEN ARROW (BLAKES CAMP)

Gold, Silver

Location. The Golden Arrow District lies on a plain just W. of the Kawich Range and E. of the desert in central Nye Co. Blakes Camp is 4 m. W. of Golden Arrow. Golden Arrow is 50 m. by road E. of Tonopah which is on the T. & G. R. R.

History. Blakes Camp was discovered in 1905, in which year the prospects at Golden Arrow were being explored. The production of the camp has been small. A few tons were shipped from the Cotter property to the West End Mill at Tonopah in 1916.

Geology. The country rock at Blakes Camp is Tertiary rhyolite, according to Ball, and the crushed and more or less iron and manganese-stained rhyolite in a fault zone pans free gold. At Golden Arrow, according to Ferguson, andesite has been faulted into contact with rhyolite and the ore occurs along this fault and in veins which are associated with it, principally in the rhyolite. The wall rock near the fissures is silicified. The ore consists of quartz veinlets in which pyrite is the chief metallic mineral. Gold occurs native in a finely divided state and alloyed with silver, and in some of the ore the silver values are said to exceed the gold.

Bibliography. MR1905 272 MR1916 I 491
USGS Kawich topographic map.
Ball285 66. Ball308 110.
Ferguson, H. G., "The Golden Arrow, Clifford and Ellendale Districts, Nye County, Nev.," USGS B 640F (1916) 113-121.
Hill507 220.
StuartNMR 91.
WeedMH 1208 Goldfield Blue Bell M. Co.

GOLD PARK see JACKSON
GOLD REED see KAWICH
GOLDYKE see FAIRPLAY

HANNAPAH (VOLCANO, SILVERZONE)

Silver, Gold

Location. The Hannapah District is located at Hannapah in the low hills at the S. end of the Monitor Range in central Nye Co. Tonopah on the T. & G. R. R. is 18 m. W.

History. The Hannapah Mine was discovered in 1902, and the district was prospected and explored for several years thereafter. In 1908 and 1909, the Silver Glance Mine shipped ore; and a shipment was made from the district in 1914. In 1915, a discovery at Volcano in the S. part of the district led to a brief boom and the shipment of a little ore. The name Silverzone has recently been applied to the Hannapah District. Ore shipments were made to the Belmont Mill at Tonopah from the Richardson property in 1922.

Geology. The ore deposits of the Hannapah District are veins in Tertiary eruptives. At Volcano, the country rock is rhyolite. The ore contains free gold and silver sulphide in a quartz gangue, and some secondary hornsilver is present in the vein and in the hanging wall rhyolite.

Bibliography. MR1905 272 MR1909 I 420 MR1914 I 699
MR1908 I 499 MR1915 I 650
Hill507 222.
Lincoln, F. C., "Volcano, A New Nevada Strike," E&MJ 100 (1915) 73.
WeedMH 1922 Hannapah Mine. Hannapah D. Ex. Ms. Co.
1266 Nevada Central Ms. Co
1336 Silverzone Ex. M. Co. Silverzone Ms. Co.
Silverzone Mohawk Co.

HOT CREEK see TYBO
IRWIN CANYON see TROY

JACKSON (GOLD PARK)

Gold, Silver, (Copper, Lead)

Location. The Jackson District is on the border of Lander and Nye Cos., the principal mines being in Nye Co. The mines are situated in Gold Park, a mountain basin about 2 m. in diameter on the W. slope of the Shoshone Range with altitude ranging from 6,500 to 8,500 ft. Gold Park is 44 m. by road S. E. of Austin, which is on the N. C. R. R. and 34 m. from Ledlie, its shipping point on the same railroad. *

* This description is based upon data kindly furnished by Mr. R. B. Todd, including a report on the Star of the West Group by Prof. W. O. Crosby.

History. The district was discovered in 1880 by Frank Bradley and others. In 1893, 3 claims and a mill site were patented and the property sold to the Nevada M. Co. which erected a stamp mill and operated for quite a period. Litigation arose in 1911, and Robert B. Todd took over the property in 1919, and organized the Star of the West M. Co. to operate it. The new company completed its 50-ton mill in 1921, and made a trial run.

Production. The Nevada M. Co. is credited with a production of from $500,000 to $1,000,000.

Geology. Granite porphyry forms the ridges to the N., N. W., and N. E. of Gold Park basin. To the S. W., is a ridge of andesite separated from the granite porphyry by a half-mile belt of Paleozoic sediments which are probably of Lower Carboniferous age. The floor and the S. E. rim of the basin are composed of rhyolite. The sedimentary rocks were formed first, intruded after a long interval by the granite porphyry, which was followed in turn by the andesite and the rhyolite.

Ore. The principal orebodies in the Jackson District are shoots in fissure veins in the andesite which forms the S. W. wall of Gold Park. The surface ores are oxidized gold-silver ores in which the gold occurs in a finely divided state. The deep ores contain pyrite, chalcopyrite, galena, and a trace of blende.

Mines. The Star of the West M. Co. was incorporated in Delaware in 1919 with a capital stock of 1,000,000 shares of $1.00 par value, of which there are 500,000 outstanding. The home office is in the Gazette Building, Reno, Nevada. R. B. Todd is Pres.; M. W. O'Boyle, V. P.; G. M. Todd, Sec.; with R. G. Withers and C. M. O'Boyle additional Directors. The company owns 15 mining claims and 3 mill sites. The property is opend by 4,000 ft. of workings and equipped with a 50-ton amalgamation and concentrating mill.

Bibliography. MR1910 I 520 MR1911 I 682 MR191 I 800 MR1918 I 829 WeedMH 1345

JEFFERSON CANYON

Silver, Gold

Location. The Jefferson Canyon District is located in Jefferson Canyon on the W. flank of the Toquima Range. It is 70 m. by road N. of Tonopah which is on the T. & G. R. R., and adjoins the Round Mountain District on the N. E.

History. The district was discovered in 1866 but did not become active till 5 yrs. later when a test lot of ore sent to Austin yielded $28,800. The principal mines, the Jefferson and the Prussian, changed hands in 1873 and each had a 10-stamp mill erected in 1874. In 1875 and 1876, the recorded partial production amounted to $292,944; and it said the two mills produced $1,000,000. C. J. Kanrohat discovered the Sierra Nevada Mine in 1873. A mill was constructed upon this property by New York people in 1919, but shut down on acount of ligitation. S. H. Brady & Co. reconstructed the mill and equipped it for flotation in 1917, producing some flotation concentrate containing gold and silver in 1918, but the mill proved unsuccessful.

Geology. The country rock consists of Ordovician sedimentary rocks intruded by Cretaceous granite and capped by Tertiary volcanic rocks, according to Ferguson. Packard states that the Jefferson and Prussian vein lies between a rhyolite hanging wall and a slate foot wall. Veins also occur in the rhyolite, where gold values predominate, while in the contact vein the principal values are in silver. At water level, silver sulphides and sulphantimonides were found.

Bibliography.

B1876 420	SMN1871-2 106-7	MR1905 272
1867	SMN1873-4 70-72	MR1909 I 422
R1874 281	SMN1875-6 105-6	MR1917 I 288
R1875 138	SMN1877-8 91	MR1918 I 252

USGS Tonopah topographic map.

Ferguson, H. G., "The Round Mountain District, Nevada," USGS B 725 I (1921) 387.

Hill507 222.

Packard, G. A., "Jefferson Canyon, Nevada," M&SP 99 (1909) 26.

Spurr208 90-93 Toquima Range.

StuartNMR 84.

WeedMH 1230 Jefferson G. & S. M. Co. 1321 Round Mt. M. Co.

JETT

Silver, Lead, (Zinc)

Location. The Jett District is located in Jett Canyon on the E. flank of the Toyabe Range. Millers on the T. & G. R. R. is 45 m. S.

History. The Jett District was discovered by John Davenport in 1876 and became quite active in 1880. Considerable ore was shipped to Eureka and smelted. The Gibraltar S. Ms. Co., of which W. J. Loring, 80 Maiden Lane, New York City, is Pres., has been carrying on active exploration work in the district recently.

Geology. According to Mines Handbook, the Gibraltar vein occupies a basal fault cutting an andesite dike or stock and consists of highly silicified andesite carrying values in silver, lead, and zinc. Whitehill states that the veins in the district contain antimony as well as lead and zinc and that they are found in slate and limestone; while Thompson & West confirm the antimony but say that the veins are found between slate and porphyry.

Bibliography. SMN1875-6 106. USGS Tonopah topographic map.

Thompson & West 517.

WeedMH 1205 Gibraltar Silver Hill M. Co. 1206 Gibraltar S. Ms. Co. 1321 Round Mt. M. Co.

JOHNNIE

Gold, Silver, Lead, Placer Gold

Location. The Johnnie District is located at Johnnie at the N. W. end of the Spring Mountain Range in S. E. Nye Co. Death Valley Jct. on the T. & G. R. R. is 25 m. S. W.

History. The district was active in 1905 and development work was in progress in 1907. In 1908, the Johnnie 16-Nissen-stamp mill was put in operation on ore from the Johnnie Mine; and in 1909, ore from the Crown Point property was treated in a 1-stamp Kendall mill. The Johnnie Mine and Mill continued to operated till 1914, In 1915, they changed hands, and have been operated continuously since. Lead-silver ore was shipped from the Black Jack in 1917. The Eureka Johnnie installed a 2-stamp mill in 1917 which operated up to 1919. In 1921, placer gold was discovered in the Johnnie District and a small boom ensued. The placer gold is recovered mainly by dry washing.

Geology. At the Johnnie Mine there is a gold quartz vein in quartzite and limestone, according to Mines Handbook. The placer pay dirt lies on bedrock, according to Labbe, and is distributed rather evenly through a foot of slightly rounded gravels in a soil-like matrix. The channels spilt and reunite and are of very variable width. The gold is coarse and angular, with fragments of quartz attached, and sometimes enclosed in pyrites. Black sand consisting of pyroxene, titaniferous iron, and pyrites, accompanies the gold.

Mines. The **Johnnie Mine** is the only property recently active in the district. It consists of 20 claims, developed by an incline shaft 1100 ft. deep and 2 m. of workings, and equipped with hoist, compressor, and 16-stamp amalgamating mill. A. P. Johnson, 207 O. T. Johnson Bldg., Los Angeles, is owner.

Bibliography. MR1905 116 | MR1911 I 689 | MR1916 I 491
MR1907 I 374 | MR1912 I 806 | MR1917 I 287
MR1908 I 497 | MR1913 I 834 | MR1918 I 250
MR1909 I 420-1 | MR1914 I 700 | MR1919 I 403
MR1910 I 525 | MR1915 I 645 | MR1921 I 389
USGS Furnace Creek topographic map.
Ball308 150. Hill507 222.
Labbe, C., "The Placers of the Johnnie District, Nevada," E&MJ 112 (1921) 895-6.
WeedMH 1232 Johnnie Mine.

KAWICH (GOLD REED)
Gold

Location. The Kawich or Gold Reed District is located S. of Kawich on the E. side of the Kawich Range. Goldfield on the T. & G. R. R. and the T. & T. R. R. in 54 m. W.

History. The Kawich District was discovered in 1904 and a small boom took place the following spring but soon subsided. The first locations were made on the property of the Gold Reed M. Co.

Geology. The country rocks are monzonite porphyry and rhyolite, according to Ball. Free gold occurs with limonite in silicified monzonite porphyry. Dendrites of manganese oxide are associated with the gold and gypsum is present in the ore. In depth auriferous pyrite of low grade is found. According to a letter from E. S. Giles, cinnabar also occurs with the gold.

Bibliography. MR1904 199-200 | MR1905 272
USGS Kawich topographic map.
Ball285 67-8 Ball 308 108-113. Hill507 222.
Spurr, J. E., "Genetic Relations of Western Nevada Ores," T AIME 36 (1906) 382-3.
Spurr208 181 Kawich Range. (1906) 382-3. StuartNMR 92.

LEE (BIG DUNE)
Gold, Copper

Location. The Lee or Big Dune District is situated W. of the Big Dune in the Amargosa Desert in S. Nye Co. on the California border. It is just N. W. of Leeland which is on the T. & T. R. R.

Geology. According to Ball, pyrite and hematite occur in quartz veins on the small Cambrian hills to the N. W. of Big Dune; and these minerals with malachite in the quartz veins of the large Cambrian inlier to the S. W. of the dune.

Bibliography. USGS Furnace Creek topographic map.
Ball308 175. Hill507 220.

LODI (MAMMOTH, MARBLE, ELLSWORTH)

Silver, Gold, Lead, Copper (Tungsten)

Location. The Lodi District is situated in the neighborhood of Lodi in the Mammoth Range in N. W. Nye Co. near the Churchill Co. boundary line. Lodi is 45 m. N. N. E. of Luning which is on the S. P. R. R. The district is sometimes subdivided into the Mammoth District on the E. and the Lodi District on the W. The Bruner District adjoins it on the N.

History. The district was discovered by an Indian who revealed its location to some miners who organized it as the Mammoth District in December, 1863. The town of Ellsworth was started shortly afterwards. A 10-stamp mill was erected in the district in 1871 which treated not only the local ores but also ores from the Union and Belmont Districts. The mines first discovered did not keep up their production, but in 1874 Welch and Kirkpatrick discovered ore in the W. part of the district and the Lodi District was organized in 1875. The principal mine in this section was the Illinois Mine which put down a 1000-ft. shaft and produced considerable ore. Activities in the district practically ceased in 1880, and were not revived till 1905, since which year small shipments of ore have been made annually. The Illinois Mine was purchased by the Lodi Ms. Co. in 1906 and a small smelter erected in 1908. This proved unsuccessful, and an experimental concentrator was erected in 1919. In 1921, the Illinois Mine was sold to the Illinois-Nevada Ms. Corp. Old tailings at Ellsworth were retreated in a 30-ton cyanide plant in 1916. In 1921, a rich gold strike was made by Hughes and Hatterly 2 m. S. of the Illinois Mine.

Geology. The lodes in the E. part of the district are in porphyritic granite and gold values predominate in them. The Illinois lode is in limestone near a granite contact and its ore is argentiferous galena carrying gold. Huebnerite occurs at Ellsworth in quartz veins cutting granite, which were worked for silver in the early days, according to Hess. Fluorite, biotite, orthoclase and gold are associated with the huebnerite.

Mines. The **Illinois-Nevada Ms. Corp.** is developing the Illinois Mine and has purchased the Black Eagle 20-stamp cyanide mill at Rawhide which will be moved to Lodi. L. H. Carver is Pres.-Mgr. and F. H. Lerchen, Supt. The **Goldfield Blue Bell M. Co.** owns 6 patented claims 5 m. S. of Lodi which have been leased to F. H. Lerchen. D. S. Johnson is Pres; B. W. Ward, Treas.; and J. M. Hiskey, Sec.; the mine office being at Austin. This company also owns properties in the Goldfield, Golden Arrow, and Berlin Districts. **The Tonopah-Brohilco Ms. Corp.** has a capital stock of 2,500,000 shares of 10 cts. par value of which half remain in the treasury. G. E. Porter, Fallon, is Pres.; G. W. Hoffman is V. P.; C. R. Olson, Cons. Engr.; and G. W. Forbes, Sec.-Treas. The property is located 2 m. S. W. of the Illinois Mine and consists of the Silver Leaf Group of 7 claims developed by a 100-ft. shaft, and the Black Reef Group of 5 claims developed by a 65-ft. shaft.

Bibliography.

B1867 319, 419	SMN1875-6 109	MR1910 I 525, 529
R1868 99-100	SMN1877-8 87-8	MR1914 I 700
R1871 182-3	MR1905 272	MR1916 I 491
SMN1866 58	MR1906 299	MR1917 I 287.
SMN1867-8 60-2	MR1907 I 372	MR1918 I 250
SMN1869-70 88-9	MR1908 I 497, 724-5, Tungsten	MR1919 I 403-4
SMN1871-2 108		MR1920 I 329
SMN1873-4 78	MR1909 I 421	MR1921 I 389

USGS Tonopah topographic map.

Hill507 222. Thompson & West 523, 525.

WeedMH 1153 Brohilco S. Corp. 1209-9 Goldfield Blue Bell M. Co. 1240 Lodi Ms. Co.

MAMMOTH see LODI

MANHATTAN

Gold, Placer Gold, Silver, Arsenic, Rhyolite Pebbles

Location. The Manhattan District is at Manhattan in the S. part of the Toquima Range. Manhattan is 45 m. by road N. of Tonopah which is on the T. & G. R. R. It is situated in Manhattan Canyon on the W. side of the range at an altitude of 6,905 ft.; while Bald Mt. to the N. reaches a height of 9,275 ft.. The old Belmont District adjoins the Manhattan District on the N. E.

History. Manhattan was discovered by John C. Humphrey in 1905 and a rush of prospectors into the district occurred that summer and again the following winter. Placer mining was inaugurated the following year, and was of particular importance from 1909 to 1915. In 1916, rich ore was found upon the lower levels of the White Caps Mine and led to another boom. In 1912, the Associated Mlg. Co. treated the ore of the White Caps Mine in a 75-ton mill which it had erected; shutting down the mine and mill when the oxidized ore was exhausted. In 1915, the White Caps M. Co., took over the White Caps mine and the Associated Mill; and in 1917, reconstructed the mill, adding a roasting furnace. Litigation with the Manhattan Glory was settled in 1918; but considerable difficulty has been experienced in devising a milling system adapted to the base arsenical ores of the White Caps Mine. Since 1914, artificial pebbles for tube mills have been manufactured in the district from silicified rhyolite.

Production. From 1906 to 1921, the Manhattan District produced 375,292 tons of ore containing $4,112,607 in gold and 76,855 ozs. silver, valued in all at $4,160,921, according to Mineral Resources of the U. S. Geol. Survey.

Geology. The country rocks of the Manhattan District consist of Paleozoic sediments cut by Cretaceous granodiorite on the S. and capped by Tertiary eruptives on the N. The Paleozoic rocks are mainly schists with included lenses of quartzite and beds of limestone. They have been compressed into close folds in part overturned toward the N., according to Ferguson, and the prinçipal anticline has been cut off obliquely by a reverse fault. The beds are further disturbed by a large number of small normal faults belonging to two series. The Tertiary eruptives consist mainly of rhyolite breccias but include lake bed deposits and andesite.

Veins. The ore deposits of the Manhattan District include veins in the Tertiary eruptives, veins in the Paleozoic sediments, stockworks in the

Paleozoic schist, replacement deposits in the Paleozoic limetsone, and placers. Small veins in the Tertiary lavas contain iron-stained quartz with minor amounts of calcite which has been largely replaced by quartz and more rarely by fluorite and adularia, according to Ferguson. Minute specks of free gold are visible on the surface of pyrite grains which have been altered to limonite. Small veins of white glassy quartz occur in the Paleozoic rocks. Minor amounts of calcite are present and occasional patches of sulphides including pyrite, galena, tetrahedrite, and chalcopyrite; while tourmaline is present in several instances and feldspar in one. The silver and gold content of these veins is low.

Stockworks. Stockwork deposits in the Paleozoic schist have made the principal production of the camp. They consist of innumerable little veinlets in a fairly well-defined zone in the schist. These stringers contain free gold in a gangue of quartz and adularia. The adjacent schist is pyritized in places.

Replacements. In the lower part of the Paleozoic series there is a mineralized bed of limestone which contains ore in the form of large replacement deposits closely connected with faults of the earlier series, according to Ferguson. The gangue minerals of the White Caps mine are calcite, quartz, and less commonly dolomite, fluorite and leverrierite, (a hydrous aluminum silicate). The metallic minerals are arsenopyrite, pyrite, stibnite, realgar, and orpiment; and, in the upper part of the mine, cinnabar. Visible gold is absent; the arsenopyrite, realgar, and orpiment being auriferous, while the stibnite is barren. In the mines to the W. of the White Caps on this same bed of limestone, free gold occurs, flourite and leverrierite become more common and adularia is present.

Placers. Placer gold has been mined from patches of old gravel on the sides of the gulch, from deep gulch gravels, and from the surface wash and shallow stream gravels near the lode outcrops. The gold is usually arborescent and but slightly abraded while the larger pieces contain quartz. The particles decrease in size and increase in fineness down the gulch. The gold is accompanied by barite and magnetite and by minor amounts of psilomelane, cinnabar, limonite, pyrite, and fluorite.

White Caps Mine. The White Caps M. Co. is the only company at present active. This company was incorporated in Nevada in 1915 with a capital stock of 2,000,000 shares of 10 cts. par value of which 1,915,300 have been issued. The stock is listed on the New York Curb and San Francisco Stock Exchange. The home office is at Tonopah. J. G. Kirchen is Pres.; H. R. Cooke, V. P.; A. G. Raycraft, Sec.-Treas.; with C. J. Blumenthal and W. L. Taylor, additional Directors. W. L. Taylor is Supt. at Manhattan. The property consists of 7 claims developed by an 800-ft. shaft and equipped with hoist, 2 compresors, pump, and 100-ton mill.

Bibliography of Manhattan District.

General

R1867 420
R1868 111
MR1905 272,116
MR1906 299
MR1907 I 374,398
MR1908 I 498
MR1911 I 689
MR1912 I 886-7
MR1913 I 834-5
MR1914 I 700-1
MR1915 I 645-6
MR1916 I 491-2
MR1918 I 250-1
MR1919 I 19 Arsenic, 404
MR1920 I 329-330
MR1921 I 390
MR1914 II 567 Pebbles
MR1916 II 210

MR1909 I 421 MR1917 I 287-8 MR1917 II 229 Pebbles
MR1910 I 525-6 MR1918 II 1185 Pebbles
Davis, 967-71.

Dynan, J. L., "The White Caps Mine, Manhattan, Nevada," M&SP 113 (1916) 884-5.

Jones, C. C., "Notes on Manhattan Placers, Nye County, Nevada," E&MJ 88 (1909) 101-4.

M&SP, Occasional Correspondent, "Developments at Manhattan," 103 (1911) 230.

Nash, P., "Manhattan, Nevada," (Abstract from the Tonopah Miner), M&SP 111 (1915) 523.

Rice, C. T., "The Manhattan Mining District, Nevada," E&MJ 82 (1906) 581-4.

StuartNMR 11, 86-8.

WeedMH 1146 Big Pine M. Co.
1147 Black Mammoth Cons. M. Co.
1166 Commercial Ms. & Mlg. Co.
1184 Dexter-Union Ms. Co.
1247 Mammoth G. M. Co.
Manhattan Big Four M. Co.
Manhattan Cons. Ms. Dev. Co.
Manhattan Dexter M. Co.
Manhattan Glory M. Co.
Manhattan Mustang M. Co.
1249 Manhattan Red Top M. Co.
Manhattan Sunrise M. Co.
Manhattan Union Amalg. Ms. Synd.
1316 Robust Group.
1326 Seyler-Humphrey G. M. Co.
1374 White Caps Ex. Mc. Co.
1375 White Caps M. Co.
1376 William Patrick M. Co.
1382 Zanzibar M. Co. of Nevada

Geology

Emmons, W. H., & Garrey, G. H., "Notes on the Manhattan District," USGS B 303 (1907) 84-92.

Ferguson, H. G., "Placer Deposits of the Manhattan District, Nevada," USGS B 640J (1917) 163-193; Abstract, Wash. Acad. Sci. J7 (1917) 266.

——— "The Limestone Ores of Manhattan," EG 16 (1921) 1-36.

Hill507 222-3.

Jenney, W. P., "Geology of the Manhattan District, Nevada," E&MJ 88 (1909) 82-3.

Spurr208 90-93 Toquima Range.

Wherry, E. T., "A Peculiar Intergrowth of Phosphate and Silicate Minerals," Wash. Acad. Sci., J. 6 (1916) 105-8.

Metallurgy

Kennedy, J. C., "Manhattan Ore Mililng Co's. Mill, Nev.," M&EW 38 (1913) 859-861.

——— "The Associated Mill, Manhattan," M&SP 105 (1912) 108-9.

Kirchen, J. G., "Solving the Ore Treatment Problem at White Caps Mine," E&MJ 104 (1917) 905-7; Discussion, W. S. Palmer, E&MJ 107 (1919) 123-4.

Palmer, W. S., "Formation of Quicklime in Roasting Ores from Manhattan, Nev.," E&MJ 104 (1917) 525-6.

Quinn, P. J., "Treatment at the Big Pine, Manhattan," M&SP 111 (1915) 320.

Miscellaneous

USGS Tonopah topographic map.

MARBLE see LODI

MILLETT (NORTH TWIN RIVER)

Silver, Gold, Lead, Copper

Location. The Millett District is situated just W. of Millett, 45 m. S. of

Austin which is on the N. C. R. R. and 105 m. N. of Tonopah which is on the T. & G. R. R. The district lies on and near Summit Creek on the E. flank of the Toyabe Mts. N. of the Twin River District.

History. The North Twin River District was organized in 1863. The principal early mine, the Buckeye , was discovered in 1865, and shipped ore to Austin to be milled. Interest in the district revived in 1905, and small productions were made from 1906 to 1916. The Nevada National property operated a 5-stamp mill from 1911 to 1913.

Geology. According to Hague, the Buckeye vein is at a limestone-slate junction and is irregular in dip and width containing pockets of ore consisting of argentiferous galena, blende, and silver sulphides.

Bibliography. B1867 414-5, 431 SMN1867-8 65-6 SMN1871-2 106
SMN1866 59 SMN1869-70 89 SMN1873-4 78
USGE40th 3 (1870) 342,383. USGS Tonopah topographic map.
Spurr208 93-7 Toyable Range.

MONTE CRISTO see TOLICHA

MOREY

Silver, Gold, Lead

Location. The Morey District is located W. of Morey in the Hot Creek Range.

History. The district was discovered in 1866 by a party of prospectors from Austin. The ores were first shipped to Austin, but in 1872 the Morey M. Co. erected a 10-stamp mill which treated the ores of the camp for several years. Activities in the district ceased about 1876. In 1921, the Airshaft and Smuggler properties produced silver ore containing a little gold and lead.

Geology. According to Raymond, perpendicular silver-bearing veins from 3 ft. to 4 ft. in width occur in granite.

Bibliography. B1867 425-6 R1873 232 SMN1869-70 91
R1868 112 R1874 281 SMN1871-2 110
R1869 142-3 SMN1866 62 SMN1873-4 73
R1870 130-1 SMN1867-8 76-7 SMN1875-6 107
R1872 173 MR1921 I 390

NORTHUMBERLAND

Silver

Location. The Northumberland District is located in the Toquima Range W. of Northumberland in N. Nye Co.

History. The district was organized in 1866. A little ore was shipped to Austin; and in 1868 a 10-stamp mill was erected by the Quintero Co. but only operated a short time. The camp became inactive in 1870. In 1917, silver ore was produced in the district.

Geology. The ore deposits are veins in granite porphyry, according to Raymond, and the silver occurs as sulphantimonates such as ruby silver and as hornsilver, while azurite and malachite are sometimes encountered.

Bibliography. B1867 404,423 SMN1866 63-4 SMN1869-70 89
R1868 84-5 SMN1867-8 73 MR1917 I 288

OAK SPRING

Gold, Silver, Copper, Chrysocolla, (Molybdenum, Tungsten)

Location. The Oak Spring District is located at Oak Spring on the E. flank of the Belted Range near its S. end. Caliente on the U. P. R. R. is 80 m. E. N. E.

History. A number of prospects were under development in the Oak Spring District in 1905. In 1917, copper ore containing a little silver was shipped from the Horseshoe claim.

Geology. South of Oak Spring, a granite stock has been intruded into Pennsylvania limestone and sends numerous apophyses into it, according to Ball. Pegmatitic quartz veins occur in the granite with values in gold and to a less extent in silver. The sulphides which are sparingly present in these veins are pyrite, chalcopyrite, galena, and blende; and from them hematite, limonite, malachite, azurite, and cerussite have been derived. A vein in the limestone near Oak Springs contains malachite, chrysocolla, jaspery and clear quartz, manganese dioxide, cerussite, and a hydrated ferric tellurite. Further S., veins in limestone carry chrysocolla, brochantite, manganese dioxide, quartz, and calcite. This chrysocolla has been mined to a limited extent and sold as "turquoise." Hess mentions the occurrence of scheelite accompanied by powellite derived from molybdenite in a prospect S. of Oak Springs.

Bibliography. MR1908 I 725 MR1917 I 288.
Ball285 70-1 Ball308 44, 45, 122-130. Hill507 233.

PHILADELPHIA see BELMONT.
PHONOLITE see BRUNER
PIONEER see BULLFROG
RAILROAD VALLEY MARSH see BUTTERFIELD MARSH

REVEILLE

Silver, Lead, Copper, Gold

Location. The Reveille District is situated at Reveille in the Reveille Range. Tonopah on the T. & G. G. R. R. is 70 m. W. The Arrowhead District adjoins the Reveille District on the N.

History. The district was discovered by Indian Jim in 1866. He showed it to Arnold, Monroe and Fairchild who organized the district and named it in honor of the Reese River Reveille of Austin. In 1867, a 5-stamp mill was erected W. of the mines, and in 1869 a 10-stamp mill, but they only operated a short time. In 1875, the Gila Silver M. Co. acquired the principal properties and reconstructed the 10-stamp mill, which ran intermittently for 4 yrs. and produced about $1,500,000 in bullion. The camp was abandoned in 1880. In 1904, interest in the district revived and an irregular production has been kept up since.

Geology. The Reveille Range is composed of Paleozoic quartzite and limestone which have been folded into a monocline on its W. flank, according to Ball, and are capped by flows of Tertiary rhyolite and basalt on the E. The Gila Mine is situated in the Paleozoic sedimentary rocks near the rhyolite contact. The ore occurs in quartz veins and stringers in brecciated zones and contains malachite, azurite, cerussite, and galena as ore minerals, with quartz and gypsum as gangues. The rich ore of the early days contained hornsilver.

Bibliography.

B1867	402, 425, 431	SMN1871-2	110-1	MR1910 I 527
R1868	107-9, 112	SMN1873-4	74	MR1911 I 690
R1869	140-2	SMN1875-6	109-110	MR1915 I 646
R1870	131	SMN1877-8	91	MR1916 I 492
R1874	35	MR1904	199-200	MR1917 I 288
R1875	177-9	MR1905	272	MR1918 I 251
SMN1866	63	MR1906	298	MR1919 I 404
SMN1867-8	77-8	MR1908 I	498	MR1920 I 330
SMN1869-70	92			MR1921 I 390

Ball308 114-7.
Hill507 223. Spurr208 161-3 Reveille Range. Thompson & West 526.
WeedMH 1242 Lousiana Cons. M. Co. 1342 Southwestern Nevada Ms. Co.

ROUND MOUNTAIN

Gold, Placer Gold, Silver, Lead, Tungsten

Location. The Round Mountain District is located at Round Mt. on the W. flank of the Toquima Range. Tonopah on the T. & G. R. R. is 60 m. S. The camp has an elevation of 6,313 ft. and Round Mt. rises to a height of 2,825 ft. The Jefferson Canyon District which adjoins Round Mountain on the N. E. is sometimes considered as a section of the Round Mountain District. The Belmont District adjoins Round Mountain on the S. E.

History. The district was discovered by Louis D. Gordon in 1906, and two companies began the exploration of lode properties. In 1907, Thomas Wilson discovered placer gravel and worked it by dry washing. The Round Mountain M. Co. has been the principal producer. It erected a mill in 1907; and later brought in water for hydraulicking the placer ground, first from Shoshone and Jefferson Canyons in the same range, and finally in 1915 from Jett Canyon in the Toyabe Range. Tungsten deposits were discovered E. of Round Mountain in 1907. A mill was erected in 1911, put into operation in 1912, and continued to produce for several years.

Production. According to L. D. Gordon, the production of the Round Mountain District from the time of its discovery to the close of 1922 is as follows:

Round Mountain Mining Co., Actual	$4,680,554.26
Fairview Round Mountain Mines Co., Actual: from 1917	752,386.03
Fairview Round Mountain Mines Co., Approx: prior to 1917	50,000.00
Daisy, Approximate	100,000.00
Blue Jacket, Approximate	100,000.00
Sphinx, Approximate	50,000.00
All others, Approximate	50,000.00
	$5,782,940.29

From 1906 to 1920, according to Mineral Resources of the U. S. Geol. Survey, the Round Mountain District produced 491,532 tons of ore containing $5,048,876 in gold, 188,398 ozs, silver and 3,202 lbs. lead, valued in all at $5,188,740.

Geology. The Paleozoic sediments of the Toquima Range have been intruded to the E. of Round Mt. by Cretaceous granite; while at Round Mt. these rocks and Tertiary lake bed deposits have been intruded by rhyolite. The gold deposits of Round Mountain occur in the Tertiary rhyolite, and are classified by Ferguson as primary and secondary veins. The primary veins are very narrow and their filling consists of quartz with minor adularia and alunite, free gold alloyed with silver, a little auriferous pyrite, and a very little realgar. The secondary veins are later fissures which sometimes follow and sometimes cross the primary veins, and contain iron and manganese oxides and gold. The placer gravel consists largely of unsorted, angular, rhyolite wash; although rough stratification occurs out in the valley to the W. The tungsten deposits E. of Round Mountain are narrow veins in the Cretaceous granite. The gangue consists of quartz, fluorite, and muscovite. Huebnerite is the principal metallic mineral with which occurs a little tetrahedrite.

Mines. Two companies are at present operating in the Round Mountain District,—the Fairview Round Mountain Ms. Co. and the Round M. Co. These companies are closely related, having the same Pres., Sec.-Treas., and Supt., the same home office at 1011 First Nat. Bank Bldg., San Francisco, and the Round Mountain M. Co. owning 200,000 shares of Fairview stock. L. D. Gordon is Pres. of the **Fairview Round Mountain Ms. Co.;** F. Hamilton, V. P.; H. G. Mayer, Sec.-Treas.; with H. R. Cooke and J. R. Davis, additional Directors; and G. Berry, Gen. Supt. The company was incorporated in Nevada in 1906 with a capital stock of 1,000,000 shares of $1 par value of which 963,471 have been issued. The stock is listed on the San Francisco Stock Exchange. The property consists of 12 claims developed by a 500-ft. tunnel and 270 ft. shaft with 1 m. of workings and equipped with hoist, compressor, and 6-stamp 50-ton amalgamating mill. **The Round Mountain M. Co.** was incorporated in Nevada in 1906 and has a capital stock of 1,500,000 shares of $1 par value of which 1,400,000 are outstanding. The stock is listed on the San Francisco Stock Exchange and New York Curb. J. R. Davis is V. P., and H. R. Cooke and H. J. Hagenbarth additional Directors, the other officers being the same as for the Fairview Co. The property consists of 2,009 acres and placer rights to 216 acres, (exclusive of property in Jett and Jefferson Canyons), developed by a 1,000-ft. incline shaft and 5 m. of workings, and equipped with 2 hoists, compressor, electric power, 180-ton mill, 6 m. pipe line from Jefferson and Shoshone Canyons, 9 m. pipe line from Jett Canyon, and 2 monitors.

Bibliography.

MR1905 272	MR1911 I 690, 943 Tungsten	MR1916 I 492-3
MR1906 298-9	MR1912 I 807-8, 991	MR1917 I 288-9
MR1907 I 373-4	MR1913 I 835, 356 Tungsten	MR1918 I 250-1
MR1908 I 498,499,724 Tungsten	MR1914 I 701	MR1919 I 404-5
MR1909 I 421,560	MR1915 I 647, 825 Tungsten	MR1920 I 330-1
MR1910 I 527		MR1921 I 390-1

USGS Tonopah topographic map.

Brown, G. C., "Round Mountain Tungsten Mine," SLMR Aug. 15, 1911, 16

Davis 971-2.

Ferguson, H. G., "The Round Mountain District, Nevada," USGS B 825 I (1921) 383-406.

Hill507 223.

Loftus, J. P. "Round Mountain, Its Mines and Its History," AMC P 12th Ann. Sess. (1909) 445-8; Abstract, M&SP 99 (1909) 568.

Packard, G. A., "Round Mountain Camp, Nevada," E&MJ 83 (1907) 150-1.

——— "Round Mountain, Nevada," M&SP 96 (1908) 807-9.

Ransome, F. L., "Round Mountain, Nevada," USGS B 380 (1909) 44-7.

Spurr208 90-93 Toquima Range.

StuartNMR 11.

WeedMH 1200 Fairview Ex. M. Co., Fairview Round Mountain Mc. Co.
1321 Round Mountain Homestake M. Co., Round Mountain M. Co.

ROYSTON see SAN ANTONE

SAN ANTONE (SAN ANTONIO, ROYSTON)

Silver, Gold, Lead, Copper

Location. The old San Antonio District was of very large size and of indefinite extent. The San Antone District of today embraces a part of the San Antonia Mountains in the neighborhood of the Liberty Mine, together with a spur of the Monte Cristo Range across the Big Smoky Valley to the W., in W. Nye Co., on the Esmeralda Co. border. The W. section of the district has recently been called the Royston District and adjoins the Crow Springs District on the E. Tonopah on the T. & G. R. R. is 28 m. S. E. of the Royston section and 20 m. S. of the Liberty Mine.

History. The San Antone District was discovered by Robles, Fisk, and others in 1863. The Liberty Mine which was the most important mine in the district was discovered at this time. A 10-stamp mill was erected at San Antone station or Indian Springs, 12 m. to the N., in 1865. It was run for a year and then torn down and moved elsewhere. A 4-stamp mill was erected in 1867 but only operated for a year. Under more recent operations by the Tonopah Liberty M. Co. and later by J. G. Lindsay as sole owner, two mills were built at the Liberty Mine. The first was a concentrating mill not adapted to the ore and the second a cyanide plant without sufficient filtration capacity. Attention was attracted to the Royston section in 1921 by the discovery of rich silver ore on the Betts Lease on ground belonging to the Hudson M. & Mlg. Co. Some $20,000 worth of ore was taken from a hole 24 ft. deep in 6 wks. time and a small boom ensued.

Geology. The vein at the Liberty Mine is in porphyry, according to Raymond, and the vein filling is clay, porphyry, and quartz. Hornsilver and native silver are found in the quartz. A letter from J. G. Lindsay says the country rock is silicified volcanic ash, and that gold values are low and stephanite occurs in the richest ore.

Bibliography. B1867 319, 402, 418 — SMN1871-2 107 — MR1916 I 493
R1868 103-7, 112 — SMN1873-4 78 — MR1917 I 289
SMN1866 61-2 — SMN1875-6 109 — MR1919 I 406
SMN1867-8 68-9 — MR1910 I 527 — MR1920 I 331
SMN1869-70 89 — MR1921 I 391
USGS Tonopah topographic map.
M&SP, Editorial, "Royston or San Antone," 123 (1921) 804-5; Discussion, "Royston," M&SP 123 (1921) 917-8, E. C. Watson, R. H. Stretch, W. F. Korf; M&SP 124 (1922) 114-5, A. C. North.
Thompson & West 518.
WeedMH 1225 Hudson Leasing Co. — Hudson M. & Mlg. Co.
1240 Liberty Group — 1347 Super Six M. Co.

SAN ANTONIO see SAN ANTONE

SILVER BEND see BELMONT

SILVERBOW

Silver, Gold

Location. The Silverbow District is located at Silverbow on the W. slope of the Kawich Range in central Nye Co. Goldfield on the T. & G. R. R. and the T. & T. R. R. is 45 m. W.

History. The Silverbow District was discovered in 1904, explored in 1905, and made its first shipments in 1906. In 1913, a 2-stamp mill was erected in the district. In 1920, the Blue Horse M. Co., of which J. W. Plant of Tonopah is Pres., was active and treated its ore in a 20-ton Gibson mill, shutting down in 1921.

Geology. The ore deposits of the Silverbow District are quartz veins in Tertiary rhyolite which is either silicified or kaolinized in their vicinity. The quartz has been deposited in faults, in breccias, and in joint cracks, according to Ball. It is stained by iron and occasionally by malachite. Specks of stephanite, ruby silver, and cerargyrite are present in the quartz. Gold occurs free. Crushed rhyolite sometimes constitutes ore, and more rarely limonite stringers in the country rock.

Bibliography. MR1905 272 MR1906 298 MR1907 I 374 MR1908 I 499 MR1910 I 527 MR1911 I 690 MR1913 I 835 MR1914 I 702 MR1920 I 331 MR1921 I 391

USGS Kawich topographic map.

Ball285 65-6. Ball308 109-110.

Hill507 223-4.

StuartNMR 92.

WeedMH 1149 Blue Horse M. Co. 1329 Silver Bow D. M. Co.

SILVERZONE see HANNAPAH

SPANISH BELT see BELMONT

STONEWALL

Silver, Gold

Location. The Stonewall District is located at Stonewall Spring on the N. slope of Stonewall Mt. Goldfield on the T. & G. R. R. and the T. & T. R. R. is 17 m. N. N. W.

History. The district was prospected as early as 1905. Small shipments were made in 1911 and 1915. The Yellow Tiger Cons. M. Co. of Goldfield is now carrying on active development work.

Geology. The formations of Stonewall Mt. beginning with the oldest are as follows: Cambrian limestone, post-Jurassic granitoid igneous rocks, earlier rhyolite, quartz-syenite and quartz-monzonite porphyry, later rhyolite, and basalt. Quartz veins and stringers are common in the earlier rhyolite and intruded quartz-syenite and quartz-monzonite near Stonewall Spring. The most prominent vein follows the great fault scarp which is 800 to 1,000 ft. in height on the N. end of Stonewall Mt. just S. of the spring. To the E. of the spring Cambrian limestone occurs. Heavy limonite stains and light azurite stains together with occasional pyrite are found in the large vein.

Mines. The **Yellow Tiger Cons. M. Co.** has a capital stock of 4,500,000 shares of 10 cents par value, of which 1,764,176 shares remain in the treasury. The stock is listed on the San Francisco Stock Exchange. H. E. Clark of Rochester, N. Y. is Pres., and G. M. Bettles of Goldfield is Mgr. The company owns 350 acres S. of the Goldfield Cons. Ms. Co. property at Goldfield but is doing very little work there at present. It acquired the Sterlag Group of 6 claims in the Stonewall District for a stock consideration and is running a long development tunnel on this property at the present time.

Bibliography. MR1911 I 691 MR1915 I 647. USGS Lida topographic map.

Ball285 70.

Ball308 28, 32, 33, 46-9, 83-9, 133, 136.

Hill507 224.

WeedMH 1380 Yellow Tiger M. Co.

TOLICHA (MONTE CRISTO)

Gold, Silver

Location. The Tolicha District is located in the neighborhood of Monte Cristo Springs on the N. slopes of Tolicha Peak and of Quartz Mountain in S. W. Nye Co. Bonnie Clare on the T. & T. R. R. is 20 m. W., but the district is more commonly reached from Goldfield which lies 50 m. to the N. W. by way of the Gold Crater road.

History. The district was prospected as early as 1905, but did not attract particular attention until Jordon and Yaiser found rich ore in 1917. Then George Wingfield explored the Life Preserver Group under lease and bond; later, E. Harney became interested and a Gibson mill was installed; and more recently the claims were taken over by Tonopah interests. The Landmark Group was acquired by C. E. Knox and associates of Tonopah; and the Southey property on Quartz Mountain is to be developed by Salt Lake people. Two carloads of ore have been shipped from the district.

Geology. The country rock is Tertiary rhyolite. This rock has been kaolinized or silicified around Monte Cristo Springs and at Quartz Mountain camp, according to Ball. Veins in the altered rhyolite consisting of quartz and silicified rhyolite stained with limonite carry gold and a little silver.

Bibliography. USGS Kawich topographic map. Ball308. 141-2. Hill507 *223*.

TONOPAH

Silver, Gold, Lead, Copper

Location. The Tonopah District is at Tonopah on the T. & G. R. R. in Nye Co. close to the Esmeralda Co. boundary line. The Divide District in Esmeralda Co. adjoins the Tonopah District on the S. The name "Tonopah," is of Indian origin meaning "water brush."

Discovery. James L. Butler discovered the Tonopah District on May 17th, 1900, while on a trip from Belmont to Southern Klondike. Butler took his samples to an old assayer in Southern Klondike named Frank Hicks and offered him a quarter interest in the discovery for an assay, but Hicks replied that he would not give a dollar for a thousand tons of such stuff and threw the samples on the dump. On the return trip to Belmont, Butler collected some more samples from the same ledge, and turned them over to Tasker L. Oddie of Austin, who happened to be in Belmont, with the promise of a quarter interest for an assay. Oddie took the samples to William C. Gayhart, an Austin assayer and offered him an interest for an assay. Gayhart ran the samples and obtained results of several hundred dollars per ton, whereupon Oddie dispatched an Indian runner with the news to Butler on his ranch near Belmont. Butler had the reputation of being the laziest white man in the world, and in this instance he lived up to his reputation, for although the news of his discovery spread abroad and others attempted to find and locate it, he made no move until August. Then he went to Belmont and arranged with Oddie and Wilse Brougher, the recorder of Nye Co. to take up the ground, which they did on August 27th, 1900. The story of the discovery of Tonopah would not be complete without further mention of Frank Hicks, the Southern Klondyke assayer. Upon learning of the value of the strike, Hicks fished the original samples out of the dump, assayed them, reported results, and claimed a one-quarter interest in the property; and although he had no moral or legal right to it, Butler generously gave him a one-thirty-second-interest.

History. After they had located all the outcrops, Oddie and Brougher took out about one ton of ore for which they received $600; which was the total production of the camp for the year 1900. The claims were divided into small sections and leased on a 25% royalty basis during 1901, and the leasers took out some $4,000,000 worth of ore that year, although they were unable to ship much of it on account of lack of transportation and smelting facilities. In 1902, the property was sold to the Tonopah M. Co., which began active development at once, and in 1904 constructed a narrow-gauge railroad 60 m. in length which connected Tonopah with the Carson and Colorado branch of the Southern Pacific System at Sodaville. With the completion of this rail connection accomplished, the Tonopah M. Co. began making larger shipments, and in 1905 e larged the gauge of its railroad to standard to conform to a simil change on the branch of the Southern Pacific. In the meantime, prospecting disclosed that beneath the capping of barren andesite surrounding the original claims, there existed veins just as valuable as those which outcropped in the ground of the Tonopah M. Co. This discovery greatly enlarged the possibilities of the camp and afforded an opportunity for new companies to obtain good holdings. No ore was milled till 1906 when the Desert Power & Milling Co. constructed a 100-stamp cyanide mill at Millers on the railroad 13 m. W. of Tonopah, to treat the ores of the Tonopah M. Co. In the following year, the Tonopah Belmont Dev. Co. built a 60-stamp mill at Millers to treat its ore, and the Montana-Tonopah M. Co. constructed a 40-stamp mill of the all-sliming type at Tonopah. The Tonopah Extension mill was put in operation with 30 stamps in 1910, and later enlarged to 50 stamps; and the West End mill began operating in 1911,—both of these mills being located at Tonopah. In 1912, the Tonopah Belmont mill of 60 stamps and the MacNamara mill of 10 stamps were constructed at Tonopah.

Production. Tonopah is the most important producer of siliceous silver-gold ore in the United States. Its silver production has only been exceeded of recent years by camps whose principal product was copper or lead from which the silver was extracted as a by-product. In 1921, Tonopah had four mines on the list of the 25 larger producers of silver in the United States as published by the United States Geological Survey. Although its gold production is a little less than a quarter of its silver production, it is great enough to make Tonopah the second largest gold producer in Nevada, being exceeded only by Jarbidge. The annual productions of Tonopah from its discovery in 1900 to the year 1921 are shown in the following table.

PRODUCTION OF THE TONOPAH DISTRICT

Year	Mines	Tons	Gold $	Silver Ozs.	Copper Lbs.	Lead Lbs.	Total $
1900 (1)		1					600
1901 (2)		2,534	202,035	623,516			576,145
1902 (2)		11,258	547,030	2,434,453			1,764,271
1903 (3)		9,055	522,576	2,548,430			1,898,782
1904 (4)	10	22,703	386,526	2,119,942			1,594,893
1905 (4)	12	91,651	1,206,345	5,369,439			4,449,486
1906 (4)	9	106,491	1,304,677	5,697,928			5,122,289
1907 (4)	10	214,608	1,183,628	5,370,891	5,939	195,508	4,739,966
1908 (4)	13	273,176	1,624,491	7,172,396			5,425,861
1909 (4)	9	278,743	1,400,361	7,872,967	1,784	1,488	5,494,600
1910 (4)	18	365,139	2,303,702	10,422,869	942	6,902	7,932,475
1911 (4)	19	404,375	2,366,495	10,868,268			8,126,677
1912 (4)	17	479,421	2,223,878	10,144,987	14	700	8,463,079
1913 (4)	32	574,542	2,613,843	11,563,437	1,150	9,001	9,598,733
1914 (4)	23	531,278	2,648,833	11,388,452	2,284	924	8,946,987
1915 (4)	22	516,337	2,228,983	10,171,374			7,385,870
1916 (4)	21	455,140	1,941,441	8,734,726			7,688,891
1917 (4)	20	470,122	1,539,529	7,068,737			7,364,168
1918 (4)	24	501,190	1,287,745	5,929,920			7,217,665
1919 (4)	14	268,658	771,804	3,568,875			4,768,944
1920 (4)	19	387,489	1,056,986	4,816,055			6,306,486
1921 (4)	25	367,909	999,094	4,623,901			5,622,995
1900-1921		6,351,820	$30,360,903	138,511,563			$120,490,863

(1) Spurr, J. E., USGS PP 42, (1905), 26.
(2) Director of Mint, AR 1902, (1903), Production of the Precious Metals in the United States, 164.
(3) Unpublished information from U. S. Geol. Survey.
(4) "Mineral Resources of the United States," U. S. Geol. Survey.

Dividends. The following dividends have been paid by Tonopah companies up to 1922, according to a table prepared by Miss N. E. Preece, Geol. Dept., International Smelting Co., Salt Lake City:

Company	Dividends
Jim Butler	$ 1,151,073
MacNamara	40,213
Montana-Tonopah	504,997
Rescue-Eula	35,539
Tonopah Belmont	10,643,059
Tonopah Midway	250,000
Tonopah Mining	15,525,000
Tonopah Extension	3,087,437
West End Consolidated	1,809,214
Total	$33,046,532

Geology. The rocks of Tonopah are Tertiary volcanics. They consist mainly of surface flows as interpreted by Burgess; while intrusives are also important according to Spurr; and they include a lake deposit of volcanic tuff. Complex faulting has occured, but the volcanic rocks lie in general nearly flat. The following table showing the geologic formations at Tonopah is adapted from Bastin and Laney.

Approximate Stratigraphic Succession	Relative Age and Origin According to Burgess	Relative Age and Origin According to Spurr
Brougher Dacite and Oddie Rhyolite	9. Surface flows and conduits through which the lava rose	11. Intrusive as volcanic necks
Basalt	8. Surface flow	10. Surface flow
Tonopah Rhyolite of Spurr in northern part of area	6b. Surface flows	8b. Intrusive
Siebert Tuff	7. Lake deposits of volcanic material	9. Lake deposits of volcanic material
Tonopah Rhyolite of Spurr in S. part of area	6a. "Surface flows of rhyolites and breccias."	8a. Intrusive, surface flows and tuffs
Fraction Dacite Breccia	Not separately considered	7. Tuffs and mud flows
Heller Dacite		6. Volcanic necks and possibly some surface flows
Midway or "Later" Andesite	5. Surface flow	5. Surface flow
Mizpah Trachyte or "Earlier Andesite"	4. Surface flows	1b. Surface flow
West End or "Upper" Rhyolite	3. Surface flows	4. Intrusive sheet
Montana Breccia	(Not specifically mentioned, but evidently classed as surface volcanic rocks.)	3. Intrusive sheet
Glassy Trachyte		1a. Basal phase of Mizpah Trachyte flow
Sandgrass or "Calcitic" Andesite	2. Surface flow	2. Flat-lying intrusive
Lower Rhyolite	1. Complex of tuffs, breccias, and flows	8c. Intrusive with "autobrecciation"

Orebodies. The ore deposits are typical replacement veins. The wall rock has been replaced along sheeted zones, and every degree of replacement may be observed from slightly altered rock to heavy sulphide ore. The principal production of Tonopah has come from veins lying wholly within the Mizpah Trachyte, although recently an increasing proportion of the ore has been mined from veins with one or both wall-rocks of some other formation,—notably the West End Rhyolite. The Midway Andesite is younger than these veins and forms the capping. The depth of oxidation is variable, depending upon the degree of fracturing, slight oxidation having been observed at as great a depth as 1,170 ft. on the Murray vein.

Ore. Quartz is the most abundant gangue mineral in the Tonopah ores. The pink material, which occurs as a gangue in many ores is a microscopic intergrowth of quartz with a pink carbonate or mixture of carbonates, containing calcium, magnesium, iron, and manganese. From their studies of primary ores, Bastin and Laney found that in general quartz and sphalerite were deposited early in the process of mineralization; galena, chalcopyrite, pyrargyrite, polybasite, argentite, elecrtum, and carbonates forming afterwards; while at a still later period galena and to a lesser degree sphalerite, quartz, and polyasite showed a tendency to be replaced by argentite, chalcopyrite, carbonates, and electrum. Arsenopyrite has been recognized among the primary ore minerals, and argyrodite is probably also present. Wolframite has been noted in a number of instances but belongs to a later period of mineralization than that of the silver ores. The secondary minerals present in the Tonopah ores include gold, silver, pyrite, chalcopyrite, argentite, polybasite, pyrargyrite, cerargyrite, embolite, iodobromite, iodyrite, quartz, hyaline silicia, hydrous oxides of iron, oxides of maganese, calcite, mixed carbonates, malachite calamine, kaolin, dahllite, barite, and gysum. Selenium occurs in the Tonopah ores in an undetermined mineral combination, perhaps with gold and silver in electrum.

Mines. The most important mines in the camp of Tonopah today are those of the Tonopah Belmont Dev., the West End Con., the Tonopah Ex.; and the Tonopah Mining Co., of Nevada. These mines ranked respectively sixth, ninth, twelfth, and sixteenth among the silver mines of the United States in 1921. They are also important producers of gold, and in 1921 the Tonopah Belmont and West End mines ranked second and third respectively among Nevada gold mines. These four mines are briefly described in succeeding paragraphs. Other Tonopah companies which appear on the Nevada State Mine Inspector's active list for 1922 are:

After All Mines Co., M. E. Albert, Sec.
Buckeye Belmont Mining Co., operated by Tonopah Belmont.
Gypsy Queen Mining Co., C. F. Wittenberg, Supt.
Halifax Tonopah Mining Co., Controlled by West End Cons.
Jim Butler Tonopah Mining Co., L. R. Robins, Supt.
MacNamara Mining & Milling Co., E. C. Simpson, Supt., (Operated by leasers).
Montana-Tonopah Mines Co., M. E. Albert, Sec., (Operated by leasers).
Rescue-Eula Cons. Mining Co., Harry D. Moore, Supt.
Tonopah Midway Cons. Mining Co., W. J. Douglas, Supt.
Tonopah North Star Tunnel & Development Co., A. E. Lowe, Supt.
Tonopah Oriental Mining Co., E. C. Simpson, Supt.
Umatilla Tonopah Mining Co., A. H. Haskell, Resident Agent.

Tonopah Belmont. The Tonopah Belmont Dev. Co. is incorporated in New Jersey with a capital stock of 1,500,000 shares of $1.00 par value. The stock is listed on the Philadelphia Stock Exchange and traded in on the New York and Boston Curbs. The home office is at 500 Bullitt Bldg., Philadelphia. C. A. Heller is Pres.; W. M. Potts, V.-P.; J. K. Kitto, Sec.-Treas.; and C. B. Taylor, General Counsel. The mine office is at Tonopah. F. Bradshaw is Gen.-Mgr.; A. Silver, Supt. of Milling; L. R. Robins, Supt. of Tonopah Belmont Mine; F. Steen, Auditor, Herman Dauth, Eng., W. H. Royston, Mill Supt. In addition to the Tonopah Belmont mine, the company controls the Belmont Surf Inlet Mines, Ltd., of British Columbia; the Belmont Shawmut Mining Company of California; and the Belmont Wagner Mining Company of Colorado. The Tonopah Belmont Mine consists of 11 claims aggregating 161 acres and containing 38½ m. of workings. It adjoins the Tonopah Mining Co. Mine on the E. The property is developed by 2 vertical, three-compartment shafts, 1,127 ft. and 1,718 ft. in depth. The company operates a 60-stamp cyanide mill and a cyanide plant for the treatment of concentrates.

Tonopah Extension. The Tonopah Extension Mining Co., is incorporated in Arizona with an authorized capital of 2,000,000 shares of $1.00 par value of which 1,392,800 are outstanding. The stock is listed on the San Francisco Stock Exchange and traded in on the New York curb. The home office is at 30 Church Street, New York City. M. R. Ward is Pres.; W. A. Mitchell, V.-P.; W. G. Benham, Sec.-Treas.; and J. G. Love, Asst.-Sec.-Treas.; John G. Kirchen is General Manager. The company own 670 acres of mineral land with 33 miles of underground workings. The property adjoins that of the Tonopah Mining Co., on the W. and that of the West End Cons. Mining Co. on the N. It is developed by 3 main working shafts, the No. 2, McKane, and Victor, which are 1,440, 1,650, and 2,000 feet deep respectively. The mine equipment includes a 300 horse-power hoist at the Victor Shaft, a 150 horse-power hoist at the No. 2 Shaft and a 75 horse-power hoist at the McKane Shaft; duplex, quintuplex, and turbine pumps, and 3 compressors. There is a tramway from the mine to the 50-stamp cyanide mill. All machinery is electrically driven and there is an auxiliary power plant to provide power in the event of the failure of the hydro-electric supply. The Victor Shaft is to be sunk to 2,125 feet and crosscuts run to the veins. Two vertical triplex pumps with a capacity of 1,200 gals. per minute have recently been installed in the McKane Shaft which is to be used for westward exporation.

West End Consolidated. The West End Cons. M. Co. is incorporated in Arizona with an authorized capital of 2,000,000 shares of $5.00 par value, of which 1,788,486 shares are outstanding. The stock is listed on the San Francisco Stock Exchange and traded in on the New York curb. The executive offices are in the Syndicate Bldg., Oakland, Calif. F. M. Smith is Pres.-Treas.; J. F. Carlton, V.-P.; G. C. Ellis, Sec.; J. W. Sherwin, Gen. Mgr., and J. R. Blair, Auditor. The mine office is at Tonopah. F. C. Ninnis is Mill Supt.; H. D. Budelman, Mine Supt.; Otto Falk, Chief Clerk and Purchasing Agent; and C. E. Waldner, Mining Eng. The company owns 10 patented and 7 unpatented mining claims aggregating 184 acres adjoining the property of the Tonopah Ext. M. Co. on the S. and that of the Tonopah M. Co. of Nev. on the W. It controls the Halifax Tonopah M. Co. and has a lease on the Jim Butler ground at Tonopah, and also controls the Santa Rosa M. Co. in Inyo County, Cal. and the Mabel mine in the Garfield District of Mineral County. The properties at Searles Lake, California and Opoteca, Honduras, Central America, formerly held by the company have been transferred to the West End Chem. Co. and the West End Opoteca Ms. Co. respectively. The Tonopah property is developed by 3 shafts, the deepest being 1,212 ft., and has 14 m. of underground workings. The ore is milled in a cyanide plant with a capacity of 225 tons per day, which treats West End, Jim Butler and Halifax ores.

Tonopah Mining. The Tonopah M. Co. of Nev. is incorporated in Delaware with an authorized capital of 1,000,000 shares of $1.00 par value, all of which are outstanding. The stock is listed on the Philadelphia and San Francisco Stock Exchanges and traded in on the New York and Boston curbs. The general office is at 572 Bullitt Bldg., Philadelphia. J. H. Whiteman is Pres.; C. R. Miller, V. P.; and P. S. Bickmore, Sec.-Treas. The mine office is at Tonopah. W. H. Blackburn is Gen. Supt.-Cons. Eng.; H. A. Johnson, Supt.; and T. A. Frazier, Chief Clerk. The Tonopah M. Co. owns 160 acres of mineral land at Tonopah adjoining the Tonopah Extension and West End ground on the E. and the Tonopah Belmont claims on the W. The company also owns controlling interests on the T. & G. R. R., Esmeralda Power Co., Tonopah Placers Co. in Colorado, Eden M. Co. and Tonopah Nicaraugua Co. in Nicaragua, and Mandy M. Co. in Manitoba, Canada. Recently, the company has engaged in the development of the Crystalline-Alabama properties at Jamestown, Cal.; has taken over the Ajax mine in the Cripple Creek District, Col., under the incorporation of Tonopah Ajax M. Co.; and has done a limited amount of work on the Lombard mine at Idaho Springs, Col. The Tonopah property is developed to a depth of 1,500 ft. It has 3 main working shafts, and 46 miles of underground workings. The ore is treated in a 100-stamp cyanide mill located at Millers, where a little dump and custom ore is also milled.

BIBLIOGRAPHY OF THE TONOPAH DISTRICT

General

MR1903 183	MR1909 I 422-4	MR1916 I 493-5
MR1904 198	MR1910 I 527-9	MR1917 I 289-293
MR1905 116,271-2	MR1911 I 691-6	MR1918 I 252-6
MR1906 122,298	MR1912 I 808-812	MR1919 I 406-9
MR1907 I 375-8, 385-39	MR1913 I 835-8	MR1920 I 331-3
MR1908 I 499-501	MR1914 I 702-6	MR1921 140
	MR1915 I 647-650	Selenium, 391-3

Bradshaw, F., "Operation of the Tonopah Belmont Mine", M&SP 106 (1913) 730-1.

———"Tonopah in 1914", E&MJ 99 (1915) 154-5.

Carenter, J. A., "Effect of the War upon Tonopah's Mining", E&MJ 106 (1918) 601-4.

Easton, S. A., "Notes on Tonopah, Nevada", E&MJ 73 (1902) 697.

E&MJ, "Some Notes on Tonopah," Special Correspondence, 79, (1905) 1084-5.

———"Prosperous Year at Tonopah", 95 (1913) 126-7.

Haas, H., "The Tonopah District in 1910", EM&J 91 (1911) 125-6.

Johnson, A. T., "Mining in the Tonopah District," P AMC, 12th Ann. Sess. (1909) 412-7.

Knapp, S. A., "Tonopah, Nevada," MS&P 82 (1901) 231.

Lakes, A., "Tonopah Mining Camp," M&M 24 (1904) 479-481.

Martin, A. H., "Montana-Tonopah Mine and Mill", E&MW Mar. 12, 1910.

———"Tonopah Extension Mine and Mill," Pacific Miner, Oct., 1910.

M&SP, "Tonopah," Staff Correspondence, 83 (1901) 192-4.

———"The Discovery and Development of Tonopah," 90 (1905) 8-9.

———"The Mines of Tonopah Nevada," 90 (1905) 182-3.

———"Tonopah, Nevada, and its Development", 91 (1905) 10-11.

———"The Discovery of Tonopah", Staff Correspondence, 92 (1906) 56-7.

"Tonopah and Goldfield and their Rapid Development," 90 (1905) 94.5.

———"Tonopah Mines and Mining", Editorial, 104 (1912) 749.

———"Tonopah," Editorial, 110 (1915) 750-1.

WeedMH 1127-1374 Describes 33 Tonopah companies.

Geology

Balliet, L. W., "The Geology of Tonopah, Nevada," M&EW 40 (1914) 837-841.

Bastin, E. S., & Laney, F. B., "The Genesis of Ores at Tonopah, Nevada," USGS PP 104 (1918); Abstract in M&SP 118 (1919) 287-9.

Burgess, J. A., "The Geology of the Producing Part of the Tonopah Mining District," EG 4 (1909) 681-712.

——— "The Halogen Salts of Silver and Associated Minerals at Tonopah, Nevada," EG 6 (1911) 13-21; Correction in EG 12 (1917) 593.

Butler, B. S., "Potash in Certain Copper and Gold Ores," USGS B 620J (1915) 234.

Durner, F. W., "The Great Tonopah Mineral Belt of Nevada," M&EW 40 (1914) 290-1.

Eakle, A. S., "The Minerals of Tonopah, Nevada," U. of Cal. Dept. of Geol. B 7 (1917) 1-20.

Hill507 224.

Jenney, W. P., "Geological and Physical Condition of Tonopah Mines," M&SP 99 (1909) 685-6.

——— "The Geological and Physical Condition of Tonopah Mines," E&MJ 89, (1910) 29-30.

Knopf, Adolph, "The Divide Silver District, Nevada," USGS B 715K (1921). 162-4.

Kraus, E. H., & Cook, C. W., "Die Kristallformen des Jodyrits von Tonopah, Nevada," Centralbl. Mineralogie 13 (1913) 385-6.

Lakes, A., "The Tonopah Volcanoes," M&M 26 (1906) 554.

Locke, A., "Four Famous Districts Compared," E&MJ 92 (1911) 505-6.

——— "The Geology of the Tonopah Mining District," B AIME 62 (1912) 217-226; T AIME 43 (1913) 157-166; Abstract M&SP 103 (1911) 523-5; Editorial 511; M&EW 35 (1911) 1271-2.

Rice, C. T., "Tonopah and its Geology," E&MJ 91 (1911) 966-970.

Rogers, A. F., "Dahllite (Podolite) from Tonopah, Nevada," AJS 4th Series (1912) 475.

Spurr, J. E., "Ore Deposits of Tonopah and Neighboring Districts," Nevada," USGS B 213 (1903) 81-87.

——— "The Ore Deposits of Tonopah, Nevada, (Preliminary Report)," USGS B 219, (1903).

——— "The Ore Deposits of Tonopah, Nevada," E&MJ 76 (1903) 769-770.

——— "Preliminary Report on the Ore Deposits of Tonopah, Nevada," USGS B 225 (1904) 88-110.

——— "Faulting at Tonopah, Nevada," Abstract, Science, NS 19 (1904) 921-2.

——— "Developments at Tonopah, Nevada during 1904," USGS B 260 (1905) 140-4; Abstract M&SP 90 (1905) 369--370.

——— "Geology of the Tonopah Mining District, Nevada," USGS PP 42 (1905); Abstract in E&MJ 80 (1905) 922-3; Review by W. Lindgren EG 10 (1906) 698-9; Abstract in M&SP 91 (1905) 360-1, 381-2,

——— "Tonopah Mining District," Franklin Inst. J 160 (1905) 1-20.

——— "Genetic Relations of the Western Nevada Ores," T AIME 36 (1906) 372-381.

——— "Tonopah Geology," M&SP 102, (1911) 560-2; Abstract Report on Montana-Tonopah M. Co.; Editorial, 550.

——— "Geology and Ore Deposition at Tonopah, Nevada," EG 10 (1915) 713-769; Editorial review in M&SP 112 (1916) 498-9.

Steidle, E., "Unusual Gas in a Metal Mine," M&SP 112 (1916) 368.

Mining

Dynan, J. L., "Operating Conditions at Tonopah Extension Mine," M&Met No. 170 (1921) 26-8.

E&MJ, "A New Sampling Plant at Tonopah," 82 (1906) 57.

——— "Variable Capacity Compression," 92 (1911) 632.

——— "Mine Track Switches," 92 (1911) 881.

——— "Headframe for a Prospect Shaft," 92 (1911) 980.

——— "A Safety Hook for Hoisting Buckets," 92 (1911) 1025.

——— "Saving Oil in Mines," 91 (1911) 1147.

——— "Air Cylinders for Changing Cages," 93 (1912) 253.

——— "Triangular Timbering in Tonopah Belmont Mine," 93 (1912) 931.
——— "Automatic Car Return," 101 (1916) 1112-3.
Fox, J. M., "Diamond Drilling at Tonopah," M&SP 99 (1909) 262-4.
M&SP, "New Hoisting Equipment at the Tonopah Extension Mine," 121 (1920) 277-8.
Rice, C. T., "Mining at Tonopah," E&MJ 82 (1906) 156-7.
——— "Mining at Tonopah," E&MJ 82 (1906) 199.
——— "Tonopah-Belmont Surface Plant," E&MJ 91 (1911) 853-7; Editorial 841-2.
——— "Present Conditions at Tonopah Mines," E&MJ 92 (1911) 17-21..
——— "Notes on Mining at Tonopah," E&MJ 92 (1911) 1184-5.
Semple, C. C., "Pipe-Straightening Device," E&MJ 101 (1916) 353-4.
———"Straightening Mine Rails," E&MJ 101 (1916) 437.
——— "A Steel Collaring Machine," E&MJ 101 (1916) 984.

Metallurgy

Barbour, P. E., "Some Screen Analyses from Tonopah," E&MJ 97 (1914) 166.
Bosqui, F. L., "Milling Versus Smelting in the Treatment of Tonopah-Goldfield Ores," M&SP 92 (1906) 217; Also in "Recent Cyanide Practice," (1907) 33-6.
Carpenter, J. A., "Continuous Agitation at the West End Mill, Tonopah," M&SP 106 (1913) 646-651.
——— "Operation of the West End Mill, Tonopah," M&SP 107 (1913) 191-2, (From annual report).
———"Slime Agitation and Solution Replacement Methods at the West End Mill, Tonopah, Nev.," T AIME 92 (1916) 82-94.
Clyne, C. B., "Concrete Work at New Belmont Mill," E&MJ 92 (1911) 1050.
Daman, A. C., "Automatic Ore Sampler," (at Millers), E&MJ 103 (1917) 188.
E&MJ, "Montana-Tonopah Stamp and Cyanide Mill," 85 (1908) 959-962.
——— "New Mill of Tonopah Extension Mining Co.," 89 (1910) 1066-7.
——— "Finger Chutes at Mills," (at Tonopah), 92 (1911) 581.
——— "Filter for Slime Samples" (at Tonopah), 92 (1911) 931.
——— "Tonopah Orehouses," 91 (1911) 1048.
——— "Tonopah Milling Practice," Editorial, 94 (1912) 194.
——— "Vacuum-Filter Improvement at Tonopah," 96 (1913) 1074.
Hart, J. P., & Williams, H. L., "West End Consolidated Mill, Tonopah," E&MJ 94 (1912) 163-6.
Jones, A. H., "Heating Cyanide Solutions," Discussion, M&SP 104 (1912) 176.
——— "Precipitate Melting at the New Belmont Mill," E&MJ 95 (1913) 1197-8,
——— "The Tonopah Plant of the Belmont Milling Co.," B AIME 104 (1915); T AIME 52 (1916) 95-122; Abstracts M&SP 111 (1915) 324, 391.
Kirchen, J. G., "Tonopah Extension Mill," M&SP 100 (1910) 522-3.
——— "Comparison of High and Low-Grade Cyanide," E&MJ 110 (1920) 614.
Martin, A. H., "Hundred-Stamp Desert Mill at Millers, Nevada," E&MW May 1, 1909.
——— "Mill of Tonopah Mining Company," MS, Feb. 24, 1910.
McKenzie, C. S., "Solution Meter at the Belmont Mill," E&MJ 96 (1913) 165-6.
Megraw, H. A., "Silver Cyanidation at Tonopah," E&MJ 95 (1913) 413-7, 455-9, 503-6; Discussion, 720, 767-8, 912, 1109, 1110, 1255; 96 (1913) 31, 75-7, 441-4; Editorial 534.
Olson, C. R., "A Solution Clarifier," E&MJ 93 (1912) 158.
Parsons, A. R., "Desert Mill, Millers, Nevada," M&SP 95 (1907) 494-500; Also in "More Recent Cyanide Practice," (1910) 9-20.

Rice, C. T., "Tonopah Mining Company's Mill," E&MJ 91 (1911), 1212-6.
——— "Patching Vanner Belts," E&MJ 92 (1911) 8.
——— "Tonopah-Belmont Cyanide Plant," E&MJ 92 (1911) 111-5; Discussion 244, 1070.
——— "Ore Sorting at Tonopah," E&MJ 92 (1911) 761-2.
——— "Tonopah-Belmont Orehouse," E&MJ 92 (1911) 836-7.
——— "The Tonopah Extension Mill," E&MJ 92 (1911) 1085-7.
Rotherham, G. H., "Milling Plant of the Montana-Tonopah Mining Company," M&SP 97 (1908) 324-7; Also in "more Recent Cyanide Practice," (1910) 201-8.
Von Bernewitz, M. W., "Metallurgy at Tonopah," M&SP 105 (1912) 828-9.
——— "The MacNamara Mill, Tonopah," M&SP 106 (1913) 182-3.
——— "Concentration of Dissolved Metals in Slime Ponds," M&SP 106 (1913) 145.
Worcester, S. A., "The Tonopah Mining Co.'s Mill," E&MJ 80 (1905) 682-4.
Young, G. J., "Precipitate Smelting at Tonopah," E&MJ 108 (1919) 892.

Miscellaneous

Averill, V., "Details and Cost of a Mine Model," E&MJ 108, (1919) 824-5.
Butson, H., "The Strike at Tonopah," Discussion, M&SP 122 (1920) 833.
E&MJ, "Map of the Tonopah Mining District," 89, (1910) 101.
——— "Apex Litigation at Tonopah," Special Correspondence, 99 (1915) 660-1.
——— "Mine Models in Tonopah Apex Litigation," 99 (1915) 850.
——— "Drilling Contest at Tonopah," 104 (1917) 315.
——— "Tonopah Belmont's Expansion," 104 (1917) 342.
——— Editorial, "Insurance of Employes," 120 (1920) 363.
M&SP, "The Strike at Tonopah," Editorial, 123 (1921) 281-2.
——— "Strike at Tonopah," 119 (1919) 733-4.
Searles, R. M., "Apexes and Anticlines," M&SP 117 (1918) 43-4.
USGS Tonopah Topographic Map and Tonopah Special Mining Map.

TRAPPMANS

Silver, Gold

Location. The Trappmans District is situated on Pahute Mesa in S. central Nye Co. Trappmans Camp is 40 miles by road E. S. E. of Goldfield which is on the T. & G. R. R. and the T. & T. R. R. The district adjoins the Wilson District on the S.

History. The Trappmans District was discovered by Trappman and Gabbard in 1904, worked for a short time, and abandoned.

Geology. Quartz veins of three generations occur in Cretaceous granite, according to Ball. The ore is oxidized and the ratio of the silver to the gold values is as four to one. The principal original sulphide was pyrite with which occurred some galena. Silver chloride and native silver have been observed. In one instance calcite occurred with quartz as the gangue.

Bibliography. USGS Kawich topographic map.
Ball285 68-9. Ball308 131-9. Hill507 224.

TROY (IRWIN CANYON)

Silver, Gold

Location. The Troy District is located in Troy and Irwin Canyons on the W. flank of the Grant Range at its S. end in N.E. Nye Co. The district lies about 30 m. S. of Currant post office.

History. The district was discovered by A. Beaty in 1867, the town of Troy was laid out in 1869, and the mines in Troy Canyon purchased by an English company. This company erected a 20-stamp mill and furnaces. but shut down in 1872. Claims in Irwin Canyon were prospected in 1905 and 1912.

Geology. In the Troy District, Ordovician sedimentary rocks consisting of shaly limestones overlain by white quartzite which in turn is overlain by massive bedded limestones, have been intruded by a large stock of quartz-monzonite, according to Hill. At the Troy mine there are ore deposits in the shales in the form of lenses of white quartz, with sphalerite, pyrite, galena, and possibly chalcopyrite, that carry silver and gold. Ore deposits also occur in the quartz-monzonite in the form of narrow quartz veins sparingly mineralized with pyrite and minor amounts of sphalerite and galena and said to carry gold and a little silver.

Bibliography. R1869 177-8 SMN1867-8 78 SMN1873-4 76-7
R1872 174 SMN1871-2 112-3 SMN1875-6 108
R1873 227, 232 MR1914 I 706
Hill648 138-144.
Spurr208 68-76 Quinn Canyon and Grant Ranges.
Thompson & West 526-7.

TWIN RIVER

Silver

Location. The Twin River District is situated on the E. flank of the Toyabe Range in N. E. Nye Co. It adjoins the Millett District on the S. The principal mines are in Ophir Canyon, about 50 m. S. of Austin which is on the N. C. R. R. and 100 m. N. of Tonopah which is on the T. & G. R. R.

History. The Twin River District was discovered by S. Boulerond and a party of Frenchmen in 1863, and its principal mine, the Murphy Mine, was located the following year. The Twin River M. Co. purchased the Murphy Mine and built a 20-stamp mill to treat its ores. This mill operated in 1867 and 1868 and produced $750,000 in bullion, but it paid no dividends and the stockholders became discouraged and permitted it to go into bankruptcy. The property passed through several hands and finally in 1917 came into those of the Nevada Ophir M. Co. which began work on the ground in 1918.

Geology. According to Hague, the ore deposit at the Murphy Mine is a quartz vein in slate. The ore occurs as an irregular band from 1 to 4 ft. thick in the vein quartz. The silver-bearing minerals are tetrahedrite,—with which is sometimes associated ruby silver,—galena, and considerable quantities of sphalerite. Pyrite is frequently associated with these minerals.

Mines. The Nevada Ophir M. Co. was incorporated in Utah in 1917 with a capital stock of 1,000,000 shares of 10 cents par value, of which 500,000 are outstanding. W. E. Trent is Pres.; C. A. Higgins, Sec., with E. E. Rich and F. V. Sullivan additional Directors. The home office is at 25 Broad St., New York City. The company owns 8 claims covering the old Murphy Mine, developed by 500-foot and 300-foot shafts and a 700-foot tunnel; and equipped with power plant, hoist, and 20-stamp mill.

Bibliography. B1867 402, 415-7, 431 R1873 242 SMN1871-2 106
R1868 101-3 SMN1866 60-61 SMN1873-4 78
R1872 172-3 SMN1867-8 66-8 SMN1875-6 107
SMN1869-70 89

USGE40th 3 (1870) 343, 383-393. USGS Tonopah topographic map.
Spurr208 93-7 Toyabe Range.
Thompson & West 526.
WeedMH 1282 Nevada Ophir M. Co.

TYBO (HOT CREEK)

Silver, Lead, Gold, Copper, (Antimony, Manganese)

Location. The Tybo District is located at Tybo and Hot Creek in the Hot Creek Range. The district is about 70 m. N. E. of Tonopah which is on the T. & G. R. R. The N. section of the district near Hot Creek is sometimes considered as a separate district and called the Hot Creek District.

History. The Hot Creek section of the Tybo District was discovered in 1865 by an Indian who showed it to white prospectors in 1866. By 1868, there were 2 10-stamp mills in operation. The Tybo section was discovered in 1870 by Gally and Gillette. A smelter was erected in 1874, and the Tybo Co. built another furnace in 1875, and also erected a 20-stamp mill. For a number of years, Tybo produced large quantities of silver and lead with less gold and was the most prosperous district in Nye Co. The Louisiana Cons. M. Co. attempted to rejuvenate the Tybo District recently. A 75-ton concentrator was built by this company in 1917 and operated in 1918. In 1919, a flotation plant and smelter were installed which operated in 1920. These operations did not meet with success and the company was forced to shut down.

Geology. The Tybo ore deposit is a fissure vein at a limestone-rhyolite contact, according to Mines Handbook, and carries values in silver, lead, and gold. Manganese ore occurs in narrow veins along faults in travertine of Pleistocene or Recent age on the Surprise claim 2 m. N. of Tybo, according to Pardee and Jones. Stibnite occurs in quartz veins in the Hot Springs section.

Bibliography.

B1867 402, 424	SMN1869-70 90-1	MR1914 I 706
R1868 109, 111	SMN1871-2 108-9	MR1915 I 650
R1873 232	SMN1873-4 73-4	MR1917 I 293
R1874 280-1	SMN1875-6 104-5	MR1918 I 256
R1875 172, 174-5	SMN1877-8 91-2	MR1919 I 409
SMN1866 61		MR1920 I 333
SMN1867-8 73-6		MR1921 I 393

Pardee, J. T., & Jones, E. L., Jr., "Deposits of Manganese Ore in Nevada," USGS B 710F (1920) 234.
Schrader624 199.
Spurr208 84-7 Hot Creek Range.
StuartNMR 7.
Thompson & West 527.
WeedMH 1242 Louisiana Cons. M. Co.

UNION (BERLIN)

Mercury, Gold, Silver, Lead, Copper

Location. The Union District is situated S. of Ione on the W. slope of the Shoshone Range. The mercury deposits lie a few miles S. E. of Ione, while the precious metal deposits are further S. in the neighborhood of Berlin and Grantsville. Berlin is about 45 m. by road N. E. of Luning which is on the S. P. R. R. and 60 m. S. W. of Austin which is on the N. C. R. R. The elevation of Berlin is 6,683 ft. and Mt. Berlin, the highest point in the district, reaches an altitude of 9,081 ft.

History. The Union District was discovered by P. A. Haven in 1863, and the town of Ione and camp of Grantsville established the same year. Upon petition of the miners of the district, Nye Co. was separated from Esmeralda Co. in 1864 and Ione designated as the county seat. The Knickerbocker 20-stamp mill and the Pioneer 10-stamp mill were erected, but the district failed to meet expectations, depression ensued, and the county seat was moved to Belmont in 1867. In 1870, there was a brief revival of mining activity in the district due to the construction of a mill at Ellsworth in the neighboring Mammoth District; and again in 1877 when the Alexander Co. erected a 20-stamp mill at Grantsville which was enlarged to 40 stamps in 1880. In 1905, the Nevada Co. was operating a 30-stamp mill at the Berlin Mine. This property was acquired by the Goldfield Blue Bell M. Co. in 1908. A 40-ton cyanide mill to treat the old tailings of the Berlin Mine was installed in 1912 and made the principal production of the district until recently when the Alexander Mine at Grantsville was temporarily reopened by E. M. Bray. Mercury ore was discovered in the N. part of the district in 1907, and the Shoshone Quicksilver Co. began production with a small test retort in 1908. In 1909, the Mercury M. Co. developed a property and in 1910 erected a 25-ton Scott furnace and commenced to produce. The Nevada Cinnabar Co. acquired the Shoshone Mine in 1912, built a 50-ton Scott furnace in 1913, and started production in 1914. The Mercury M. Co. and Nevada Cinnabar Co. continued to make a considerable although irregular production up to 1920.

Production. The Union District is said to have produced about $1,000,000 in gold and silver up to 1880. The Mercury M. Co. has produced over 6,000 flasks of mercury, and the Nevada Cinnabar Co. is credited with a production of 5,000 flasks.

Geology. The country rocks of the Union District are Carboniferous sediments and Tertiary eruptives. In the vicinity of the mercury deposits, according to Ransome, the crest of the Shoshone Range is composed of rhyolite which extends part way down the W. slope. Near the mines, Carboniferous limestone, shale and conglomerate, have been faulted up into contact with the Tertiary rhyolite, while to the W. they are in contact with a broad belt of ancient greenstones. The ore on the property of the Mercury M. Co. occurs in and near a volcanic neck which is filled with an agglomerate consisting of angular fragments of rhyolite and sedimentary rocks. The ore in this agglomerate usually consists of altered rhyolite fragments impregnated with cinnabar; while that in the enclosing rocks consists of cinnabar filling cracks and crevices in shattered limestone and more rarely in shale in the limestone. The country rock at the Berlin Mine is andesite, according to Daggett. The ore deposit is a quartz vein which has been intricately faulted. About 2% of sulphides occur in the quartz including iron, copper, lead, zinc, and antimony sulphides. The proportion of silver to gold by weight is from 7 to 12 silver to 1 gold. The vein is a filled fissure with the filling frozen to the walls. The ore at the Alexander Mine is said to be a sulphide replacement in alaskite and limestone.

Bibliography. B1867 419-429
R1868 100,112
R1871 182-3
SMN1866 57-8
SMN1867-8 62-5
MR1907 I 372, 683 Mercury
MR1908 I 496, 692 "
MR1909 I 419, 555 "
MR1910 I 529, 702 "
MR1911 I 696, 908 9 "

SMN1869-70 89	MR1912 I 812,	945	Mercury
SMN1871-2 107	MR1913 I 838,	208	"
SMN1873-4 78	MR1914 I 706,	328	"
SMN1875-6 108-9	MR1915 I	272	"
SMN1877-8 88-9	MR1916 I	767	"
MR1905 272	MR1917 I 293,	389, 418	"
MR1906 298	MR1918 I	164-5	"
	MR1919 I	173	"
	MR1920 I	432	"
	MR1921 I 393		

USGS Tonopah topographic map.
Bancroft 269.
Daggett, E., "The Extraordinary Faulting at the Berlin Mine, Nevada," T AIME 38 (1908) 297-309; Abstract E&MJ 83 (1907) 617-621.
Davis 960-2.
Davis, W. C., "Quicksilver in Nevada," M&SP 99 (1909) 663.
Hill507 220, 225.
StuartNMR 85-6.
Thompson & West 523.
WeedMH 1208 Goldfield Blue Bell M. Co. 1254 Mercury M. Co.

VOLCANO see HANNAPAH

WASHINGTON

Silver, Lead

Location. The Washington District lies S. of Washington in N. E. Nye Co. on the border of Lander Co. It adjoins the Kingston District on the S. W. Austin on the N. C. R. R. is 28 m. N. N.E. The mines are located on San Pedro, Cottonwood and San Juan Creeks on the W. flank of the Toyabe Range. San Juan has an elevation of 8,075 ft. and San Juan Peak rises above it to a height of 10,982 ft.

History. The Washington District was organized in 1863 and for a few years was a very active camp. In recent years a little work has been done at the Warner Mine or Washington property, silver ore having been shipped in 1918 and 1919.

Geology. The ore deposits occur in Paleozoic shales, with some interbedded quartzite and limestone, whose age is thought to be Ordovician. At the Warner Mine there is a quartzite breccia recemented with quartz and iron-stained at the surface which is associated with a fracture cutting lime-shale and quartzite country rocks. The ore minerals are argentite, galena, tetrahedrite, and their alteration products. At the St. Louis and Richmond Mine the vein is parallel to the bedding of the sedimentary rocks occurring in a limestone interstratified with shales. The minerals present are quartz, siderite, pyrite, sphalerite, galena, and arsenopyrite.

Bibliography. B1867 413 SMN1866 97, 102 SMN 1869-70 89 USGE40th 3 (1870) 339-340, 347
Hill507 216. Hill594 114-121, 123-5. Spurr208 93-7 Toyable Range.

WELLINGTON

Gold, Silver

Location. The Wellington District is at Wellington in the foothills at the S. end of the Cactus Range, 11 m. S. of Cactus Springs. It adjoins the Antelope District on the S. W. and the Gold Crater District on the N. E. Wellington is 20 m. E. of Cuprite which is on the T. & T. R. R.

History. The Wellington District was discovered in August, 1904, and abandoned a few years later.

Geology. The country rock in the Wellington District is the same Tertiary rhyolite in which the ore deposits of the Cactus Springe and Antelope Springs Districts occur, but in the Wellington District the rhyolite is cut by dikes of andesite, and both rocks are in turn cut by the veins. The quartz veins have been brecciated and the interstices filled with calcite and limonite. The ore is a free-milling gold ore containing a little silver, the gold being associated with limonite.

Bibliography.

USGS Kawich topographic map.

Ball285 68. Ball308 46, 48, 95. Hill507 225.

WILLOW CREEK

Gold, Silver

Location. The Willow Creek District is located in the foothills on the N. W. side of the Quinn Canyon Mts. near the S. end of Railroad Valley in N. E. Nye Co. Willow Creek is about 90 m. S. W. of Ely which is on the N. N. R. R. and about an equal distance E. of Tonopah which is on the T. & G. R. R.

History. The district was discovered by Sampson and Jenkins in 1911, and their discovery purchased by George Wingfield in 1913. In that same year, excitement was occasioned by the discovery of free gold in the Melbourn vein by Papas and Blackwell. The Willow Creek M. Co. produced gold and silver ore in 1913 and 1914, and there was one producer of gold ore in 1917, while the Gold Springs M. Co. operated a new 5-stamp amalgamation and concentration mill in 1921.

Geology. The lodes of the Willow Creek District are found in yellowish and dark-gray shaly limestones which grade into massive dark-gray limestones and are overlain by white quartzite, according to Hill. These Paleozoic sediments have been intruded by quartz-monzonite, andesite porphyry, and alaskite porphyry, and capped by rhyolite. Two types of ore deposits occur. At the Melbourn and other properties, quartz veins carrying free gold closely associated with talc and calcite are found. The primary minerals appear to have been pyrite and chalcopyrite. The other type of ore deposit is represented by the Wingfield Mine, and consists of replacements in limestone along bedding planes near fissures. The ore contains copper pitch ore, limonite, and copper carbonate and silicate. It is valuable chiefly for its silver which occurs native and as hornsilver, but also carries minor values in gold and copper.

Bibliography. MR1913 I 838 MR1914 I 706 MR1921 I 393
MR1917 I 293

Hill648 144-151.

Spurr208 68-76 Quinn Canyon and Grant Ranges.

WeedMH 1215 GGold Springs M. Co.

History. The State Prison site was leased from Abram Curry in 1862 and

WILSONS

Silver, Gold

Location. The Wilsons District is situated on the N. slope of O'Donnell Mt. which forms part of Pahute Mesa in S. central Nye Co. Wilsons Camp is 38 m. by road E. S. E. of Goldfield which is on the T. & G. R. R. and the T. & T. R. R. The district adjoins the Antelope Springs District on the S. E. and the Trappmans District on the N.

History. The Wilsons District was discovered in 1904, worked for a short time and abandoned.

Geology. Steeply dipping quartz veins with a general N. E. strike cut Tertiary rhyolite and andesite, according to Ball. The quartz is stained by limonite and more rarely by malachite, and the silver values are five or six times those of the gold.

Bibliography.

USGS Kawich topographic map.

Ball285 69. Ball308 131-7, 139. Hill507 224.

ORMSBY COUNTY

CARSON CITY

Sandstone

Location. The Carson City sandstone quarry is located at the State Prison E. of Carson which is on the V. & T. R. R.

History. The State Prison site was leased from Abram Curry in 1862 and purchased by the State of Nevada in 1864. A considerable amount of sandstone has been quarried by prisoners and used for building purposes in Reno as well as in Carson City and at the State Prison itself.

Geology. The sandstone is yellowish gray in color, rather heavily bedded, and the beds are nearly horizontal. It is of Pleistocene age and contains well-preserved footprints of extinct animals. The sandstone was derived from granite by weathering. According to Reid, the beds were formed in an ancient lake, while Smith believes the rock to have been cemented by the action of a neighboring hot spring.

Bibliography. SMN1866 19 SMN1871-2 116 MR1891 462
SMN1867-8 20 MR1913 II 1375

USGS Carson topographic map.

Blake, W. P., "The Carson City Ichnolites," Science 4 (1884) 273-6.

Reid, J. A., "Preliminary Report on the Building Stones of Nevada," Univ. of Nev., Dept. Geol. & M., B 1 (1904).

Smith, W. S. T., "Origin of the Sandstone at the State Prison near Carson City, Nev.," Abstract, GSA B 23 (1912) 73.

Thompson & West 545-6.

DELAWARE (SULLIVAN)

Copper, Gold, Silver, Iron

Location. The Delaware District is located in the neighborhood of Brunswick Canyon in E. Ormsby Co. The V. & T. R. R. lies to the N. of the district, and the Carson River describes a curve about it on the W. and N. The district was formerly known as the Sullivan District. It adjoins the old Eldorado coal district on the W.

History. The district was discovered in 1860 and a large number of locations made but little work done. Interest in the region revived recently, and a little copper ore containing gold and silver was shipped from 1909 to 1915. The Bessemer Mine contains a large deposit of iron ore from which a trial shipment was made in 1907, and commercial shipments were made to San Francisco in 1919 and 1920. Lessees were working on the silver-gold lodes of the Comstock Ex. M. Co. from 1921 to 1923.

Mines. E. B. Yerington of Carson City is Sec. of the **Bessemer Cons. M. Co.** The **Comstock Ex. M. Co.** was incorporated in Nevada in 1921 with a capital stock of 1,500,000 shares of which one-half is in the treasury. A. F. Eske is Pres.; E. L. Drappo, V. P.; F. J. DeLongchamps, Reno, Sec.-

Treas.; with H. W. Noble, Directors. The company owns 12 claims on which are 2 fissures in andesite one carrying principally silver and the other gold, developed by 2 shafts and tunnel.

Bibliography. SMN1866 18 | MR1907 I 378 Iron | MR1909 I 424 | MR1910 I 530 | MR1911 I 696 | MR1914 I 707 | MR1915I 650 | MR1920 I 333 Iron | MR1921 I 393

USGS Carson topographic map. StuartNMR 149-9.
WeedMH1167 Comstock Ex. M. Co. 1365 United M. Co.

EAGLE VALLEY see VOLTAIRE
ELDORADO CANYON see LYON COUNTY
SULLIVAN see DELAWARE

VOLTAIRE (EAGLE VALLEY, WASHOE AND EAGLE VALLEY) ORMSBY COUNTY

Silver, Gold, Copper, Graphite, (Arsenic)

Location. The Voltaire District is located in the foothills of the Sierra Nevada Range W. of Eagle Valley in W. Ormsby Co. It lies immediately W. of Carson City which is on the V. & T. R. R. Carson City has an elevation 4,675 ft. and Snow Valley Peak to the W. reaches an altitude of 9,214 ft. The district is sometimes called the Washoe and Eagle Valley District.

History. The district was discovered in 1859 and considerable prospecting performed without opening any mine of importanre. In 1865 and 1866 some work was done upon the Athens Mine and a few tons of ore treated at the Carson Mill. A renewal of interest in the district took place in 1874. The Voltaire Mine was in operation in 1880. The Athens Mine made some test runs in a 3-stamp mill in 1910 and 1911. Recently, the Panama Canal M. Co. has been active and has shipped a number of lots of silver ore. Graphite was mined in the district as early as 1903 and production has been continued up to the present year.

Geology. The country rocks of the district are Triassic schists intruded by Cretaceous granodiorite. At the Panama Canal Mine, silver-bearing quartz veins occur in the granodiorite. Amorphous graphite occurs in graphic shale which has been intruded by basalt. A lens of arsenopyrite in these schists has been opened on the Raffetto property, according an unpublished report by F. C. Lincoln.

Mines. The Panama Canal M. Co. was incorporated in Washington in 1917 with a capital stock of 1,000,000 shares of 10 cts. par value. The home office is at 104 Holly St., Bellingham, Wash. L. Neber is Pres.; H. Schupp, V. P.; A. C. Muller, Sec.-Treas., and G. H. Cress, Mgr. The company owns 12 claims aggregating 224 acres and developed by a 550-ft. tunnel. The Carson Black Lead Co. of Oakland, Cal., operates the Chedic graphic mine 5 m. S. W. of Carson City.

Biblography B1867 325 | R1868 76 | SMN1866 18-19 | SMN1871-R 116-7 | SMN1873-4 81-3 | MR1903 1122, 1125 Graphite | MR1910 II 909 " | MR1911 II 1092 " | MR1912 II 1067 " | MR1913 II 204 " | MR1915 II 90-1 | MR1916 II 57 | MR1917 II 117 | MR1918 II 240 | MR1919 II 312 | MR1920 II 84 | MR1921 II 9

USGS Carson topographic map. StuartNMR 146, 148.

Thompson & West 537.
WeedMH 1162 Carson Free Gold M. & M, Co. 1300 Panama Canal M. Co.

WASHOE AND EAGLE VALLEY see VOLTAIRE
AMERICAN see SPRING VALLEY

PERSHING COUNTY

ANTELOPE (Cedar)

Silver, Gold, Lead, Copper, Antimony, (Tin, Arsenic)

Location. The Antelope District is located in the neighborhood of Antelope Spring and of Cedar Spring at the N. end of the Trinity Range in N. Pershing Co. Imlay on the S. P. R. R. is 25 m. E. The Rosebud District adjoins the Antelope District on the N. W., and the Farrell District lies to the S. W.

History. The Nevada Superior Ms. Co. began shipping in 1906 and installed a 75-ton concentrating plant in 1911 which operated up to 1913. In 1915, the Antelope Spring M. Co. built a 25-ton concentrator, which was enlarged to a 50-ton flotation plant in 1917. The Antelope Spring Mine shipped some antimony ore in 1916. From 1916 to 1919, the Majuba Hill Group was under option to the Mason Valley Ms. Co. which shipped some 2,000 tons of copper ore containing a little silver and did considerable development work. Tin ore was found on this property in 1917. The claims now belong to the Majuba Silver, Tin & Copper Co., of which G. S. Brown, 315 Nixon Bldg., Reno, is Pres.

Geology. The country rock is Triassic slate intruded by dikes of rhyolite porphyry. The Majuba Hill fissure vein is in tourmalinized rhyolite breccia and rhyolite porphyry, according to Knopf. The copper ore consists of chalcocite, cuprite, chrysocolla, arsenopyrite, tourmaline and some fluorite; and the primary copper mineral was probably chalcopyrite. The tin ore forms an isolated shoot in the same fissure and consists of an intimate intergrowth of cassiterite, tourmaline, and quartz. Arsenopyrite occurs in narrow veins in Triassic slate near Majuba Hill, according to an unpublished report by F. C. Lincoln. A little argentiferour galena is present with the arsenopyrite.

Bibliography. MR1905 269 MR1912 I 796 MR1916 I 481
MR1907 360 MR1913 I 826 MR1917 I 278; 65 Tin
MR1908 490 MR1914 I 684 MR1918 I 240; 26 Tin
MR1910 I 516 MR1915 I 635 MR1919 I 410
MR1911 I 678 MR1920 I 333
Hill507 212.
WeedMH 1132 Antelope Spring M. Co. 1247 Majuba Fresno Silver M. Co. Majuba Silver, Tin & Copper Co. 1287 Nevada Superior Ms. Co.

ANTELOPE SPRINGS see RELIEF
ARABIA see TRINITY

BLACK KNOB

Antimony

Location. The Black Knob District is located in the neighborhood of Black Knob which is at the N. end of the Humboldt Lake Range. The Sacramento District adjoins it on the N., the Willard District on the W., the Relief District on the E., and the Rochester District on the N. E.

History. The Sutherland antimony mine in the Black Knob District was the principal shipper of antimony in Nevada during the World War, and is said to have produced more antimony than any other mine in the United States.

Geology. The antimony veins of the Black Knob District occur in Jurassic calcareous shale a short distance from rhyolite flows, according to a letter from J. T. Reid.

Bibliography. MR1916 I 724 MR1917 I 657.

BUENA VISTA (UNIONVILLE)

Silver, Gold, Lead, Copper, Antimony

Location. The Buena Vista District is located on the E. slope of the Humboldt Range in central Pershing Co. It includes Buena Vista Canyon; where the town of Unionville is located, on the N.; Jackson Canyon, and Cottonwood Canyon on the S. Unionville is about 25 m. by road S. of Mill City which is on the S. P. R. R. The Star District adjoins the Buena Vista on the N., the Indian District on the S., and the Echo District on the W.

History. The Buena Vista District was organized in 1861, and its principal mine, the Arizona, was discovered by Graves and Kelly the following year. Unionville was founded in 1861, and when Humboldt Co. was organized in 1862, it became the county seat, retaining that position till 1873. The Arizona Mine was purchased by Fall & Temple in 1866. The Pioneer 8-stamp mill was the first to be built and several others were erected later. Litigation between the Arizona and the adjoining Silver Mine led to a consolidation in 1870. The property was shut down in 1880; but reopened in 1906 and in 1911 a cyanide plant was erected. In 1919, this mine was purchased by the Arizona Ms. Co. of which E. S. Van Dyck is Pres. and was closely allied with the Universal Silver Co.; of which he was also Pres., passing finally into the hands of the Cons. M. & Refining Co. of which Van Dyck is now Pres. The Black Warrior antimony Mine in Jackson Canyon was worked in 1874 and 1875 and reopened recently, producing considerable antimony from 1915 to 1917.

Production. Ransome estimates the production of the Arizona and Wheeler Mines at about $3,000,000, mostly in silver, produced from approximately 80,000 tons of ore. About 95% of this production was from the Arizona Mine.

Geology. The country rocks of the Buena Vista District are the Koipato and Star Peak formations of Triassic age. The Koipato consists mainly of rhyolitic flows with some andesitic lavas, associated with tuffs, conglomerates, grits, and limestones, according to Ransome. The Star Peak formation which overlies this consists of limestones, quartzites, and slates. The Arizona lode is a bedded vein in Star Peak limestone about 25 ft. above Koipato rhyolite. The deposit is spoon-shaped, being horizontal in places and having dips up to 30 degrees ,and has been disturbed by small normal faults. The gangue consists of banded quartz with rare calcite and the ore minerals are pyrite, galena, sphalerite, tetahedrite, and argentite. A letter from M. E. Pratt adds silver sulphantimonite, and scheelite to this list and states that there are other horizons of mineralization in the limestone. The Black Warrior antimony deposit in Jackson Canyon is a narrow vein in Koipato rhyolite with a filling of quartz and stibnite. At the Manoa Mine in Cottonwood Canyon, Koipato rhyolite is cut by a basalt dike, and the ore occurs in a shear zone along and near this dike. The ore is mainly a replacement of crushed rhyolite by galena, sphalerite, tetrahedrite, and possibly some silver sulphantimonite, Ransome states, in a gangue of barite and quartz.

Mines. The only company active in the Buena Vista District at present is the Cons. M. & Refining Co., which holds 53 claims aggregating 910 acres and including all the old mines of importance. E. S. Van Dyck is Pres. and Gen. Mgr., and M. E. Pratt, Gen. Supt. The main office of the company is at Unionville; business office, Suite D, Cronin Bldg., Sacramento, Cal., and transfer office, Room 614, 79 Milk St., Boston, Mass.

Bibliography. R1868 121-3 | SMN1866 52 | MR1906 295
R1869 187-190 | SMN1867-8 36-7 | MR1907 I 360
R1870 134-7, 141 | SMN1869-70 19-23 | MR1908 I 491
R1871 205-8, 216-7 | SMN1871-2 58-9 | MR1909 I 413-4
R1872 153-5 | SMN1873-4 49-8 | MR1911 I 518
R1873 210-2 | SMN1875-6 64 | MR1911 I 678
R1874 257-260 | SMN1877-8 64-5 | MR1917 I 280
R1875 185-7, 447 | MR1918 I 242
MR1919 I 411
USGE40th 3 (1870) 308-314 | MR1920 I 334
MR1921 I 395

Bancroft 263.
Borzynski, F., "The Arizona Silver Mines Mill," M&SP 122 (1921) 717-8.
Hill507 214.
Ransome414 7, 10-11, 38-41.
StuartNMR 120.
WeedMH 1133 Antimony Syndicate. 1135 Arizona Ex. Silver Ms. Co.
1175 Cons. Silvers, Inc. 1262 Myra Divide M. Co.
1334 Silver Reef Ms. Co. 1364 Unionville M. Co.
1365 Universal Silver Co.

CEDAR see ANTELOPE
CENTRAL see MILL CITY
CHAFEY see SIERRA

COPPER VALLEY (RAGGED TOP)
Tungsten, Copper

Location. The Copper Valley District is located in the neighborhood of Ragged Top Mountain in S. W. Pershing Co. extending into Churchill Co. The district is 10 m. W. of Toulon, a siding on the S. P. R. R.

History. Copper claims were located and patented in this district many years ago. Tungsten was discovered by E. J. Mackedon and others early in 1916, and the claims sold to the Chicago-Nevada Tungsten Co., a subsidiary of H. M. Byllesby & Co., of Chicago. The deposits were rapidly explored and developed, and a mill was erected at Toulon which was operated in 1916 and 1917. During 1918, the mill was operated as a custom mill by the Humboldt County Tungsten Ms. & Mills Co. This mill was recently acquired by the Toulon Arsenic Co. and turned into a white arsenic plant.

Geology. The tungsten ore occurs in masses of limestone which are included in quartz-diorite and cut by sodic aplite and other dikes connected with the quartz-diorite and also by latites that have welled up to form the S. end of the range, according to Hess and Larsen. The ore consists of garnet, calcite, quartz, and scheelite, with a little epidote, chloropal, halloysite, pyrite, and chalcopyrite. The greater part of the sheelite is in particles not over 1-50th of an inch in diameter. The copper deposits lie to the S. and are within the quartz-diorite close to intrusive dikes of latite, according to a letter from J. T. Reid.

Bibliography. MR1916 I 792-3.

Hess, F. L., and Larsen, E. S., "Contact-Metamorphic Tungsten Deposits of the U. S.," USGS B 725D (1921) 289-290.

M&EW, "New Concentrator for the Beeson Tungsten Property, Nevada," 45 (1916) 496.

——— "Caterpillar Haulage at Nevada Mines," 45 (1916) 1112.

WeedMH 1164 Chicago-Nevada Tungsten Co.
1226 Humboldt County Tungsten Ms. & Mills Co.

DUN GLEN see SIERRA

ECHO (RYE PATCH)

Silver, Gold, Copper, Lead, Tungsten

Location. The Echo District is situated on the W. flank of the Humboldt Range in central Pershing Co. It includes Echo Canyon on the N., Panther Canyon, Rye Patch Canyon, Rocky Canyon, and Wright Canyon on the S. The Rye Patch Mine in Rye Patch Canyon lies 4 m. E. of Rye Patch, a station on the S. P. R. R. The Imlay District adjoins the Echo District on the N., the Buena Vista and Indian Districts on the E., and the Sacramento District on the S.

History. The Echo District was organized in 1862 and the Rye Patch Mine located in 1864. An English company purchased the mine in 1869, shipping ore to its mill in Reno that year, and building a 10-stamp mill at Rye Patch station in 1870. In 1873, the mine and mill were taken over by the Rye Patch Mining Co. of San Francisco, but the ore gave out the following year. A little work has been done by lessees since, and from 1919 to 1921, the American M. and Exp. Co. prospected the property without success. The tailings dump is now being cyanided in a 50-ton mill by the Sam O'Connell Mining Syndicate. The La Toska Mine of the Humboldt Cons. Mines Co. in Wrights Canyon produced a little gold-silver ore in 1916 and has had a large amount of exploration work performed upon it. Some tungsten ore was produced in Wrights Canyon in 1918.

Production. The Rye Patch Mine is said to have produced more than $1,000,000 in silver in the early days.

Geology. The Rye Patch Mine is in the black limestone which constiuttes the basal member of the Middle Triassic Star Peak formation near is junction with the underlying Koipato formation. A diabase dike cuts this limestone in the E. part of the mine; while in depth an apophysis of granite has been found in diamond drill holes, and probably comes from the Cretaceous granite which outcrops in Rocky Canyon to the S., according to a report by F. C. Lincoln. The fissure system is complicated and irregular. The filling of the fissures consists of brecciated fragments of the limestone wall rock, quartz, and a minor amount of calcite. The ore minerals occur scattered through the quartz and consist of tetrahedrite, galena, and sphalerite. All the ore minerals are argentiferous, but the tetrahedrite is richest, containing 10% silver by weight. A small amount of gold is present. The La Toska vein has a limestone hanging wall and rhyolite foot wall, according to Weed, and carries free gold and silver minerals. The Wrights Canyon tungsten deposit is in a contact-metamorphosed limestone block in granite. Scheelite is present, and a vein of chloropal was observed, according to Hess and Larsen.

Bibliography. R1869 191-2 SMN1866 49-50 MR1910 I 517
R1870 134 SMN1867-8 37-8 MR1911 I 678
R1871 141, 213, 216 SMN1869-70 18-19 MR1913 I 826
R1872 156-7 SMN1873-4 51 MR1914 I 684
R1873 214 SMN1875-6 67 MR1916 I 481
R1874 264 SMN1877-8 67 MR1917 I 944 Tungsten
MR1918 I 241
MR1919 I 411
USGE40th 3 (1870) 314. MR1920 I 334

Hess, F. L., & Larsen, E. S., "Contact-Metamorphic Tungsten Deposits of the U. S.," USGS B 725D (1921) 295.

Hill507 214. Ransome414. Thompson & West 450.

WeedMH 1131 American M. & Exp. Co. 1226 Humboldt Cons. Ms. Co.
1258 La Toska M. Co. 1323 Rye Patch Ms. Co.

ELDORADO see IMLAY

FARRELL (Stone House)
Gold

Location. The Farrell District is located at Farrell and Stone House. The Seven Troughs District adjoins it on the S., the Antelope District on the N. E., and the Placerites District on the N. W.

History. The district was discovered by L. H. Egbert in 1863, according to information furnished by J. T. Reid, and was known as the Stone House District. In 1908, many locations were made and the district was organized as the Farrell District. The principal mine in the district is the Wild Cat, owned by the P. N. Marker estate of Lovelock. A shipment of rich ore was made from this mine in 1922.

Geology. Gold ore occurs in seams and lenses in Tertiary rhyolite.

Bibliography. StuartNMR123.

FITTING see SPRING VALLEY

GERLACH see WASHOE COUNTY

GOLDBANKS
Mercury, (Gold, Silver)

Location. The Goldbanks District is at Goldbanks in Grass Valley in E. Pershing Co. Winnemucca on the S. P. R. R. and the W. P. R. R. is 40 mi. N.

History. The gold deposits of Goldbanks were discovered in 1907 and many leasers were at work up to 1908, when the boom subsided. The Goldbanks mercury mine is situated 4 m. W. of the Goldbanks Merger Ms. Co. gold property. It was developed in 1913 and a 10-pipe retort furnace built which with a D retort was operated the following year. In 1915 a 50-ton Herreshoff furnace was erected which was operated until 1917 when it was put out of commission by a fire.

Geology. According to Yale, the gold ore is found in quartz-porphyry associated with iron pyrites and contains low silver values. A letter from W. G. Adamson states that the gold lodes are replacement veins along fissures in andesite and that silver ore has recently been found upon the property. The mercury ore consists of small stringers and fine disseminations of cinnabar, forming a large irregular lode in silicified rhyolite, according to Ransome. Adamson's letter says the cinnabar occurs diseminated through a silicified agglomerate composed of quartzite, sandstone, and rhyolite at a rhyolite-quartzite contact.

Bibliography.

MR1907 I 362	MR1914 I 328 Mercury	MR1917 I 416-7 Mercery
MR1908 I 489	MR1915 I 272 Mercury	MR1918 I 164 Mercury
MR1913 I 208 Mercury	MR1916 I 767 Mercury	MR1919 I 172 Mercury

Ransome414 48.

HAYSTACK
Gold

Location. The Haystack District is located at Haystack Peak in N. Pershing Co. on the Humboldt Co. border. Jungo on the W. P. R. R. is 7 m. N.

History. The Haystack property operated its 5-stamp mill in 1915 and produced gold bullion. The Haystack Ms. Co. of which P. Egoscue is Pres. was incorporated in 1914 and installed machinery in 1919.

Geology. Free-milling gold ore occurs in fissure veins in granite and quartzite, according to Mines Handbook.

Bibliography. MR1915 I 635. WeedMH 1222 Haystack Ms. Co., Inc.

HUMBOLDT see IMLAY

IMLAY (Humboldt, Prince Royal, Eldorado)
Silver, Gold, Lead, Copper, Surphur, (Mercury)

Location. The Imlay District is in central Pershing Co. at the N. end of the Humboldt Range on its W. flank. It includes Prince Royal Canyon on the N., Humboldt Canyon where old Humboldt City was located, and Eldorado Canyon on the S. The district is about 6 m. S. of Imlay and 4 m. E. of Humboldt, both of which are on the S. P. R. R. Star Peak to the E. rises to a height of nearly 10,000 ft. The N. section of the district was sometimes known as the Prince Royal District, the central section as the Humboldt District, and the S. section as the Eldorado District. The Imlay District adjoins the Star District on the W. and the Echo District on the N.

History. The Humboldt District was organized in 1860. Humboldt City was founded and had a population of 500 in 1863. The mines did not prove successful and the district was soon deserted. The Imlay Mine became active in 1907, was equipped with a 10-stamp amalgamation and concentrating mill, and continued operations up to 1918. Sulphur was mined near Humboldt in 1869; and a mercury deposit in Eldorado Canyon was explored in 1906.

Geology. At the Imlay Mine, gold and silver values occur in fissure veins and lead and copper are present in the concentrates. The sulphur at Humboldt occurs with gypsum in the cones of extinct hot springs, according to Day. At the Ruby Cinnabar Mine in Eldorado Canyon, cinnabar is irregularly distributed through fractured limestone in a nearly horizontal mass and is associated with pyrite or marcasite, according to Ransome.

Bibliography.

B1866 28	SMN1875-6 66	MR1910 I 517
R1868 124, 125	MR1883-4 865	MR1911 I 678
R1869 192 Sulphur	Sulphur	MR1912 I 797
R1873 214	MR1915 268	MR1914 I 684
R1874 264	MR1906 495 Mercury	MR1915 I 636
SMN1866 48-9	MR1907 I 360	MR1916 I 481;
SMN1869-70 24-5	MR1908 I 490	II 410-1 Sulphur
Sulphur		MR1918 I 241

SMN1871-2 52-5 MR1909 I 412 MR1919 I 410
SMN1875-4 52-3
Hill507 213.
Ransome414 10, 45-6.
Russell, I. C., "Sulphur Deposits in Utah and Nevada," N. Y. Acad. Sci. T 1 (1882) 168-175.
Thompson & West 451, 454.
WeedMH 1131 American M. & Exploration Co.

INDIAN

Silver, Gold, Placer Gold

Location. The Indian District is located at Indian Canyon on the E. flank of the Humboldt Range in central Pershing Co. The Buena Vista District adjoins it on the N., the Spring Valley District on the S., and the Echo District on the W.

History. The district was organized in 1861. The Moonlight Mine was the only mine of importance discovered in the district. This mine produced $100,000 in silver from surface workings. Placer ground was worked in 1875 and 1876.

Bibliography. R1870 137 R1875 187 SMN1873-4 49-50
R1873 212 SMN1866 52 SMN1875-6 63-4
R1874 260-1 SMN1871-2 59

IRON HAT

Lead, Silver, (Copper)

Location. The Iron Hat District is located on the E. slope of the Sonoma Range in E. Pershing Co. It is about 20mi. S. of Valmy on the S. P. R. R.

History. The Iron Hat Corporation has a 2-compartment shaft 200 ft. deep on its property, and shipped several car-loads of lead-silver ore obtained from development in 1910.

Geology. Nearly vertical bedded vains in limestone contain lead, silver, and copper. These veins are from 2 ft. to 20 ft. in width, averaging about 4 ft., according to a letter from J. T. Reid.

Bibliography. MR1910 517.

JERSEY

Silver, Lead, Mercury

Location. The Jersey District is located at Jersey on the Pershing and Lander Co. boundary line. Jersey is 32 m. by road W. of Watts, a station on the N. C. R. R. and 43 m. S.W. of Battle Mountain which is on the S. P. R. R.

History. The Jersey District was discovered by A. S. Trimble in 1874. In 1875, 500 tons of ore were shipped to Oreana to be smelted. A small smelting furnace erected in the district proved unsuccessful for lack of flux. Considerable silver-lead ore was shipped from the Jersey Valley Mine by Abel & Son, from 1880 to 1910. The Jersey Valley Mines Co. made an unsuccessful attempt to concentrate the ore. The Rex Pershing Co. shipped a little silver-lead ore from the Jersey Valley Mine in 1921. In 1918, the Quicksilver Ms. Co. acquired the Ruby mercury mine, erected a retort furnace, and made small productions in 1919 and 1920.

Geology. The country rocks of the district are quartzite and porphyry, according to Whitehill, and the ore in the Jersey vein is argentiferous galena with carbonates of lead.

Bibliography. R1874 261 SMN1873-4 52 MR1918 I 164 Mercury
R1875 188 SMN1875-6 82-3 MR1919 I 172 Mercury
MR1920 I 432 Mercury
MR1921 I 394 Murcery

Thompson & West 474.
WeedMH 1230 Jersey Valley Ms. Co. 1309 Quicksilver Ms. Co.

JUNIPER RANGE
Tungsten, (Copper, Silver, Gold)

Location. The Juniper Range District is in the Juniper Range on the W. part of the border between Pershing and Churchill Cos. It adjoins the Jessup District on the N.W. and is about 20 m. N.W. of Huxley, a station on the S. P. R. R.; and about 40 m. S.W. of Lovelock which is on the same railroad.

Geology. The S. end of the range is composed of quartz-diorite which is instrusive into calcareous shales and limestones which are perhaps of Jurassic age, according to Hess and Larsen, Contact-metamorphism has altered these sedimentary rocks to tactites in which scheelite and copper minerals are present. At the Mines Development Co.'s copper property, veins are said to occur beside porphyry dikes intrusive into monzonite and to contain chalcocite, bornite, and tetrahedrite carrying small silver and gold values.

Bibliography. Hess, F. L., & Larsen, E. S., "Contact-Metamorphic Tungsten Deposits of the U. S.", USGS B 725D (1921) 285-6.
Weed MH 1257 Mines Development Co. of Nevada.

KENNEDY
Gold, Silver

Location. The Kennedy District is located at Kennedy in E. Pershing Co. Kennedy is situated at the E. base of Granite or Cinnabar Mt.. 45 m. by road S.E. of Mill City, which is on the S. P. R. R. and 55 m. S. of Winnemucca, which is on the S. P. R. R. and W. P. R . R. The Kennedy District is S.W. of the Goldbanks District.

History. A rumor that gold had been discovered by a Chinaman in Cinnabar Creek led to an influx of prospectors, and Charles E. Kennedy located the Imperial Mine in 1893. The Gold Note Mine was located at about the same time by the Lawler Bros. Two small mills were erected in the district, and in 1895 the 20-stamp Imperial mill was built and operated for a year. In 1901, this mill was taken over by the Wynn-Lasher Syndicate which added a leaching plant and operated the mill up to 1905. The district has made small and irregular shipments since that time. A mill built in 1914 to try out the Roylance electro-chemical amalgamation process produced a little bullion in 1917.

Production. Klopstock estimates the early production of the camp as follows:

Years	Shipped	Milled
1893-1896	$175,000	$100,000
1901-1905	60,000	150,000

Geology. Granite Mountain is composed of granodiorite which is intrustive into Triassic volcanic rocks which may contain some sedimentary beds, according to Ramsome. At the Gold Note Mine the country rock is rhyolite intruded irregularly by sheets and dikes of basalt, and altered andesite is associated with the rhyolite in some parts of the workings.

The vein dips flatly following the bedding of the rhyolite and in some places traverses basalt. The vein filling is mainly quartz and pyrite with which in the ore occur galena, sphalerite, tetrahedrite, and a little chalcopyrite.

Bibliography. MR1905 269 | MR1909 I 412 | MR1917 I 279
MR1906 295 | MR1910 I 517 | MR1918 I 241
MR1907 I 362 | MR1911 I 678 | MR1918 I 241
MR1918 I 490 | MR1914 I 684 | MR1921 I 394

USGE40th 1 (1878) 83-4, 278-9, 665; 2 (1877) 691.
Hill507 213.
Klopstock, P, "The Kennedy Mining District, Nevada," B AIME June 1913; Abstract, M&EW 39 (1913) 63-5.
Ransome414 52-5.
Roylance, L. St. D., Communication, "New Plant for Kennedy, Nevada," M&EW 41 (1914) 773.

LAKE see CHURCHILL COUNTY

LORING (Willard)
Gold, Silver

Location. The Loring District is situated in Coal Canyon in the Humboldt Range. It is 6 m. E. of Kodak, a station on the S. P. R. R. and 8 m. N. E. of Lovelock on the same railroad. The Sacramento District lies to the N. E. and the Muttleberry District to the S. W.

History. The Loring District was discovered by J. H. Downey in 1915 and was first called Willard. The Sheepherder was the principal mine and lessees built a 30-ton Gibson mill at Kodak to treat its ores. In 1921, lessees shipped gold ore and bullion from the Treasure Hill Mine.

Geology. The country rocks of the district are limestones, rhyolite, and basalt dikes, according to a letter from J. T. Reid. Ore occurs in crevices in limestones at and near the contact with the underlying rhyolite, and mineralization is also present in brecciated rhyolite close to basalt dikes. The formation is disturbed by numerous step faults.

Bibliography. MR1915 I 637-8 | MR1920 I 333 | MR1921 I 394

E&MJ, Special Correspondence, "The New Camp of Willard," 99 (1915) 1088-9.

WeedMH 1232 Jose-DavisM. & M. Co. 1241 Loring Nev. Ms. Co.
1241 Loring Treas. Hill M. Co. 1243 Lovelock Willard M. Co.
1280 Nev. Honey Bee M. Co. 1281 Nevada Loring Ms. Co.
1292 New Willard G. Ms. Co. 1327 Sheepherder Mine.
1367 Victory G. Ms. Co. 1376 Willard Ms Co.

MILL CITY (Central)
Tungsten, Silver, Copper

Location. The Mill City District, formerly known as the Central District, is located in the Eugene Mts. in N. Pershing Co. on the Humboldt Co. border. The tungsten deposits are 7 m. N. W. of Mill City, which is on the S. P. R. R.

History. L. Vary discovered the Central District in 1856, but it was not organized until 1861, at which time the Fifty-Six Mine was located. About 1869, the Golden Age built a 2-stamp mill which operated for a while; and in 1872 a 4-stamp mill and Stetefeldt furnace were erected but this mill was soon shut down by litigation and burned in 1876. Ore from the Central District was worked at the Wise Reduction Plant in Winnemucca in the early days. Considerable ore was also reduced at the Clark Mill on the Humboldt river, according to a letter from J. T. Reid, including ore from the Blackbird Mine which later erected a mill of its own. The district made a small bullion production in 1908, and copper ore was shipped from the Fifty-Six Mine in 1917. A large amount of tungsten ore was shipped from the district in 1917 and 1918. The Nevada Humboldt Tungsten Ms. Co. and the Pacific Tungsten Co. both built mills in the district which went into operation late in 1918 and shut down in 1919 owing to the drop in the price of tungsten.

Geology. The sedimentary rocks at the tungsten deposit are mainly slates which grade into hornfels, with some interbedded quartzite and some comparatively thin layers of limestone, according to Hess and Larsen. These rocks strike N. with the range and for the most part dip steeply into it. They have been intruded by several kinds of porphyritic rocks in small dikes or sills, and later by a large body of granodiorite with associated aplite, pegmatite, and porphyritic dikes. The orebodies are parts of the limestone beds near the granodiorite contact which have been replaced by garnet and epidote, considerable quartz and calcite, and some sulphides, zeolites, and scheelite.

Bibliography R1868 123-4 SMN1869-70 26-7.
R1870 133-4, 141 SMN1871-2 54
R1871 212-3, 216-7 SMN1873-4 52
R1872 156 SMN1875-6 62
R1873 213 MR1908 I 490
R1874 264 MR1917 I 278 Copper; 943-4 Tungsten
R1875 190 MR1918 I 974 Tungsten

Hess, F. S., & Larsen, E. S., "Contact-Metamorphic Tungsten Deposits of the U. S.", USGS B 725D (1921) 295-300.

Thompson & West 450.

WeddMH 1226 Humboldt County Tungsten Ms. & Mills Co.
1255 Mill City Tungsten Co.
1280 Nevada Humboldt Tungsten Ms. Co.
1299 Pacific Tungsten Co.
1316 Richland Ms. Co.
1378 Wyoming-Nevada Co.

MINERAL BASIN

Silver, Antimony, (Iron, Pumice)

Location. The Mineral Basin District is situated in S. Pershing Co. on the Churchill Co. border. Lovelock on the S. P. R. R. is 25 m. N. W. The Relief District adjoins the Mineral Basin District on the N.

History. Rich silver ore was discovered and worked by J. T. Culver in the eighties, according to a letter from J. T. Reid. About 1884, George Senn and associates of San Francisco developed antimony ore and erected a small crucible furnace to treat it but only produced a little antimony. During the World War, Causten and Shelby shipped about 400 tons of antimony ore averaging 35% from this district. Extensive iron deposits in the Mineral Basin District are owned by the Central Pacific R. R. and by J. T. Reid and associates.

Geology. The Culver silver vein is a narrow vein in limestone with a 45-degree dip. Lead and antimony are present with the silver. Pumice of good grade is said by Reid to occur in the district.

MUTTLEBERRY

Gypsum, Silver, Lead, (Copper)

Location. The Muttleberry District is located in Muttleberry Canyon in the Humboldt Lake Range. Lovelock on the S. P. R. R. is 9 m. W. N. W. The Loring District lies to the N. E.

History. Gypsum was shipped from the district from 1891 to 1893; and later was shipped to the mill of the Western Gypsum Co. at Reno; but the quarries have now been idle for many years. About 1910, the Old Tiger Mine shipped silver-lead-antimony ore. Recently this mine was equipped with a Gibson mill and concentrator and shipped a little silver concentrate in 1919; but the mill did not give satisfaction and has since been removed.

Geology. The country rocks of the district are slates, limestones, and quartzites of Triassic and Jurassic age, capped in places by Tertiary volcanic rocks, according to Jones. The gypsum occurs as a bed from 100 ft. to 200 ft. thick in the Triassic limestone, which has been faulted transversely so as to produce a number of outcrops extending for some 3 m. N. of Muttleberry Canyon. The gypsum bed contains many thin bands of limestone. The Old Tiger lode is a flat vein in limestone, according to a letter from J. T. Reid. It contains silver-lead ore with a little gold, and antimony is present in spots. To the S. E. of this property are some copper claims in diorite.

Bibliography.

MR1902 907 Gypsum   MR1905 1108 Gypsum   MR1913 II 367 Gypsum
MR1903 1039 "   MR1910 II 727 "   MR1919 I 410
MR1904 1044 "   MR1912 II 648 "

Jones, J. C., "Nevada" in "Gypsum Deposits of the U. S.", USGS B 697 (1920) 146-9.

Louderback, G. D., "Gypsum Deposits in Nevada," USGS B 223 (1904) 112-117.

NIGHTINGALE

Tungsten

Location. The Nightingale District lies in the Nightingale Range E. of Winnemucca Lake, upon the boundary line between Pershing and Washoe Cos. The nearest railroad is the Susanville Branch of the S. P. R. R., 17 m. to the S. W.

History. Tungsten ore was mined from the Ranson property on the E. side of the range by Joe Bean and associates in 1918 and sent to Toulon for treatment. About 80 tons from the Cowles property on the W. side of the range was milled that same year.

Geology. On the E. flank of the Nightingale Range is a narrow belt of sediments standing on edge with quartz-monzonite forming the valley to the E. and the ridge to the W. A limestone bed 16 ft. thick contains the ore, according to Hess and Larsen. The ore consists of epidote, quartz, laumontite, with smaller quantities of calcite and a little actinolite, but no garnet. The valuable mineral is scheelite which occurs in particles which range from misroscopic dimensions to the size of a walnut. On the W. flank of the range, limestone and shale have been intruded by the quartz-monzonite; and ore occurs which is similar to that on the E. flank except that it carries considerable garnet and sphene, and diopside and albite are also present.

Bibliography. MR1918 I 979.

VSGS Granite Range topographic map.

Hess, F. L., and Larsen, E. S., "Contact-Metamorphic Tungsten Deposits of the U. S.", USGS B 725D (1921) 282-5.

ORO FINO see SIERRA

PLACERITES (Rabbit Hole)

Placer Gold, Copper

Location. The Placerites District is situated at Rabbit Hole on the W. P. R. R., is 15 m. N.E. The Placerites District adjoins the Farrell District on the N. W. and is a short distance S. of the Sulphur District in Humboldt Co. which is also sometimes called the Rabbit Hole District.

History. According to notes furnished by J. T. Reid, the district was discovered in the fifties but was not worked at that time on account of lack of water. Placer operations were carried on in the nineties, first by carting the gravel to a spring, and later by bringing in a ditch from Cow Canyon. Dry washing was attempted a few years ago but met with failure. Considerable work has been done upon a copper mine owned by the Olson Bros. of Salt Lake City.

Geology. The placers occupy positions in narrow ravines at and near the tops of hills. The gravels are shallow and rest upon false bedrock. The copper deposit is said to be a replacement in granite.

PRINCE ROYAL see IMLAY
RABBIT HOLE see PLACERITES
RAGGED TOP see COPPER VALLEY

RELIEF (Antelope Springs)

Silver, Gold, Mercury, Antimony

Location. The Relief District is situated at Relief on the S. slope of Buffalo Peak in S. Pershing Co. It lies to the N. of Antelope Springs, and Lovelock on the S. P. R. R. is 22 m. I. W. The Rochester District adjoins it on the N., the Black Knob District on the W., and the Mineral Basin District on the S.

History. The Relief District was discovered in 1869 by P. Andrews and A. J. Knapp. The principal early mine was the Batavia & Pacific which shipped high-grade silver ore and operated a mill up to 1874. The mine has been worked intermittently since the early days and is now owned by the Rochester Treasure M. Co. Antimony deposits were discovered early in the history of the camp and have been worked to a slight extent. The Antelope Springs mercury deposit lying E. of the silver mine was explored in 1913. Two 12-pipe retort furnaces began treating its mercury ore in 1914 and operated up to 1917.

Production. Estimates as to the production of the old Relief silver mine range from $200,000 to $2,500,000, according to Schrader.

Geology. The Relief silver lode is an argentiferous quartz vein dipping at 55 degrees. It is in limestone belonging to the Triassic Star Peak formation. This limestone is underlain by rhyolite and overlain by andesite, according to Schrader. The valuable constituents in the ore are silver chloride, silver bromide, native silver and probably also argentite. The mercury deposits are residual remnants left on the crests and ridges of hills of soft shale, according to Ransome. The ore consists of cinnabar dispersed through masses of shattered limestone conglomerate or breccia.

Bibliography.

R1871 216	SMN1869-70 17-18	MR1913 I 208 Mercery
R1872 156	Silver & Antimony	MR1914 I 328 "
R1873 212	SMN1871-2 55	MR1915 I 272 "
R1874 261	SMN1873-4 54	MR1916 I 767 "
R1875 187	MR1883-4 643 Antimony	MR1917 I 417 "

Schrader, F. C., "The Rochester Mining District, Nevada," USGS B 580M (1914) 361-4 Mercury, Silver.

WeedMH 1320 Rochester Treasure M. Co.

ROCHESTER

Silver, Gold, Lead, Copper

Location. The Rochester District is located at Rochester and Packard in central Pershing Co. It lies just N. of Coal Canyon in the Humboldt Range. The town of Rochester is 9 m. E. of Oreana, which is on the S. P. R. R. Packard is 2½ m. S. of Rochester. The Spring Valley District adjoins the Rochester District on the N. E. and E., the Sacramento District on the N. W. and W., and the Relief District on the S.

History. The Rochester District was discovered and named in the sixties by prospectors from Rochester, New York. Although known and prospected from that early time, its importance was not suspected until Joseph Nenzel and others relocated the ground in 1911. Large bodies of ore were discovered in 1912, and a mining boom ensued in 1913. From that date to the present, the camp has been one of the important silver producers of Nevada.

PRODUCTION OF THE ROCHESTER DISTRICT
(According to Mineral Resources of the United States, U. S. Geol. Survey.)

Year	Mines	Placers	Tons	Gold $	Silver Ozs.	Copper Lbs.	Lead Lbs.	Total $
1912	5		144	491	6,850			4,704
1913	32	2	16,152	53,816	701,407	135		477,487
1914	24	8	14,499	79,143	621,939	885	7,278	423,478
1915	20	5	26,667	105,407	663,791	422	22,716	443,091
1916	11	5	67,992	81,228	816,620	268	22,046	620,151
1917	19	2	88,371	153,572	799,871	646	14,521	814,091
1918	20	1	95,747	177,295	810,975	4,215	42,530	992,330
1919	14	1	103,662	117,899	667,161	958		865,297
1920	9	1	75,048	130,796	620,047	2,342	460	807,115
1921			87,628	176,144	667,084	351		843,273
Total 1912-1921			575,910	1,075,791	6,375,745	9,871	109,551	$6,291,027

Geology. The country rock of the Rochester District is the Koipato formation of Lower Triassic age. This is a volcanic complex consisting mainly of rhyolite but including quartz-latite, dacite, andesite, altered quartz-prophyry and greenstone, together with some non-volcanic sediments among which limestones predominate with smaller amounts of shales and quartzites. The orebodies occur chiefly in the rhyolite, according to Schrader. There are two lode systems, N. E.-S. W. and N.-S., the former being the most important. The lodes vary in width up to 40 feet and in length up to 400 feet and are quite irregular. The ore consists mainly of antimonial silver and gold-bearing deposits in altered, silicified, and replaced rhyolite with some small fissure-filling quartz veins and stringers. Quartz is the principal gangue mineral.

Mines. **The Rochester Silver Corp.** owns the Rochester Mines, Rochester Merger, and Rochester Elda Fina properties and operates a 160-ton cyanide plant. O. W. Jones is Pres. and F. M. Manson is Chairman of the Executive Board, with home office at Room 310 Reno Nat. Bldg., Reno. The company is incorporated in Nevada with a capital stock of 2,000,000 shares of $1 par value with 550,000 shares in the treasury. The property is now being operated under lease to E. R. Bennett. **The Nevada Packard Ms. Co.** operates a mine and a 125-ton cynide mill at Packard. F. Margrave is the receiver in charge of the property and the post office address is Lower Rochester. The only other property at present active is the **Rochester Lincoln Hill G. & S. M. Co.** which has been carrying on development work with a small force for the past 3 years. W. H. Stott is Sec., with home office at Lower Rochester and John J. Munzer at Pittsburgh is Pres. The company is incorporated in Nevada with a capital stock of 1,000,000 shares at $1 par value.

Bibliography. R1868 125, R1869 192, R1870 137, MR1912 I 798-9, MR1913 I 827, MR1914 I 685-6, MR1915 I 636-7, MR1916 I 481-2, MR1917 I 279, MR1918 I 241, MR1919 I 410-411, MR1920 I 333, MR1921 I 394

USGE40th 2 (1877) 713

USGS Rochester Special Topographic map. ...

Bradley, F. D., "Filter Adjustments at Packard Mill at Rochester, Nev.," E&MJ 106 (1918) 207-9.

Bunker, C. R., "What a Nevada Man Thinks of the Rochester District," M&EW 43 (1915) 431-5.

C&ME, "Cyaniding by Continuous Decantation at Two Nevada Mills," 14 (1916) 437-440.

Cutler, H. C., "Rochester, Nevada's Newest Boom Camp," E&MJ 95 (1913) 355-6.

Damon, A. C., "A Mechanical Disintegrator," E&MJ 101 (1916) 944.

————"The Rochester Cyanide Mill," E&MJ 103 (1917) 133-5.

Emminger, W. G., "Sampling in a Mill," Discussion, M&SP 122 (1921) 152.

Freitag, K., "Nevada Packard Reduction Plant of Rochester, Nev.," M&EW 43 (1915)) 847-8.

Jones, J. C., "Geology of Rochester, Nevada," M&SP 106 (1913) 737-8.

M&EW, "Mill of the Nevada Packard Mines Company," 45 (1916) 707.

M&SP, Editorial 106 (1913) 134.

————"Rochester Mining District, Nevada," 106 (1913) 994.

Schrader, F. C., "The Rochester Mining District, Nevada," USGS B 580 1914.

————"Ore Deposits of the Rochester District, Nevada," (Abstract), Wash. Acad. Sci. J 6 (1916) 518-9.

Thompson, H. G., "Construction and Operation of Nevada Packard Mill," M&SP 113 (1916) 377-384.

Todd, R. B., "The Nevada Packard Mill," E&MJ 101 (1916) 247-8.

WeedMH 1137 Associated Ms. Dev. Co. 1205 General Ms. Co. of Nev. 1240 Lincoln Hill M. & M. Co. 1264 Nenzel Crown Point M. Co. 1282 Nevada Packard Ms. Co. 1299 Packard Ex. Ms. Co. 1300 Packard S. Ms. Co. 1314 Reorg. United Ms. Co. 1316 Rochester Buck & Charley Ms. Co. 1317 Rochester Home Trail Ms. Co.

Rochester Lincoln Cons. Co.	Rochester Lincoln Hill G. & S. M. Co.
1318 Rochester Raven Ms. Co.	Rochester Silver CoCrp.

Wood, G. W., "The Rochester Mill, Nevada," M&SP 111 (1915) 317-320.

ROSEBUD

Silver, Gold, Copper, Lead, Placer Gold

Location. The Rosebud District is located at Rosebud in the Kamma Mt. in N. Pershing Co. on the Humboldt Co. border. Rosebud is 10 m. S. E. of Sulphur, which is on the W. P. R. R. The Sulphur District adjoins the Rosebud District on the N. W., and the Antelope District adjoins it on the S. E.

History. Silver ore was discovered near Rosebud in 1906 and a short-lived boom ensued. Production began in 1908 and has continued on a small scale down to the present time. Placers were first worked in 1911 and have added slightly to the production of the district.

Production. From 1908 to 1920, the Rosebud District produced $10,428 in gold, 70,479 ounces of silver, 18,722 lbs. of copper, and 4,633 lbs. of lead, valued in all at $59,473, according to Mineral Resources of the U. S. Geol. Survey.

Geology. The country rock is Tertiary rhyolite. The ore deposits are mineralized zones in the rhyolite containing a little pyrite disseminated through the rhyolite and a few small irregular stringers of quartz. The valuable constituent of the ore is argentite, according to Ransome.

Bibliography.

MR1906 295	MR1911 I 680	MR1917 I 279
MR1907 I 362	MR1912 I 799	MR1918 I 241
MR1908 I 490	MR1913 I 827	MR1919 I 411
MR1909 I 413	MR1914 I 686	MR1920 I 333
MR1910 I 518	MR1915 I 637	MR1921 I 394
	MR1916 I 482	

Hill507 214. Ransome414 25-6.

RYE PATCH see ECHO

SACRAMENTO

Silver, Gold, (Placer Gold)

Location. The Sacramento District is located on the W. flank of the Humboldt Range in central Pershing Co. It includes Horse, Pole, Sacramento, and Limerick Canyons and adjoins the Spring Valley District on the N.W. the Rochester District on the W., and the Echo District on the S.

History. Many claims were located in this district in the early days. The Humboldt Queen shipped a large tonnage of silver ore to the Alex Wise Mill at Winnemucca, and after that mill was destroyed by fire to a mill at Mill City. A shipment was made from this mine by the Lone Mt. M. Co. in 1907. Placer gold has recently been found in Limerick Canyon, and preparations are being made to mine it.

Geology. The country rock at the Humboldt Queen Mine is thin-bedded Star Peak limestone of Triassic age, according to Ransome. The limestone beds have been sharply and complexly folded and in some cases overturned. The veins, or vein, follow the bedding of the limestone and as they are near the surface, the same vein outcrops several times as a result of erosion cutting across its folds. The vein filling is milk-white quartz banded by thin dark seams of argillaceous material, and containing calcite, pyrite, galena, sphalerite, and tetrahedrite. At Pole Canyon, quartz occurs in fissure veins in the Triassic Koipato rhyolite, according to Schrader, and has replaced the rhyolite for considerable distances from the fissures. The ore minerals are native gold alloyed with silver, ruby silver, and cerussite; with associated pyrite, arsenopyrite, hematite, and sphalerite. Considerable sericite and a little fluorite are present in places.

Bibliography. R1868 125 R1870 137 SMN1866 50
R1869 192 R1873 212 MR1907 I 359
Ransome414 33-5
Schrader, F. C., "The Rochester Mining District, Nevada," USGS B 580M (1914) 364-7.
Thompson & West 452.

SAN JACINTO

Silver, Lead, (Arsenic)

Location. The San Jacinto District is located on the E. flank of the Trinity Range in central Pershing Co. It lies between the Antelope District on the N. and the Trinity District on the S. and is 9 m. N.W. of Rye Patch, a station on the S. P. R. R.

History. A San Francisco company shipped silver-lead ore from this district in 1876. Arsenic ore has recently been discovered in the district.

Geology. The country rock is slate and granite, according to Whitehill. Argentiferous galena and carbonate of lead occur in bunches in veins from 2 ft. to 6 ft. in width.

Bibliography. SMN1875-6 66-7.

SANTA CLARA see STAR

SEVEN TROUGHS

Gold, Silver, Copper, Lead

Location. The Seven Troughs District is located at Seven Troughs on the E. slope of Seven Troughs Mt. Lovelock on the S. P. R. R. is 32 m. S. E. The Farrell District adjoins the Seven Troughs District on the N.

History. The district was discovered in 1905 but did not attract outside attention until 1907. In 1908, the Mazuma Hills and Kindergarten Mines each erected 10-stamp mills. In 1909, the Darby Mill at Mazuma began treating custom ores and continued its operations until dismantled in 1918. The Seven Troughs M. Co. formed in 1911 included the Kindergarten property; and a 50-ton cyanide plant was added to the Kindergarten mill by the new company. Since 1918, most of the work in the camp has been done by leasers.

Production. From 1908 to 1921, the district produced 76,090 tons, containing $2,300,837 in gold, 551,317 ozs. silver, 562 lbs. copper and 757 lbs. lead, valued in all at $2,606,912 according to Mineral Resources of the U. S. Geol. Survey.

Geology. The country rocks of the Seven Troughs District are Tertiary volcanic rocks, according to Ransome, and the principal veins follow basalt dikes which cut these rocks. The vein filling consists of sugary quartz and shattered country rock, and the valuable constituent is native gold with a considerable proportion of silver, which is visible as clusters or irregular particles in most of the rich ore. Pyrite, stibnite, and a little chalcopyrite, native silver, and proustite have been observed in the ore.

Bibliography. MR1906 295 MR1911 I 680-1 MR1917 I 279
MR1907 I 360-2 MR1912 I 799 MR1918 I 241
MR1908 I 490-1 MR1913 I 827-8 MR1919 I 411
MR1909 I 413 MR1914 I 686-7 MR1920 I 334

MR1910 I 518 MR1915 I 637 MR1921 I 394-5
MR1916 I 482

Adams, W. S., "Cyanide Practice at the Darby, Nevada, Ore Reduction Co.," Mex. Min. Jour. 11 (1910) 75.

Crehore, L. W., "Modern Cyanide Mill at Mazuma, Nevada," M&EW Sept. 11, 1909.

Hill507 214. Ransome414 14-25. StuartNMR 7, 121-2.

WeedMH. 1183 Delaware M. Co. 1308 Province Ex. Gold M. Co. 1326 Seven Troughs Reorg. Ms. Co.

Young, G. J., "Co-operation Among Small Mines," E&MJ 106 (1918) 813.

SIERRA, (SUNSHINE, DUN GLEN, CHAFEY, ORO FINO)

Placer Gold, Gold, Silver, Lead, Copper, (Graphite)

Location. The Sierra District lies E. of Dun Glen in the N. part of the East Range in N.E. Pershing Co. It extends from Dun Glen Peak at the N., across Natchez Pass to Oro Fino Canyon and Rock Hill Canyon on the S. Dun Glen is 10 m. N.E. of Mill City which is on the S. P. R. R.

History. The Sierra and the Oro Fino Districts were both organized in 1863, and in 1869, the Ore Fino District was annexed to the Sierra. Dun Glen, settled in 1862, became the business center of the district. Early in the history of the camp, the Auld Lang Syne Mill was built, and a little later the Essex 10-stamp mill was erected. In 1871, a Paul process dry amalgamation mill was put up on the site of the Auld Lang Syne Mill, but did not prove a success. It was given repeated trials up to 1875 when the mill and mine shut down. The Thacker and Goodrich 10-stamp mill was erected in 1875. Chinese placer miners produced some $2,000,000 from Auburn and Barber Canyons near Chafey in the seventies, eighties, and nineties, according to a communication from J. T. Reid, while in the eighties and nineties they produced $2,000,000 or more from Rock Hill Canyon. The district gradually declined and was practically abandoned when E. S. Chafey began work there in 1908. Dun Glen was renamed Chafey, the Mayflower Mine was opened, and a 30-ton mill set to work on its ore. This led to a general revival, and the district has continued to make small productions up to the present day.

Production. The recent production of the district, from 1908 to 1921, has been 13,568 tons containing $279,082 in gold, 46,886 ozs. silver, 10,283 lbs. copper, and 59,481 lbs. lead, valued in all at $314,441, according to Mineral Resources of the U. S. Geol. Survey.

Geology. The Mayflower lode is a banded quartz vein accompanying a diabase dike which cuts obliquely across a series of volcanic flows and associated slates, while at one place the foot wall is limestone. The ore minerals are galena, pyrite, sphalerite, and native gold containing a good deal of silver and sometimes partly embedded in the galena. The Auld Lang Syne lode consists of a number of paralled quartz veins and stringers in silicified andesite near a diabase dike, according to Ransome, and pyrite and arsenopyrite are present in the ore. Mines Handbook states that the Sunshine deposit consists of fissure veins in limestone containing quartz carrying argentiferous lead carbonate.

Mines. The only company at present operating in the district is the Sunshine Ms. Co. of which O. W. Jones of Chicago is Pres. and N. Adamson is Supt. at Winnemucca.

Bibliography.

B1867 332	SMN1866 53-4	MR1911 I 678
R1868 125-6	SMN1867-8 39-40	MR1912 I 797
R1869 192	SMN1869-70 25-6	MR1913 I 481
R1870 140	SMN1871-2 55-7	MR1914 I 684
R1871 213	SMN1873-4 51	MR1915 I 635
R1872 155-6	MR1883-4 916	MR1916 I 481
R1873 213	Graphite	MR1919 I 411
R1874 263-4	MR1919 I 411	MR1920 I 334
R1875 190	MR1910 I 516-7	MR1921 I 395

USGE40th 3 (1870) 316

Hill507 212.
M&SP, "Chafey, Nevada," 97 (1908) 625-6.
Ransome414 11, 49-52, 70.
Thompson & West 451-2.
WeedMH 1156 Buena Vista Del Oro M. & M. Co.
1204 Gem Five Ms. Co. 1328 Sierra Silver Ms. Corp.
1347 Sunshine Ms. Co.

SPRING VALLEY (FITTING, AMERICAN)

Placer Gold, Gold, Silver, (Mercury)

Location. The Spring Valley District is located on the E. flank of the Humboldt Range in central Pershing Co. It includes Spring Valley Canyon in which Fitting is situated on the N., Dry Gulch, American Canyon and South American Canyon on the S. It lies S. of the Indian District and E. of the Sacramento and Rochester Districts.

History. The Eagle Mine, later known as the Bonanza King, is situated in Spring Valley Canyon and is the only important lode mine in the district. It was located in 1868, relocated in 1871, and sold to the Oakland Mill & M. Co. in 1873. This company erected a 15-stamp mill in 1874 and failed in 1875. Placer mining operations in American Canyon began about 1881 and continued till 1895. The mines were first operated by Americans and later by Chinese and are credited with a production of $10,000,000. The discovery of cinnabar in the placer workings led to prospecting for mercury lodes; and several deposits were discovered and explored in 1906. The Bonanza King Mine was operated by the Bonanza King M. Co. and lessees from 1905 to 1908 and made small productions. In 1911 the Federal Ms. Co. of Chicago built a 2,000 cu. yd. dredge in Spring Valley Canyon which operated up to 1914. Small placer productions have been made since.

Geology. The Bonanza King vein is in a dioritic dike which cuts porphyritic rhyolite belonging to the Triassic Koipato formation, according to Ransome. The vein filling is banded quartz at times slightly stained by copper carbonates. The common ore minerals are galena, pyrite, and sphalerite, while tetrahedrite occurs in the rich ore. At the Dixie mercury mine in American Canyon, cinnabar occurs in specks in a kaolinized zone in porphyritic rhyolite; while at the Nevada-Almaden, cinnabar occurs in small irregular fissures forming a vein in limestone with a hanging wall of eruptive rock. The placer gravels of American Canyon consist of subangular and imperfectly assorted wash from the adjacent hills and the gold is found at the bottom of gravel layers underlain by seams of clay. In Spring Valley Canyon the dredge worked placer ground from 20 ft. to 30 ft. thick containing many boulders and having a hard limestone bed rock, according to Walker.

Bibliography.

R1873 212	SMN1875-6 63	MR1910 I 518
R1874 260	MR1905 269	MR1911 I 681
R1875 187	MR1906 495 Mercury	MR1913 I 828
SMN1866 53	MR1907 I 359	MR1914 I 687
SMN1871-2 59	MR1908 I 489	MR1916 I 482
SMN1873-4 50	MR1909 I 411-2	
	554 Mercury	

Hill507 212. Ransome414 12, 35-8.

Schrader, F. C., "The Rochester Mining District, Nevada," USGS B 580M (1914) 360-1 Mercury, 638-371 Placer Gold.

Walker, H. G., "First Gold Dredge in Nevada," E&MJ 91 (1911) 1210-I.

STAR (SANTA CLARA)

Silver, Antimony

Location. The Star District is located at Star City in Star Canyon in the N. part of the Humboldt Range on its E. flank. Star City is 10 m. S. of Mill City, which is on the S. P. R. R. Star Peak to the W. is the highest peak in the range, reaching a height of nearly 10,000 ft. The N. section of the district in Santa Clara Canyon was sometimes called the Santa Clara District. The Buena Vista District adjoins the Star District on the S.; and the Imlay District adjoins it on the W.

History. The Star District was organized in 1861. Star City was founded and reached a population of 1,200 in 1864, although now abandoned. The growth of Star City was due to the discoverey of a bonanza close to the surface on the Sheba Mine. Several unsuccessful attempts to rejuvenate the Sheba Mine have been made within recent years. Antimony has been produced from a mine in Bloody Canyon, a few miles S. of Star Canyon, in desultory fashion for many years. In 1917, the National Antimony Co. mined some antimony ore from its property in the district and in 1918 erected an antimony oxide plant at Mill City.

Production. The Sheba Mine is said to have produced $5,000,000 in silver in the early days.

Geology. The Triassic rocks at the Sheba Mine consist of thin-bedded limestones and tuffaceous sandstones interbedded with thin flows of rhyolite and related rocks, according to Ransome. The Sheba bonanza consisted of lenses of ore following bedding planes in the limestone or between the limestone and the associated sandstones and volcanic rocks. In many cases, these seams were connected by a network of veinlets across limestone beds; and fissures were as a rule confined to the limestone. The ore consists of white quartz carrying argentiferous jamesonite, galena, sphalerite, pyrite, and tetrahedrite. A mile down Star Canyon from the Sheba Mine is a quartz vein in limestone which carries stibnite, and another stibnite deposit occurs in Bloody Canyon.

Bibliography.

B1866 33, 114, 117	R1874 264-5	SMN1877-8 65
B1867 420	R1875 191	MR1905 269
R1868 126-9, 132-3	SMN1866 50-51	MR1906 295
R1869 192-3	SMN186--8 38-9	MR1907 I 360
R1870 134, 141	SMN1869-70 23-5	MR1908 I 489, 490
R1871 141, 208-212	SMN1871-2 57	MR1915 635-6
R1872 155	SMN1873-4 50	MR1917 I 278, 657
R1873 214-6	SMN18756 63	Antimony
		MR1919 I 411

USGE40th 3 (1870) 314-6.

Hill507 214
Ransome414, 7, 10-11, 41-3.
StuartNMR 120-1.
Thompson & West 452-3, 458.
WeedMH. 1240 Little Giant G. & S. M. Co., 1284 Nevada Sheba S. M. Co.
1133 Antimony Syndicate 1304 Pioneer Exploration Co.
1327 Sheba G. & S. M. Co.

STONE HOUSE see FARRELL
SUNSHINE see SIERRA
TABLE MOUNTAIN see CHURCHILL COUNTY

TRINITY (ARABIA)

Silver, Lead, Antimony, Gold, Copper, (Tungsten)

Location. The Trinity District is located on the E. flank of the Trinity Range and includes Trinity Canyon on the S.W., Black Canyon, and the Arabia section on the N.E. The Arabia section in the neighborhood of the Montezuma Mine is sometimes considered a separate district. Oreana which is on the S. P. R. R. is 5 m. E.

History. The Trinity District was discovered by George Lovelock in 1859. A small mill was built at the Evening Star Mine which operated for a number of years and was then moved to Oreana. The district was organized in 1863. A smelter was built at Oreana to treat the ores of the Montezuma Mine in 1867. This was the first smelter in the United States to ship lead into the general market; although the little Mormon furnace at Las Vegas and the smelter at Argenta, Montana, produced lead for local consumption before this date. Torrey operated a 5-stamp mill on the Humboldt River upon ores from this district in 1873. The Evening Star was reopened and made shipments of gold-silver ore in 1914 and 1915. Slag was shipped from the Montezuma dump in 1916, and ore shipments have been made from the mine since that date; while several other properties in the district have also been active recently.

Production. According to Thompson & West, the early production was 30,000 tons which paid from $30 to $700 per ton. The production of the Montezuma and Jersey Mines is said to have been about $1,000,000 gross.

Geology. The country rock of the Trinity District consists of altered granodiorite through which are scattered irregular blocks of hornfels derived from sediments by contact-metamorphism. The sedimentary rocks may be Jurassic slates and the granodiorite of Cretaceous age. The ore deposits are fissure veins in the granodiorite and hornfels which are well-defined in the former rock but feather out in the latter, according to Knopf. At the Montezuma Mine, the principal ore mineral is argentiferous bindheimite, a hydrous antimonate of lead which sometimes fills the vein and at other times is associated with milk white quartz. Residual jamesonite is occasionally found with the bindheimite. Other minerals present are cerussite, black manganiferous oxide of antimony, gypsum, and arsenopyrite in part oxidized to scorodite, plumbojarosite, and jasper. According to Hess & Larsen, scheelite is said to occur with garnet and epidote in contact-altered sediments at Black Canyon.

Mines. The only property at present active is that of the Pershing County Ms. Co. This company was incorporated in Nevada in 1920 with a capital stock of 100,000 shares of 25 cts, par value. The office of the company is at Oreana. G. D. Cook is Pres.-Mgr.; R. M. Hardy, Sec.;

and F. M. Cook, Treas. The company owns 9 claims including the old Montezuma and Jersey Mines developed by a 120-ft incline and 1,000 ft of workings and equipped with 2 small hoists.

Bibliography R1868 129-132 SMN1867-8 33-5 MR1914 I 687
R1870 137-9 SMN1869-70 16-17 MR1915 I 637
R1873 212 SMN1871-2 50-51 MR1918 I 242
R1874 261 SMN1875-6 67 MR1919 I 410
SMN1866 55 SMN1877-8 66 MR1920 I 333
MR1921 I 393
USGE40th 3 (1870) 297-308
Hess, F. L., & Larsen, E. S., "Contact-Metamorphic Tungsten Deposits of the U. S.," USGS B 725D (1921) 292.
Knopf, A., "The Antimonial Silver-Lead Veins of the Arabia District, Nevada," USGS B 660 (1918) 240-255.
M&SP, "Antimony in Humboldt County, Nevada," 116 (1918) 127.
StuartNMR 120. Thompson & West 453.
WeedMH 1226 Humboldt Trinity G. M. & M. Co.
1302 Pershing County Ms. Co.

UNIONVILLE see BUENA VISTA

VELVET

Gold, Diatomaceous Earth, Opal

Location. The Velvet District is located on the E. flank of the Trinity Range in S. Pershing Co. It lies to the S.W. of the Trinity District, and is 10 m. W. of Lovelock, which is on the S. P. R. R.

Geology. At the Velvet Mine, gold occurs in a vein in Tertiary eruptives in much the same way as in the Seven Troughs District, according to information furnished by J. T. Reid. Some good fire opals have been found in the volcanic rocks of this district. Diatomaceous earth of good quality also occurs, and some is being shipped from the E. part of the district at the present time.

WILD HORSE

Antimony, (Lead, Silver, Arsenic, Pyrite, Copper.)

Location. The Wild Horse District is located S. of Lovelock on the E. side of the Humboldt Lake Range in S. Pershing Co. on the Churchill Co. border.

Geology. Antimony deposits in this district were worked during the World War, according to information furnished by J. T. Reid. Silver and lead occur with the antimony in places, and there is some arsenopyrite here and there. Massive pyrite occurs in limestone at the head of Lovelock Canyon; and in Dutchman Canyon silver-copper ore is found in Triassic slates and limestone near porphyry intrusions.

WILLARD see LORING

STOREY COUNTY

CHALK HILLS (PARKER AND NOE)

Diatomaceous Earth

Location. The Chalk Hills District, sometimes called the Parker and Noe District, is located in central Storey Co. adjoining the Comstock District on the N.E. It is 9 m. N.E. of Virginia City at the head of Long Valley in a fairly level basin surrounded by hills.

Geology. The hills are composed of basalt, as is also the floor of the basin, except where cut by the wash. Beneath this basalt lies a series of thick beds of rhyolitic and andesitic tuff which dips northerly at low angels. Interbedded and folded with these tuffs is a bed of diaomaceous earth from 30 ft. to 40 ft. thick The diatomaceous earth is fine-grained and white with brown stains on the joints. The first grade material is 3 ft. in thickness.

Mines. The property belongs to the Electro-Silicon Co. of 30 Cliff St., New York City. The mine was formerly worked by means of open cuts but is now operated through a tunnel. The first grade material is mined separately, some brown strains being removed from it with a hatchet before sacking for shipment. By the liberal use of the hatchet, some 40 per cent of good diatomaceous earth may be obtained from the poorer grade material. Several carloads are shipped from this property to New York annually, where the earth is used in the manufacture of electro-silicon.

USGS Carson topographic map.

Bibliography.

Hill, J. M., "Notes on the Economic Geology of the Ramsey, Talapoosa, and White Horse Mining Districts, in Lyon and Washoe Counties, Nevada," USGS B 470, (1911), 108.

COMSTOCK (Washoe, Virginia City, Gold Hill, Silver Star, Flowery)
Silver, Gold, Lead, Copper, (Stone)

Location. The Comstock District is in Storey Co. at Virginia City and Gold Hill, two towns with stations on the V. & T. R. R. In the early days the Comstock District was known as the Washoe District. It is sometimes subdivided into the Virginia City District on the N. and the Gold Hill District on the S.; the E. portion of the district on and near the Brunswick Lode is sometimes called the Silver Star District; and the extreme E. part situated in the Flowery Range is known as the Flowery District. S. of the Comstock District in Lyon Co. is the Silver City District, and to the W. in Washoe Co. lies the Jumbo District, while to the N. E. in Storey Co. the Chalk Hills diaomaceous earth district is situated. Mt. Davidson, the highest peak in the district, has an altitude of 7,870 ft.; and the Comstock Lode outcrops on its E. flank at elevations in the neighborhood of 6,550 ft., the towns of Virginia City and Gold Hill being a few hundred feet lower. Flowery Peak in the E. part of the district is 6,650 ft. high.

Discovery. The discovery of the Comstock Lode cannot be attributed to any one individual nor given any exact date, for it was the outcome of the activities of a large number of people extending over a long period of time. Abner Blackburn, a Mormon immigrant belonging to the Beatie party which erected the first house in Nevada at Genoa in 1849, discovered placer gold at the mouth of Gold Canyon near the present site of Dayton in July, 1849. The discovery was not rich enough to occasion excitement, but Blackburn informed other Mormons of it, and the little placer mining camp of Johntown grew up in Gold Canyon. The Grosh brothers found rich silver ore in the region in 1853, and it is quite possible that they located the outcrop of the Comstock Lode on Gold Hill at the head of Gold Canyon, but both brothers met with hard luck and died before they could take advantage of their find. James Fennimore, familiarly known as "Old Virginia," and who gave Virginia

City its name, together with several other placer miners from Johntown, located the Gold Hill croppings of the Comstock Lode as placer ground on January 29, 1859. They worked these croppings as a placer without realizing their true character. Henry T. P. Comstock, for whom the district is named, claimed to have been one of the original locators of Gold Hill, but his associates said that he joined them later. In June, 1859, two Irish miners named Peter O'Riley and Patrick McLaughlin, were working at the head of Six-Mile Canyon to the N. of Gold Hill. In digging a water hole, they uncovered the top of the Ophir bonanza, located it, and washed gold from it. Henry Comstock happened along that evening, and by putting up a bluff secured a place for himself and a friend on the location notice. The rich silver sulphide which occurred with the gold, both at Ophir and Gold Canyon was not recognized at first, but was looked upon as a nuisance because it interfered with the recovery of the gold. In July, 1859, Augustus Harrison carried a piece of this heavy black mineral as a curiosity to Judge James Walsh at Grass Valley, where Judge Walsh had it assayed by Melville Atwood and its real nature was discovered.

Surface Bonanzas. Judge Walsh and a friend left for the Comstock the very night that the famous assay was made, thus inaugurating the "great Washoe rush." Hundreds crossed the mountains to the Comstock that year, late as it was, while thousands followed the next year. Had it not been for the Washoe boom, there would not have been a sufficient population in Nevada in 1864 to have warranted making it a state, the abolition amendment could not have been passed, and the Civil War would have been greatly protracted. The soft wide ore in the Ophir bonanza could not be mined by the ordinary methods, and in 1860, Philip Deidesheimer invented and introduced the square-set system which has since come into world-wide use for similar ore-bodies. Almarin B. Paul experimented with the extraction of silver from the Comstock ores and in 1860 built a mill near what is now Silver City for the treatment of these ores by a modification of the old Mexican patio process which came to be known as the Washoe process. Good roads were constructed across the Sierras, and Virginia City and Gold Hill grew rapidly. In 1867, William Sharon, representative of the Bank of California, organized the Union Milling and Mining Co., controlled by what was known as the "Bank of California Ring," which exercised a predominating influence in Comstock affairs for many years to come. This company completed the V. & T. R. R. from Carson City to Gold Hill in 1869. In that same year, Adolph Sutro began work on the Sutro Tunnel, 4 m. from the Comstock lode and designed to drain it at a depth of over 1,600 ft., a work which he eventually brought to a successful conclusion despite the powerful opposition of the Bank of California Ring. By this time, the orebodies first discovered had begun to play out, enormous sums had been spent upon mining litigation, and a period of depression had set in. The bullion production which had risen to $16,000,000 in 1864 sank to less than half that amount in 1869.

Deep Bonanzas. Interest in the Comstock revived with the discovery of the Crown Point-Belcher bonanza at a depth of 900 ft. in 1871, by John P. Jones, superintendent of the Crown Point Mine. Alvinza Hayward, a member of the Bank of California Ring but working for himself in this instance, obtained advance information of this bonanza and he and Jones reaped immense fortunes as Crown Point stock rose from $2 to $1,825 per share. This bonanza produced about $60,000,000. At about this time, John W. Mackay and James G. Fair, who were associated in the Hale and Norcross Mine, decided to open the unproductive ground between the Ophir and Gould and Curry, both of which mines had contained surface bonanzas. They secured financial assistance from O'Brien and Flood, and in 1873 discovered another deep-lying bonanza, the largest ever found on the Comstock. This was known as the "Big Bonanza," was located in the Con. Virginia and California Mines, and produced about $105,000,000. Mackay, Fair, O'Brien, and Flood through their fortunate connection with the rich find were known as the "Four Bonanza Kings." The same year that the Big Bonanza was discovered saw the completion of a pipe-line bringing fresh water to the Comstock from the Sierras. This water was brought from a distance of 25 miles and passed through an inverted siphon designed by Henry Schuessler to resist the enormous head of 1720 feet, nearly twice as great as had ever been withstood before. The mines were extremely hot in depth and produced great quantities of hot, acid water, so that mining was rendered extremely difficult. In 1878, the Sutro Tunnel made connection with the Comstock Lode and immediately became of great value for drainage purposes, although many of the mines had already been sunk to greater depths and were forced to pump to the tunnel level. Great ingenuity in solving mechanical problems was displayed by the men who were called upon to design the hoists for deep mining and the pumps to handle the large volumes of hot water. The bullion production rose from less than eight million dollars in 1869 to over thirty-six million dollars in 1877 when the Big Bonanza was at its height, falling rapidly to a little more than a million dollars in 1881, when it had become exhausted.

Low-Grade Ores. After the exhaustion of the Big Bonanza, the low-grade ores left in the mines were worked systematically and the production rose to a maximum of over $1,000,000 in 1888, declining to less than $200,000 in 1899. An immense flow of hot water was struck in 1882 which drowned out the Gold Hill mines below the Sutro Tunnel level. The Virginia City mines continued pumping and a depth of 3,300 ft. was reached in the Mexican winze in 1884, but in 1886 these mines also suspended work below the Sutro Tunnel level. The Comstock Pumping Association was formed in 1898 by 28 companies in the Virginia City section and pumping resumed in 1899. Considerable ore was mined, some from as deep as 2,900 ft, before pumping was again discontinued in 1922. The bullion production during this period was irregular, ranging from under four hundred thousand dollars to under one million four hundred thousand dollars. In 1920, through the efforts of Alex Wise who had been financed by Herbert G. Humphrey, Bulkeley Wells became interested in the low-grade ores of the Gold Hill mines. Wells took over these mines and organized the United Comstock Ms. Co., a subsidiary of the Metals Exploration Co., to operate them. Gen. Wells resigned in 1923 and Roscoe H. Channing became Pres. of the Metals Exploration Co.. In 1922, Wise and Humphrey interested Thomas F. Cole and associates in the Middle Mines Group, and the Comstock Merger Mines, Inc., was formed to operate these and other groups in the Comstock District. The North End mines are also considering the mining and milling of their low-grade ores on a large scale.

Production. The recorded production of the Comstock District is given in the following table. It is probable that there was a large unrecorded production in the early days, but the amount of this cannot be even roughly estimated.

PRODUCTION OF THE COMSTOCK DISTRICT

Year		Tons	Gold $	Silver $	Total $
1859	(1)		30,000.00		30,000.00
1860	(1)	10,000	550,000.00	200,000.00	750,000.00
1861	(1)	140,000	2,500,000.00	1,000,000.00	3,500,000.00
1862	(1)	250,000	4,650,000.00	2,350,000.00	7,000,000.00
1863	(1)	450,000	4,940,000.00	7,460,000.00	12,400,000.00
1864	(1)	680,450	6,400,000.00	9,600,000.00	16,000,000.00
1865	(1)	430,745	6,133,488.00	9,700,232.00	15,833,720.00
1866	(1)	640,282	5,963,158.00	8,944,737.00	14,907,895.00
1867	(1)	462,176	5,495,443 20	8,243,164.80	13,738,608.00
1868	(1)	300,560	3,391,907.60	5,087,861.40	8,479,769.00
1869	(1)	279,584	2,962,231.20	4,443,346.80	7,405,578.00
1870	(1)	238,967	3,481,730.16	5,222,595.24	8,704,325.40
1871	(1)	409,718	4,099,811.46	6,149,717.19	10,249,528.65
1872	(1)	384,668	4,894,559.86	7,341,839.79	12,236,399.65
1873	(1)	448,301	8,668,793.40	13,003,187.13	21,671,980.53
1874	(1)	526.743	8,990,714.06	13,486,071.09	22,476,785.15
1875	(1)	546,425	10,330,208.62	15,495,312 92	25,825,521.54
1876	(1)	598,818	12,647,464.08	18,971,196.12	31,618,660.20
1877	(1)	562,519	14,520,614.68	21,780,922.02	36,301,536.70
1878	(1)	272,909	7,864,557.64	11,796,836.47	19,661,394.11
1879	(1)	178,276	2,801,394.33	4,202,091.49	7,003,485.82
1880	(1)	172,399	2,051,606.00	3,077,409.00	5,129,015.00
1881	(1)	76,049	430,248.00	645,372.00	1,075,620.00
1882	(1)	90,181	697,385.60	1,046,078.40	1,743,464.00
1883	(1)	125,914	802,539.54	1,203,809.29	2,006,348.83
1884	(1)	188,369	1,261,313.60	1,577,438.40	2,838,752.00
1885	(1)	226,147	1,729,531.25	1,415,071.04	3,144,602.29
1886	(1)	238,780	2,054,920.15	1,681,298.31	3,736,218.46
1887	(1)	223,682	2,481,176.85	2,030,053.78	4,511,230.63
1888	(1)	271,152	3,169,209.07	4,458,058.66	7,627,267.73
1889	(1)	286,144	2,590,973.32	3,358,949.95	5,949,923.27
1890	(1)	286,075	1,992,349.03	2,988,523.56	4,980,872.59
1891	(1)	188,647	1,380,857.02	2,071,285.53	3,452,142.55
1892	(1)	133,678	1,043,158.86	1,130,088.77	2,173,247.63
1893	(1)	109,780	1,123,262.54	748,841.70	1,872,104.24
1894	(1)	97,049	768,880.63	512,587.09	1,281,467.72
1895	(1)	63,558	548,873.68	365,915.79	914,789.47
1896	(1)	39,240	340,253.36	226,835.57	567,088.93
1897	(1)	17,850	223,808.63	149,205.76	373,014.39
1898	(1)	10,766	123,023.89	82,015.92	205,039.81
1899	(1)	6,780	103,006.74	68,671.16	171,677.90
1900	(1)	35,300	381,423.56	319,441.70	700,865.26
1901	(1)	56.577	746,477.00	521,032.00	1,267,509.00
1902	(1)	96,490	785,030.50	495,944.96	1,280,975.55

(1) Director of Mint, AR 1902, (1903), 157.

Production of the Comstock District Continued

Year	Tons	Gold	Silver	Copper	Lead	Total
1859-1902*	10,851,748	$148,145,385	$204,653,040			$352,798,425
1871-1902*	Mill Tailing	6,272,953†	12,176,910†			18,449,863
1903‡	36,142†	329,656	124,132			453,788
1904‡	69,972**	636,563	728,942			1,365,505
1905‡	89,484**	613,425	348,959	$3,795		966,179
1906‡	52,507**	327,766	185,763			513,529
1907‡	39,400**	254,690	107,442	543	$126	362,801
1908‡	113,908**	599,617	287,269			886,886
1909‡	92,092**	556,621	260,845		80	817,546
1910‡	108,592**	502,843	173,187	73	82	676,185
1911‡	95,894**	977,349	327,543			1,304,892
1912‡	112,001**	855,494	496,214†	218†	163†	1,352,089
1913‡	157,562**	853,564	440,903†	290†	178†	1,294,935
1914‡	103,406**	434,387	154,065†	146†		588,598
1915‡	85,144**	298,990	79,419†	93†	229†	378,731
1916‡	63,187**	483,137	188,074†	222†	763†	672,196
1917‡	59,805**	360,875	200,291			561,166
1918‡	95,051**	518,154	321,636			839,790
1919‡	50,376**	355,640	443,278†	41†	157†	799,116
1920‡	58,494**	321,982	353,971			675,953
1921‡	64,601**	324,827	263,931			588,758
Total 1859-1921	12,399,366	$164,023,917	222,315,814	$5,421	$1,778	$386,346,931

* Director of Mint, AR 1902, (1903), 157.
‡ "Mineral Resources of the United States," U. S. Geol. Survey.
† Unpublished information from U. S. Geol. Survey.
** Includes mill tailings.

Dividends. The following dividends have been paid by Comstock Mining companies according to a table prepared by Miss N. E. Preece, Geological dept., International Smelting Co., Salt Lake City.

Company	Dividends
Belcher	$15,397,200
Chollar	3,080,000
Cons. California & Virginia	78,148,000
Confidence Silver Mining	78,000
Comstock Phoenix	40,000
Crown Point	11,903,000
Gould & Curry	3,837,600
Hale & Norcross	1,850,000
Imperial Consolidated	750,000
Kentuck	1,350,000
Kossuth	10,800
Mexican	161,910
Ophir	1,856,680
Savage Gold & Silver	4,560,000
Sierra Nevada	102,500
Succor	22,800
Silver Hill	81,000
Union Consolidated	40,000
Yellow Jacket	2,184,000
TOTAL	$126,163,490

General Geology. The Comstock Lode is in the Viriginia Mountains on the E. flank of Mt. Davidson, a mountain which, according to Reid, is composed of a mass of diorite bounded upon all sides by faults. The lode occupies the great fissure made by the E. member of this block-faulting system, which is a normal fault with a movement of 3,000 ft. This movement has resulted in the shattering of the hanging wall and the production of numerous nearly vertical fissures which join the lode in depth but pinch out in height. The lode has a length of 13,000 ft., terminating by branching at both extremities, and varies in width from 100 to 1,400 ft. The strike of the lode is N. 14 degrees E. and its dip is 43 degrees easterly. The country consists of late Tertiary igneous rocks ranging from dioite on the W., which forms Mt. Davidson and the foot wall of the lode, through a hanging wall consisting mainly of hornblende-andesite, to augite-andesite on the E. Besides the Comstock Lode and its branches, there are in the Comstock District a number of cross and parallel veins including the Brunswick or Occidental Lode, a large parallel lode lying to the E. of the Comstock Lode in what is sometimes called the Silver Star District; and the Flowery Lode still farther to the E. in what is sometimes called the Flowery District.

Orebories. The country rock of the Comstock Lode has been highly altered by hydrothermal action, propylitization having effected the rocks on both sides of the lode and sericitization having occurred in the neighborhood of the principal ore channels. Hydrothermal action still continues, for great volumes of scalding sulphate water issue from the lower levels of the mines. 8he wide body of quartz and altered rock which constitutes the lode, contains rich ore shoots or "bonanzas" separated from one another by long and irregular stretches of barren or low-grade material. In the Virginia City section, the principal bonanzas occur in the vertical hanging wall fractures, while in the Gold Hill section they occur in the main lode. The location of the bonanzas appears to have been determined by N. W. and N. E. pre-mineral factures intersecting the main lode, and their size by the strength of these fractures.

Ore. According to Church, the Big Bonanza yielded 104,007,653 tons from which there was extracted an average of $93.35 per ton, a higher average than for any other bonanza. All the bonanzas taken together produced 6,350,520 tons yielding an average of $42.89 per ton. The average value of all the ore mined from 1859 to 1921, including values recovered by retreatment of tailings, is $31.16 per ton.

Ore Minerals. The typical ore of the Comtsock Lode consist of quartz and more or less calcite in which is disseminated a fine-grained mixture of sulphides. The sulphides present in practically all ores are sphalerite, galena, chalcopyrite, and pyrite. They are intercrystallized with one another and with the quartz and calcite in such a manner as to show that all were deposited contemporaneously. The common primary precious metal minerals are argentite, and gold of such a pale yellow color as to indicate a high silver content; while primary polybasite is occasionally found. Secondary ore is not found to any marked degree except within a few hundred feet of the surface, where native silver, polybasite, argentite, covellite, chalcocite, anglesite and wulfenite, all of which are clearly secondary, occur. Other ore minerals which have been observed in small quantities are hornsilver, pyrargyrite, stephanite, and sternbergite. Grant names the following secondary gangue minerals, in order of their abundance, as occurring in the Imperial mine; epsomite, gypsum, melanterite and lesser amounts of sulphates of manganese, aluminum, and copper, also kaolin and oxides of iron and manganese.

Mines. The original locations on the Comstock Lode, going in order from north to south, according to Church, were as follows:

Name	Length of Claim		Name	Length of Claim	
Utah	1,000	feet	Bullion	943.7	feet
Sierra Nevada	3,550	feet	Exechequer	400	feet
Union Con.	575	feet	Alpha	306	feet
Mexican	600	feet	Imperial Con.	466	feet
Ophir	675	feet	Challenge	90	feet
California	600	feet	Confidence	130	feet
Con. Virginia	710	feet	Yellow Jacket	953 3-4	feet
Best & Belcher	540	feet	Kentuck	93 11-12	feet
Gould & Curry	612 1-2	feet	Crown Point	541 7-12	feet
Savage	771 1-6	feet	Belcher	1,040	feet
Hale & Norcross	400	feet	Segregated Belcher	160	feet
Chollar	700	feet	Overman	1,200	feet
Potosi	700	feet	Calendonia	2,188	feet

In addition to the claims on the main Comstock Lode, there are in the Comstock District a number of properties located upon branches of the Lode, and upon parallel and cross veins.

North End Mines. A merger of the mines at the North End of the Comstock Lode for the purpose of working their low-grade ores to advantage has been proposed but not put into effect. The companies at present active in this section are the **Sierra Nevada M. Co.,** and **Union Cons. M. Co.,** of both of which Alex Wise is Supt.; and **Mexican G. and S. M. Co.,** the **Ophir M. Co.,** and the **Cons. Virginia M. Co.,** of all of which Zeb Kendall is Supt.

Comstock Merger Mines, Inc. Through the efforts of Herbert C. Humphrey and Alex Wise, Thomas F. Cole and associates were interested in the Middle Mines Group on the Comstock Lode, and in 1922 formed the Comstock Merger Mines, Inc., to operate it. This company is incorporated in Delaware with a capital stock of 2,000,000 shares, with par value of $10.00; of which 1,500,000 shares are outstanding. The Comstock Merger Mines, Inc., owns the mines from the Best and Belcher to the Exchequer inclusive, as listed above, the Caledonia at the S. end of the Lode, and also a group of claims covering 6,100 feet of the S.E. branch of the Comstock Lode in Gold Canyon. Charles V. Bob is Pres. of the company and Alex Wise, res.-mgr. The controlling interest in this company was sold in 1923 to the Goldfields American Dev. Co., Inc., a subsidiary of the Cons. Goldfields of South Africa, and the National M. Co., Ltd., both of London, England.

United Comstock. The old mines listed above from the Alpha to the Overman inclusive, were taken over in 1920 by the United Comstock M. Co.., a subsidiary of the Metals Exploration Co. The company is incorporated in Nevada with a capitalization of 7,000,000 shares of $1.00 par value. The Gold Hill Group which it has taken over are developed by 4 old shafts, the Belcher, Yellow Jacket, Knickerbocker, and Imperial. The new company has run an 8x8 adit 9,585 ft. from its portal on American Flat entering the Comstock Lode at the Knickerbocker shaft and connecting with the other shafts. This tunnel gives a maximum back of low-grade ore and old fills amounting to 650 ft. Through it the ore is drawn by electric haulage to the 2,500-ton cyanide plant of the company located near the tunnel entrance. A new town named Comstock has been built near the mill, and a 2-mile spur connects the mill with the V. & T. R. R.

Other Mines. Four companies are active in the Gold Hill section upon properties located away from the main Comstock Lode. The **Comstock-Eldorado M. & M. Co.** is incorporated in Nevada with an authorized capital of 1,500,000 shares of 10 cents par value. Its home office is at 208 Clay Peters Bldg., Reno. C. P. McColm is Pres., and C. Stoddard, Gen. Mgr. Its property is located in the S.E. part of the district near Silver City adjoining that of the Comstock S. M. Co. on the S. The **Comstock Eldorado** is treating 20 tons a day in the Overland mill of the Comstock S. M. Co., and is making plans to erect a 100-ton mill of its own. The **Comstock Silver M. C.** is incorporated in Nevada with an authorized capital of 1,500,000 shares of 10 cents par value. Its home office is at 310 Reno Natl. Bank Bldg., Reno, and its mine office at Virginia City. F. M. Manson is Pres. and G. H. Drysdale, Supt. Its property consists of 11 claims and fractions adjoining the Comstock-Eldorado on the N., and active development work is being carried on through 3 shafts. The **Nevada Canyon M. Corp.** is developing propetries in the Flowery section. Controlling interest in this company was purchased in 1923 by the same British interests which took control of the Comstock Merger Ms., Inc. The **Spearhead Gold M. Co., Reorg.** is a Goldfield company, which in 1922 acquired 2 claims a short distance to the S.W. of the properties of the Comstock-Eldorado and Comstock Silver, and in 1923 purchased the 3 adjoining claims of the Lager Beer Group. The company is incorporated in Nevada with a capital stock of 1,500,000 shares of $1.00 par value which is listed on the San Francisco Stock Exchange. Its home office is at 421 Clay Peters Bldg., Reno. A. A. Busey is Pres-Gen. Mgr., and A. A. Codd is Sec. The **Pittsburg-Comstock M. Co.** is incorporated in Nevada with a capitalization of $1,500,000. Its home office is in the Reno Natl. Bank Bldg., Reno, and its mine office at Virginia City. H. G. Humphrey is Pres. and A. McCoy, Mgr. The company is developing a group of 5 claims adjoining the United Comstock Ms. Co. on the W.

BIBLIOGRAPHY OF THE COMSTOCK DISTRICT

General

B1866 31-4, 71-88, 99-125.
B1867 341-394.
R1868 38-75.
R1869 89-116, 558, 576, 598, 600, 623, 624.
R1870 2, 93-111.
R1871 141-155.
R1872 108-119.
R1873 158-192.
R1874 8, 194-232, 470a.
R1875 145-171, 365-6, 378, 464-7.
SMN1867-8 22-32.
SMN1869-70 5-14.
SMN1871-2 118-138.
SMN1873-4 89-145.
SMN1875-6 119-151.
SMN1877-8 94-150.
MR1903 183.
MR1904 146, 200.
MR1905 116, 272-3.
MR1906 122, 299.
MR1907 I 378-381.
MR1908 I 501-2.
MR1909 I 425-6.
MR1910 I 530-1.
MR1911 I 697.
MR1912 I 813
MR1913 I 839
II 1370-3 Stone
MR1914 I 707-8.
MR1915 I 651.
MR1916 I 496.
MR1917 I 293-4.
MR1918 I 257-8.
MR1919 I 412.
MR1920 I 334-5.
MR1921 I 395.
USGSW100th AR1877 1284-5. AR1878 145-166.
Bancroft 92-149.
Blake, W. P., "Notes on the Geology and Mines of Nevada Territory," QJGS 20 (1864) 317-327.

Campbell, R. H., "Discovery of the Great Comstock Mine," Discussion, M&SP 99 (1909) 524.

Church, J. A., "The Comstock Lode; Its Formation and History," New York, John Wiley & Sons, (1879).

Colcord, R. K., "Early Comstock Days," M&SP 124 (1922) 327-330.

Davis 381-419, 997-1003.

E&MJ, "Revival on the Comstock," 91 (1911) 126-7.

———"Recent Developments on the Comstock," 98 (1914) 616-7.

———"The Comstock Lode in 1914," 99 (1915) 155-6.

———"Interest in the Comstock Revives," Editorial, 112 (1921) 362.

Goodman, J. T., "Comstock Beginnings," M&SP 99 (1909) 19-21.

Gratacap, L. P., "The Possible Revival of Virginia City, Nevada," Sci. Am. Suppl., 40 (1895) 16, 329-30.

———"The Comstock Lode, Nevada," Sci. Am. Suppl. 48 (1899) 19, 925-6.

M&SP, Editorial, "Comstock Semi-Centennial," 99 (1909) 3; Discussion G. McM. Ross, "Comstock Bullion," 149.

———"The Year on the Comstock Lode," 102 (1911) 45-6.

———Editorial, "The Comstock Meeting" 123 (1921) 48-9.

Read, T. T., "Impressions of the Comstock," M&SP 105, (1912) 244-5.

Ross, G. McM., "The Great Comstock Lode," M&SP 95 (1907) 468-9.

———"The Great Comstock Lode," in Davis 367-380.

Shinn, C. H., "The Story of the Mine as Illustrated by the Great Comstock Lode of Nevada," New York, D. Appleton & Co. (1908.)

Storms, W. H., "The Passing of the Comstock Lode," M&EW 39 (1913) 877-9, 963-6; 40 (1914) 3-6, 59-61.

Stretch, R. H., "Comtsock Lode in the Sixties," M&SP 123 (1921) 739-44.

StuartNMR 11, 33-45.

Symmes, W., "Decline and Revival of Comstock Mining," M&SP 97 (1908) 496-500, 570-576.

———"The Comstock Mines Today," M&SP 99 (1909) 24-6.

———"The Comstock Lode," E&MJ 95 (1913) 129-130.

Thompson & est 589-622.

WeedMH 1130-1366 Describes 37 Comstock companies.

Wright, W., ("Dan de Quille"), "History of the Big Bonanza," Hartford, Conn., American Pub. Co. (1877); Abstract, "Discovery of the Great Comstock Mine," M&SP 99 (1909) 22-3.

———"A History of the Comstock Silver Lode and Mines," Virginia, Nevada, (1889).

———"The Gold Belts of Nevada," E&MJ 59 (1895), 532-3.

Young, G. J., "The Position of the Comstock," E&MJ 93 (1912) 167-173.

———"History of Mining in Nevada" in Davis, 315-366.

Geology.

Barus, C., "The Electrical Activity of Ore Bodies," T AIME 13 (1885) 417-477; "Experiments Made in Some of the Mines of the Comstock," 428-435.

Bastin, E. S., "Bonanza Ores of the Comstock Lode, Virginia City, Nevada," USGS 735 (1922) 41-63.

Becker, G. F., "Notes on a New Feature in the Comstock Lode," AJS 10 (1875) 459-62.

———"A Summary of the Geology of the Comstock Lode and the Washoe District," USGS AR 2 (1882) 291-330.

———"Geology of the Comstock Lode and the Washoe District," USGS M 3 (1882).

———"The Washoe Rocks," California A. S. B 2 (1887) 93-120.

Church, J. A., "The Heat of the Comstock Mines," T AIME 7 (1897) 45.

———"The Heat of the Comstock Lode," T AIME 8 (1880) 324.

Dana, E. S., "System of Mineralogy," New York (1900) 991; Wulfenite.

Grant, W. H., "Resume of Geology of United Comstock Mines," M&Met 3 (1922) No. 191, 39-41.

Hague, A. & Iddings, J. P., "On the Development of Crystallization in the Igneous Rocks of Washoe Nevada with Notes on the Geology of the District," USGS B 17 (1885); Review by R. W. Raymond, E&MJ 40 (1885) 397-8; Review by J. A. Church, "The Geological Battle of the Comstock," E&MJ 41 (1886) 52.

Hill507 225.

King, C., "The Comstock Lode," USGE40th, 3 (1870) 9-96.

Locke, A., "Four Famous Districts Compared," E&MJ 92 (1911) 505-6.

———"The Abnormal Temperatures of the Comstock Lode," EG 7 (1912) 583-7.

Reid, J. A., "The Structure and Genesis of the Comstock Lode," Cal. Univ. Dept. Geol. B 4 (1905) 177-199; Abstract M&SP 91 (1905) 244.

Richtofen, F., Baron, "The Comstock Lode, its Character and the Probable Mode of its Continuance in Depth," San Francisco (1866).

Smith, D. T., "Vein Systems of the Comstock," E&MJ 94 (1912) 895-6.

Mining

Church, J. A., "Accidents in the Comstock Mines and their Relation to Deep Mining," T AIME 8 (1880) 84.

De Laval, C. G. P., "Pumping on the Comstock," E&MJ 79 (1905) 516-8.

Dickie, G. W., "The Men and Machinery of the Comstock," E&MJ 98 (1914) 397-8; "The Sutro Tunnel," 519-522; "The Barclay and Davey Engines," 734-8; "Hydraulic Machinery," 877-9; "The Combination Shaft," 990-4; "Pioneer Hoisting Works," 1130-4.

Eddy, L. H., "Mining Lower Levels of the Comstock Lode," E&MJ 105 (1918) 1029-1032.

E&MJ, "Unwatering the Comstock Lode," 82 (1906) 961-3.

———"The Starrett Air-Lift Pump," 83 (1907) 611-2.

———"Pumping Plant at the Wand Shaft, Virginia City, Nevada," 89 (1910) 575.

———"Fire Protection in Shafts," 103 (1917) 191.

Hague, J. D., "The Comstock Mines," USGE40th, 3 (1870) 97-192.

Hall, L. M., "Modernizing the Comstock Lode," M&SP 92 (1906) 183.

———"Comstock Drainage Problems," M&SP 99 (1909) 27-29.

Harding, J. E., "Unwatering an Inclined Shaft," E&MJ 103 (1917) 521-2.

Lord, E., "Comstock Mining and Miners," USGS M 4 (1883).

M&EW "Philip Deidesheimer, Inventor of the Square Set," 45 (1916) 193-4.

M&SP, "Unwatering the Comstock Lode," 90 (1905) 65, 73-4.

———"Pumping at the Gold Hill Mines on the Comstock," 108 (1914) 652-3.

O'Brien, W. D., "Recent Work on the Comstock," M&SP 96 (1908) 804-6.

Reigart, J. R., "Proposed Mining Methods of United Comstock Mines Company," M&Met (1922) No. 191, 41-43.

Rice, C. T., "Timbering in Swelling Ground," E&MJ 82 (1906) 306.

———"The Reopening of the Comstock," E&MJ 82 (1906) 1155-7.
———"Modern Mining on the Comstock," E&MJ 82 (1906) 1209-1211.
Sutro, A., "The Mineral Resources of the United States and the Importance and Necessity of Inaugurating a Rational System of Mining with Special Reference to the Comstock Lode and the Sutro Tunnel in Nevada," Baltimore, John Murphy & Co., (1868).
Symmes, Whitman, "Reopening the Mexican Mine, Comstock Lode," M&SP 100 (1910) 419-423.
———"Difficulties of Pumping on the Comstock Lode," M&EW 38 38 (1913) 57-9.
———"Mine-Pumping Equipment," E&MJ 103 (1917) 567-571.
USGS AR 1 (1880) 39-46, 71-2, 50-2.
USGS AR 2 (1881) xxiv-xxvi, xxxvii-xxxvii.
Walsh, A. M., "Pumping at the Comstock," M&SP 107 (1913) 305-6.
Young, G. J., "Ventilating System at the Comstock Mines, Nevada," T AIME 41 (1911) 2-57; Abstract, "Ventilation System at Comstock Mines," E&MJ 88 (1909) 1016-8.
———"Driving in Loose Ground", E&MJ 91 (1911) 161.
———"Fires in Metalliferous Mines," T AIME 44 (1913) 644-662.

Metallurgy.

Hague, A., "Chemistry of the Washoe Process," USGE40th 3 (1870) 274-293.
Hague, J. D., "The Treatment of the Comstock Ores," USGE40th 3 (1870) 193-273.
Hodges, A. D., "The Process Used at the Comstock for Refining Coppery Bullion Produced by Amalgatmating Tailings," T AIME 14 (1886) 731-757.
———"Amalgamation at the Comstock Lode, Nevada," T AIME 19 (1891) 195-231.
M&SP, Editorial, "Cyaniding at Virginia City," 123 (1921) 733-4.
Martin, A. H., "Treating Low Grade Ore at Comstock Lode," MS, Jan. 21, 1901.
Reid, W. L., "Design and Construction of United Comstock Mill," M&Met 3 (1922) No. 191, 44-7.
Rice, C. T., "The Butters Cyanide Plant, Virginia City, Nev.," E&MJ 83 (1907) 269-273.
Rothwell, R. P., "The Cost of Milling Silver Ores in Utah and Nevada," T AIME 8 (1880), "Milling Comstock Ores," 558-560.
Symmes, W., "The Mexican Mill, Virginia City, Nevada," E&MJ 94 (1912) 701-5; Discussion, J. V. N. Dorr, 969.
———"The Symmes Agitator," M&SP 107 (1913) 92-4; Discussion J. V. N. Dorr, 193; P. G. Spilsbury, 467.
Techow, W., "The Ophir Cyanidation Plant," M&SP 105 (1912) 703.
Weinig, A. J., "United Comstock Metallurgy, M&Met 3 (1922) No. 191, 47-49.
Young, G. J., "New Treatment Plant of the United Comstock Mines," E&MJ 114 (1922) 846-853.

Miscellaneous.

E&MJ, "The Scandal of the Comstock," 94 (1912) 745-7.
———"Comstock Affairs," Special Correspondence, 96 (1913) 972.
———"Comstock Developments, Editorial, 97 (1914) 1068.
———"Comstock Miners Walk Out," 110 (1920) 587.

———"Vacationing on the Comstock," Editorial, 110 (1920) 654.
M&EW, "Mr. Symmes on the Comstock Situation," Editorial, 38 (1913) 63-5.
M&SP, "View of the Comstock Lode," 90 (1905) 54-5.
———Editorial, 106 (1913) 201, 470, 505 (Symmes and Comstock Affairs).
Slosson, H. L., Jr., "Present Comstock Affairs," E&MJ 97 (1914) 1065.
USGS Carson topographic map.
Van Loan, S. E., Stories from article on Virginia City in the Saturday Evening Post, E&MJ 100 (1915) 281, 327, 450, 489, 533, 612.
Young, G. J., "Co-Operation Among Small Mines," E&MJ 106 (1918) 812.

FLOWERY see COMSTOCK
GOLD HILL see COMSTOCK
PARKER AND NOE see CHALK HILLS
SILVER STAR see COMSTOCK
VIRGINIA CITY see COMSTOCK
WASHOE see COMSTOCK

WASHOE COUNTY

BUFFALO SPRINGS

Salt (Sodium Sulphate)

Location. The Buffalo Springs Salt Works is located at Sheepshead on the W. edge of the Smoke Creek Desert. Gerlach on the W. P. R. R. is 25 m. E. N. E.

History. In 1885, these works had produced about 1500 tons of salt and were producing at the rate of 250 tons per year. In 1907, the old vats were cleaned out and small productions have been made since.

Geology. The salt comes from the Quaternary deposits formed during the desiccation of Lake Lahontan. It is obtained as brine from shallow wells and to a less degree as crusts. In addition to sodium chloride, the brine contains smaller amounts of magnesium, sodium, potassium and calcium sulphates. A short distance from the salt deposit is a deposit of sodium sulphate 8 ft. thick.

Bibliography. MR1883-4 848 MR1912 II 918 MR1913 II 300
MR1907 II 664 MR1915 II 273
USGS Granite Range topographic map.
Free, E. E., "The Topographic Features of the Desert Basins of the U. S. with Reference to the Possible Occurrence of Potash," U S. Dept. Agr. B 54 (1914) 16.
Phalen, W. C., "Salt Resources of the U. S.," USGS B 669 (1919) 145-6.
Russell, I. C., "Geological History of Lake Lahontan," USGS M 11 (1885) 232-3.

COTTONWOOD

Silver, Lead, Gold

Location. The Cottonwood District is located in the Fox Mts. N. of Pyramid Lake. The oldest mines are in Cottonwood Canyon at the N. end of the district and 7 m. N. of Pah-rum Peak, the highest point in the range; the Wild Horse Mine is 2 m. N.W. of this peak; while the Sano Mine is at the S. end of the range 5 m. N. of Pyramid Lake. Sano on the Susanville Branch of the S. P. R. R. lies a few miles to the W.

History. The mines in Cottonwood Canyon were worked in the seventies, and a 5-stamp mill was erected to treat their ores but did not meet with success. The Wild Horse Mine of the Washoe Lassen M. Co. which was operating in 1912 is equipped with a 5-stamp mill. The only company at present active in the district is the Sano Cons. Ms. Co. of which W. E.

Pruett, 406 Clay Peters Bldg., Reno, is Res. Agt., and C. A. Packard, Mgr.

Geology. The Fox Mts. are composed mainly of quartz-monzonite which is intrusive into sedimentary rocks probably of Mesozoic age, according to Hill. At Cottonwood Canyon, quartz veins occur in the quartz-monzonite carrying argentiferous galena and perhaps gray copper. At the Wild Horse Mine, ocherous quartz ores containing gold and silver are said to occur in veins in altered porphyry. At the Sano Mine, silver-lead ore containing a little gold is present in veins in the quartz-monzonite near its contact with limestone.

Bibliography. MR1907 I 382 Hill507 225
MR1908 I 503 Hill594 181

CRYSTAL PEAK see PEAVINE

DEEPHOLE

Gold

Location. The Deephole District is located at Deephole N. of the Smoke Creek Desert.

History. Gold was produced from one property in the district in 1908.

Bibliography. MR1908 I 503. USGS Granite Range topographic map. Hill507 225.

DONNELLY

Gold, Silver

Location. The Donnelly District is situated on the S.W. flank of Division Peak on the boundary line between Washoe and Humboldt Cos. Gerlach on the W. P. R. R. is 39 m. S. Division Peak rises from the Black Rock Desert, which has an elevation of 4,000 ft. to a height of 8,585 ft. above sea level, and the mines are at elevations of from 6,800 ft to 7,500 ft. The Donnelly District adjoins the Leadville District on the E. and is sometimes considered a section of that district.

History. The Donnelly Mine was discovered in 1910 by James Raser. He shipped a little high-grade gold ore and in 1911 built a 5-stamp mill, which is credited with a production of $90,000 in gold bullion. The property was operated up to 1914. In 1919, the mine was operated by lessees. The Reeder Mine to the S. is said to have made a small production with the aid of an arrastra.

Geology. The country rock of the Donnelly District consists of Cretaceous granodiorite which has been intruded into slate and quartzite which are probably of Triassic age and are capped by Tertiary andesite, according to a report by F. C. Lincoln. The ore deposits are flat, narrow, gold-quartz veins, in the granodiorite near its contact with the sedimentary rocks and andesite. The wall rock has been softened near the veins. The vein filling is white quartz and silicified granodiorite, with a minor amount of pyrite which is largely altered to limonite. The gold is disseminated in fine specks which are sometimes visible to the naked eye; and silver is present to the extent of about one-half the weight of the gold.

Bibliography MR1910 I 542 MR1912 I 815 MR1913 I 839
MR1911 I 698 MR1914 I 709
USGS Long Valley topographic map.

FLANIGAN

Bog Lime

Location. The Flanigan bog lime deposit is situated on the Susanville Branch of the S. P. R. R., 5 m. E. of Flanigan station.

History. Paul Butler located the deposit in 1919 and sold it to the Agricultural Lime & Compost Co. in 1922.

Geology. The deposit is a thin bed of unconsolidated calcareous material consisting mainly of Chara stems with a few small shells. It was formed in ancient Lake Lahontan.

Mines. The Agriculture Lime & Compost Co. of San Francisco is sacking and shipping the bog lime as a fertilizer. About 1500 tons were shipped during the past 6 mos. J. Andrews is Supt. at Flanigan.

GALENA (Washoe)

Lead, Zinc, Silver, Gold, Copper, (Arsenic)

Location. The Galena District is situated on Galena Creek W. of Washoe City, a station on the V. & T. R. R. The Jumbo District adjoins it on the E. and is sometimes considered a part of it.

History. A. J. and R. S. Hatch laid out the town of Galena and organized the Galena Mining District in 1860. A smelting furnace was erected and proved unsuccessful, and a mill met with the same fate. The mines were then abandoned and the town became a logging camp. The Nevada Commonwealth M. & M. Co. built a concentrating plant which it operated in 1906 and 1907; and the Rocky Hill M. Co. ran its mill during the latter year.

Geology. The ore in the Commonwealth vein is said to contain 0.5 per cent. copper, 4 per cent. lead, 4 per cent. zinc, 5 ozs. silver per ton, and 70 cts. in gold. According to Stretch, there is a thin seam of arsenopyrite between the wall of the lode and the galena-bearing material.

Bibliography

B1867 327.	SUN1867-8 22	MR1905 273.
R1868 76.	SMN1869-70 4	MR1906 299. 478 Zinc.
SMN1866 21-2.	SMN1871-2 142	MR1907 I 382.
	SMN1877-8 153.	MR1922 I 62 Arsenic

USGS Carson topographic map.

Hill507 226. StuartNMR 152. Thompson & West 643.

GERLACH

(Gypsum)

Location. The Gerlach gypsum deposit is located at the W. base of Luxor Peak toward the N. end of the Truckee Range on the Washoe and Pershing Co. boundary line. Gerlach on the W. P. R. R. is 12 m. N.

Geology. Triassic limestones, schists, and gypsum are exposed in contact with granodiorite on the E., according to Jones. The gypsum occurs in large lenses between a white limestone and a dark calcareous schist. The beds are nearly vertical and the gypsum stands out in terraces upon one of which hot springs have built cones of selenite. The gypsum is soft, massive, white, and crystalline and is for the most part pure, containing beds of limestone and sandy layers. In depth it changes to anhydrite.

Mines. This deposit belongs to the Pacific Portland Cement Co., Cons., which began development work upon it in 1922. A 5-mile standard-gage spur is being built from the W. P. R. R. to the point where the gypsum mill will be located, and this mill will be connected with the mine by aerial tram. A full line of the company's Empire brand gypsum products will be manufactured in the mill. The home office of the Pacific Portland

Cement Co., Cons., is in the Pacific Bldg., San Francisco. The company is now operating a gypsum quarry and mill at Mound House in Lyon County.

Bibliography. USGS Granite Range topographic map.
USGE40th 2 (1877) 808-9.
Jones, J. C., "Nevada" in "Gypsum Deposits of the U. S.," USGS B 697 (1920) 150.

JUMBO (West Comstock)
Gold, Silver.

Location. The Jumbo District is located on the W. flank of Mt. Davidson in the Virginia Range. The Galena District adjoins it on the W., and the Comstock District in Storey County on the E. The Jumbo Dstrict is sometimes considered part of the Galena District.

History. A mill was constructed in the district by the Selby Cons. M. & M. Co. in 1907. In 1909 and 1910 the Bargo M. & M. Co. operated a 10-stamp mill; and in 1910 and 1911 Fink & Mahoney operated a 5-stamp mill. There have been small productions from the district since. In 1915, the Washington-Nevada Dev. Co. constructed a 10-stamp mill.

Geology. Gold-silver ore occurs in veins in diorite.

USGS Carson topographic map

Bibliography. MR1907 I 382 MR1910 I 531 MR1914 I 709
MR1908 I 503 MR1911 I 698 MR1915 I 651
MR1909 I 426
Hill507 226. StuartNMR 152-3. WeedMH 1237 Klamath M. & M. Co.

LEADVILLE
Silver, Lead, Zinc, Gold, (Niter)

Location. The Leadville District is situated on the E. flank of Mt. Fox in the Granite Range, 38 m. N. of Gerlach, which is on the W. P. R. R. The Donnelly District adjoins it on the E. and is sometimes considered a section of the Leadville District.

History. Development work was carried out in the Leadville District in 1909 and regular annual productions were made from 1910 to date. The Tohoqua Mine was equipped with a 50-ton concentrator in 1911, which was remodeled and had a flotation plant added when the Leadville Ms. Co. took possession in 1920.

Production. Mines Handbook gives the early production of the Tohoqua Mine as $205,000; while that of the Leadville Ms. Co. for 1921 was $153,000, and for 1922, $254,600 gross.

Geology. The ore deposits are quartz veins in Tertiary diorite porphyry, according to Mines Handbook, and the principal values in the ore are lead and silver with minor ones in zinc and gold. Niter has been found in crevices in rhyolite on the W. side of the range W. of Leadville.

Mines. The Leadville Ms. Co. was incorporated in Nevada in 1920 with a capital stock of 1,500,000 shares of 10 cents par value of which 750,000 remain in the treasury. The home office is at 421 Clay Peters Bldg., Reno. A. A. Codd is Pres., F. M. Manson, V. P., and G. V. Ward, Sec.-Treas. F. A. Elliott is Supt. at Gerlach. The property consists of 3 claims; developed by 2 shafts, a 1,700 ft.(tunnel, and a 500-ft. winze; and equipped with 2-75 and 1-100 hp. semi-Diesel engines, compressor, electric locomotive, auto trucks, 7,500-ft. water line and 30,000 gal. tank, and 35-ton mill.

Bibliography. MR1909 I 426 MR1910 I 532 MR1911 I 698 MR1912 I 815 MR1913 I 839 MR1914 I 709 MR1915 I 651 MR1916 I 496 MR1917 I 294 MR1918 I 259 MR1920 I 335 MR1921 I 396

USGS Long Valley topographic map.

Gale, H. S., "Nitrate Deposits," USGS B 523 (1912) 23-5.

Hill507 225. WeedMH 1239 Leadville Ms. Co.

NIGHTINGALE see PERSHING COUNTY

OLINGHOUSE see WHITE HORSE

PEAVINE (Reno, Crystal Peak)

Copper, Silver, Gold, Lead, Placer Gold, Granite, Andesite, (Lignite, Oil Shale, Diatomaceous Earth, (Petroleum)

Location. The Peavine District is on Peavine Peak in S. W. Washoe Co. on the California border. It lies just N.W. of Reno, which is on the S. P. R. R., W. P. R. R., and V. & T. R. R. The elevation of Reno is 4,500 ft., and Peavine Peak rises to an altitude of 8,270 ft. The lignite deposits occur in the S.W. section of the district near Verdi in what was formerly known as the Crystal Peak District. The Wedekind District which adjoins the Peavine District on the E. is sometimes considered a section of the Peavine District.

History. The Peavine District was discovered in 1863. The first rail shipment across the Sierras consisted of ore from this district which was shipped to Sacramento in 1866. A smelting furnace was erected but proved unsuccessful. The district has been worked intermittently upon a small scale from the early days to the present. The Standard Metals Co. completed a 150-ton flotation plant in 1920 and ran it in 1921. Numerous attempts to find lignite deposits of commercial grade have been made. Granite and andesite have been quarried from time to time for building purposes. An unsuccessful attempt to find oil by drilling was made in 1908.

Production. From 1905 to 1921, the district produced 5,521 tons of ore containing $26,480 in gold, 44,009 ozs. silver, 424,828 lbs. copper, and 57,163 lbs. lead, valued in all at $153,450, according to Mineral Resources of the U. S. Geol. Survey.

Geology. Peavine Peak is composed of pre-Tertiary schists intruded by Cretoceous quartz-monzonite and flanked to the E. and S. by Tertiary andesite flows which in turn are overlain in places by the sediments of the Miocene Truckee Formation. The ore deposits consist of veins and lenses in the schists and quartz-monzonite and of replacement deposits in the andesite and quartz-monzonite. The Standard Metals vein is in schist cut by dikes of quartz-monzonite; and its ore consists of quartz, chalcedony, barite, rhodochrosite, pyrite, chalcopyrite, galena, sphalerite, and tetrahedrite; and carries silver and minor gold values. Diatomaceous earth, and irregular and impure seams of lignite and oil shale, are present in the Truckee beds.

Bibliography.

B1866 97 Lignite 155.
B1867 216, 316-7 315-6 Lignite.
R1868 76.
SMN1866 21, 22-3
SMN1875-6 159-160.
SMN1877-8 151-2.
MR1885 40 Coal 532 Ocher
MR1898 185 416 Granite
MR1913 I 840.
MR1913 II 1368 Granite 1370-3 Andesite.
MR1914 I 709.
MR1915 I 651.
MR1916 I 496.

SMN1866 21, 22-3 MR1905 273. MR1917 I 294.
Lignite. MR1906 299. MR1918 I 259.
SMN1867-8 22 MR1907 I 382. MR1919 I 412.
SMN1869-70 4 MR1908 I 503. MR1920 I 335.
SMN1871-2 142. MR1909 I 426. MR1921 I 396.
SMN1873-4 83 MR1912 I 815.
USGE40th 3 (1870) 296-7.
USGS Reno topographic map.
Anderson, R., "Geology and Oil Prospects of the Reno Region, Nevada," USGS B 381 (1909) 475-493.
Hill507 225-7. Hill594 184-195.
Louderback, G. D., "General Geologic Features of the Truckee Region East of the Sierra Nevada," Abstract, GSA B 19 (1908) 662-9.
StuartNMR 151-2.
Thompson & West 644.
WeedMH. 1148 Black Panther Ex. M. Co., Black Panther M. Co.
1204 Fravel-Paymaster M. Co. 1293 Nixon-Nevada M. Co.
1301 Peavine S. Corp. 1310 Red Metals Co.
1329 Silver Blossom Ms. Co. 1344 Standard Metals Co.
1365 United Metals Co. 1370 Washoe C. Co.

PYRAMID

Copper, Silver, Gold, (Lead)

Location. The Pyramid District is located S.W. of Pyramid Lake. It is a few miles S.W. of the Susanville Branch of the S. P. R. R., and Reno on the S. P. R. R., W. P. R. R. and V. & T. R. R. is 32 m. S.

History. The Pyramid District was known as early as 1860, but did not attact attention until 1876 when Dr. S. Bishop located the Monarch Mine and erected a 2-stamp mill. Since that time, the district has been active at intervals, the Franco-American or Blondin Mine having made the principal production.

Geology. The ore deposits are veins in Tertiary eruptives containing silver-bearing copper ores with a trace of gold. One property shows gold-silver-lead ore with no copper.

Bibliography SMN1875-6 158-9 MR1913 I 840 MR1916 I 496
SMN1877-8 153 MR1914 I 709 MR1918 I 259
USGS Reno topographic map.
Thompson & West 644.

RENO see PEAVINE

SAND PASS

Fuller's Earth

Location. The Sand Pass fuller's earth deposit is located in central Washoe Co., 22 m. by road N. of Sand Pass, a station on the W. P. R. R. It is situated upon a plateau which rises 500 ft. above the W. edge of the Smoke Creek Desert, and is S.W. of the Buffalo salt district.

History. The Standard Oil Co. recently prospected this clay deposit by means of trenches, pits, and borings; but the efficiency of the earth was considered too low to warrant any considerable expenditure upon it.

Fuller's Earth. The bed of fullers' earth has an areal extent of 2,100 ft. by 3,300 ft. and a thickness of approximately 60 ft. The upper surface is not a plane but appears to have the shape of a series of cones. Overlying the clay, there is a bed of lime-cemented rock and gravel and of loose gravel with an average thickness of 16 ft. The analysis of the fuller's earth is as follows:

Silica	60.22%
Alumina	15.55
Lime	2.69
Magnesia	1.11
Ferric Oxide	6.55
Combined Water	13.78
	99.90%

The moisture content is 6.21% and the weight per cu. ft. 31.3 lbs.

STEAMBOAT SPRINGS

Mercury, Sulphur

Location. The Steamboat Springs District is located at Steamboat Springs at the E. base of the Sierra Nevada Mts. in S. Washoe Co. Steamboat station on the V. & T. R. R. is at the hot springs, while the mine lies 1 1-4 m. to the W. The Steamboat Springs District adjoins the Galena District on the N.

History. The hot springs were located by Felix Monet in 1860. The mercury and sulphur deposit to the W. was discovered by Thomas Wheeler in 1875 and sold to the Nevada Quicksilver Co. in 1877. This company erected a mercury furnace and worked the mine for several years, producing sulphur as well as mercury.

Geology. The hot springs issue forth from fissures in sinter which is underlain by granodiorite. Basalt caps the granodiorite to the S.W. and the mine is at the granodiorite basalt contact. The ore consists of altered granodiorite and basalt impregnated with cinnabar and sulphur. It is of low grade.

Bibliography B1866 94 SMN1875-6 157 MR1886 644 Sulphur
B1867 307-8 MR1833-4 865 Sulphur MR1916 II 411 Sulphur
USGS Carson topographic map.

Becker, G. F., "Descriptive Geology of the Steamboat Springs District," USGS M 13 (1888) 331-353.

Hill507 226.

Jones, J. C., "The Occurrence of Stibnite at Steamboat Springs, Nevada," Science NS 35 (1912) 775-6.

———"Occurrence of Stibnite and Metastibnite at Steamboat Springs, Nevada," Abstract, GSA B 25 (1914) 126.

LeConte, J., "On Mineral Vein Formation Now in Progress at Steamboat Springs Compared with the Same at Sulphur Bank." AJS III 25 (1883) 424-8.

Lindgren, W., "Metallic Sulphides from Steamboat Springs, Nevada," Abstract, Science, NS 17 (1903) 729.

———"The Occurrence of Stibnite at Steamboat Springs, Nevada," T AIME 36 (1906) 27-31.

WASHOE see GALENA

WEST COMSTOCK see JUMBO

WEDEKIND

Silver, Lead, Gold, Zinc

Location. The Wedekind District is situated N. of the Truckee Meadows in S. Washoe Co. The Wedekind Mine is 2 1-2 m. N. of Sparks, which is on the S. P. R. R. The Wedekind District adjoins the Peavine District on the E. and is sometimes considered as a section of that district.

History. The Wedekind Mine is said to have produced $500,000 prior to 1896, according to the Mines Handbook. Old tailings were cyanided at this mine in 1911 and 1913, and productions were made in 1916 and 1917.

Geology. The ore deposit consists of irregular replacements in altered Tertiary andesite. The ore in general looks like ferruginous cemented conglomerate, according to Morris, while the richer ore contains lead sulphide and carbonate, silver sulphide and chloride, and native gold. In the neighboring Arkell Mine, sphalerite accompanies the galena.

Bibliography MR1911 I 689 MR1913 I 840 MR1917 I 294
MR1916 I 496
USGS Reno topographic map.
Hill507 226. Hill594 195-6.
Morris, H. G., "Hydrothermal Activity in the Veins at Wedekind, Nevada." E&MJ 76 (1903) 275-6.
WeedMH 1127 Adelphia M. & M. Co. 1370 Wedekind S. M. Co.

WHITE HORSE (Olinghouse)
Gold, Silver

Location. The White Horse or Olinghouse District lies immediately to the N. of Olinghouse on the E. flank of the Pah-Rah Range in S.E. Washoe Co. Olinghouse is 9 m. W. of Wadeworth on the S. P. R. R. and 7 m. N. W. of Derby on the same railroad.

History. The district was prospected in 1860 and locations made there in 1864. The Green Mt. Mines were located by Frank Free in 1874. The camp reached its greatest activity in the period from 1901 to 1903 and continued to produce up to 1921.

Production. From 1902 to 1921, the district produced 52,381 tons of ore containing $509,530 in gold and 12,956 ozs. silver, valued in all at $519,450, according to Mineral Resources of the U. S. Geol. Survey.

Geology. The country rock of the district is Tertiary andesite intruded by later andesite and by rhyolite porphyry. The ore deposits are veins and silicified zones in the older andesite, in one instance at a contact with the later intrusive andesite. The ore commonly consists of quartz, calcite, pyrite, and free gold. A minor amount of chalcopyrite was noted by Hill in one mine, and cerussite (?), argentite, and possibly telluride of gold and silver, in another.

Bibliography. MR1905 273 MR1911 I 698-9 MR1916 I 496-7
MR1906 299-300 MR1912 I 815-6 MR1917 I 294-5
MR1907 I 383 MR1913 I 840 MR1918 I 259
MR1908 I 503 MR1914 I 709 MR1919 I 412
MR1909 I 426 MR1915 I 651 MR1920 I 335
MR1910 I 531 MR1921 I 396
USGS Wadsworth topographic map.
Hill507 226.
Hill, J. M., "Notes on the Economic Geology of the Ramsey, Talapoosa, and White Horse Mining Districts in Lyon and Washoe Counties, Nev." USGS B 470 (1911) 99-106.
StuartNMR 7.

WHITE PINE COUNTY

AURUM

(Schellbourne, Schell Creek, Siegel, Queen Springs, Silver Canyon, Ruby Hill, Muncy Creek, Silver Mountain)

Silver, Lead, Covper, Gold, Manganese

Location. The Aurum District occupies the N. part of the Schell Creek Range. Schellbourne on the W. side of the range, is 18 m. S.E. of Cherry Creek, which is on the N. N. R. R. while Aurum post office, on the E. side near the S. end of the district is 42 m. by road S. S. E. of Cherry Creek. The elevations of Spring Valley on the E. of the Schell Creek Range and of Steptoe Valley on the W. are about 5,500 ft. and the crest of the range is over 9,000 ft. above sea-level. The Aurum District is sometimes subdivided into the Schell Creek of Schellbourne District, at the N.; the Siegel, formerly the Queen Springs District, next S.; the Silver Canyon District; the Ruby Hill District; and the Müncy Creek, formerly the Silver Mountain District, at the extreme S.

History. The Aurum District was discovered in 1871 and has been intermittently active up to the present time. Silver was the principal product of the early days, but lead and copper have become important of recent years.

Geology. The Schell Creek Range is composed of Cambrian quartzites, limestones, and shale, intruded by dikes and stocks of granite porphyry and capped at the N. end by Tertiary eruptives, according to Hill. Most of the ore bodies are in the limestones which overlie the quartzite, but the Signal Mine is at the top of the quartzite. The ore deposits are commonly replacements of the limestone by silver-lead ore, and in the Seigel Mine the primary ore minerals are pyrite, galena, and arsenopyrite; while the oxidized ore consists mainly of maganese and iron. Contact-metamorphic deposits containing pyrite and chalcopyrite and carrying gold and silver occur N.W. of Aurum post office. At the Siegel Mine the quartzite and a small dike of granite porphyry have been replaced by pyrrhotite and chalcopyrite.

Bibliography. R1871 200-3; R1872 168-9; R1873 229; R1874 268, 277-8; R1875 192-3; SMN1871-2 144-5; SMN1873-4 88; MR1905 273; MR1913 I 840-1; MR1914 I 710; MR1915 I 652, 654; MR1916 I 497; MR1917 I 295; MR1918 I 260; MR1919 I 413; MR1920 I 356

Hill507 226, 228. Hill648 186-196.

Pardee, J. T., & Jones, E. L., Jr., "Deposits of Manganese Ore in Nevada," USGS B 710F (1920) 212-6.

Spurr 208 38-47 Schell Creek Range.

Thompson & West 655-6, 663.

WeedMH 1127 Acme Cons. Manganese Assoc.
1131 American Standard Coal. M. Co.
1244 Lucky Deposit M. Co. 1261-2 Muncy Creek M. Co.
1375-6 White Pine C. C.

BALD MOUNTAIN

Silver, Copper, Gold, (Tungsten)

Location. The Bald Mountain District is situated in the neighborhood of Joy post office in the pass between Bald Mt. and South Bald Mt. at the S.

end of the Ruby Range. The approximate elevation of Joy is 7,400 ft; Bald Mt., 9,400 ft.; and South Bald Mt. 9,000 ft. Eureka which is on the E. N. R. R. is 40 m. S.W. of Joy.

History. The district was discovered by G. H. Foreman and others in 1869. The most important mine was the Nevada which is credited with a production of some $20,000 worth of silver chloride ore. The Copper Basin property was opened in the late seventies and shipped a little copper carbonate ore in 1905-6. Tungsten ore was discovered in the district prior to 1917.

Geology. The country rocks of the Bald Mountain District are Paleozoic limestones with some quartzite which have been intruded by quartz-monzonite porphyry. Bald Mt. is a low anticline composed of Ordovician sedimentary rocks and with its W. flank eroded. The main mass of the mountain is composed of upper Pogonip limestone, with the overlying Eureka quartzite and Lone Mt. limestone appearing to the E. A fault striking N. 60 degrees W. occurs in the pass to the S., and South Bald Mt. is made up of lower Pogonip limestone. The main body of the quartz-monzonite porphyry stock has been intruded into the Pogonip limestone along this fault. Veins of gold-bearing quartz carrying pyrite, stibnite, and chalcopyrite are present in the quartz-monzonite porphyry; replacement copper ores occur in the limestone; and copper and tungsten are found in contact metamorphic zones.

Bibliography. R1875 192 SMN1869-70 78-9 MR1914 I 710
Hess, F. L., and Larsen, E. S., "Contact-Metamorphic Tungsten Deposits of the United States," USGS B 725D (1921) 306-7.
Hill648 152-161. Thompson & West 652.

BLACK HORSE
Gold, Silver

Location. The Black Horse District is located at Black Horse on the E. flank of the Snake Range. Black Horse is 49 m. by road E. S. E. of Ely which is on the N. N. R. R. The Sacramento District adjoins the Black Horse District on the N.W. and the Osceola District on the S.W., and it is sometimes considered as a section of the latter district.

History. The Black Horse District was discovered in 1905 and produced from 1906 to 1911. The Black Horse Mine operated an arrastra in 1909 and an amalgamating mill in 1910.

Geology. The ore deposits are auriferous veins carrying a little silver, according to Hill, and the country rock consists of Paleozoic sediments cut by granite porphyry.

Bibliography. MR1905 274 MR1907 I 383 MR1910 I 534
MR1906 300 MR1909 I 428 MR1911 I 700
Hill507 206-7.
Weeks, F. B., "Geology and Mineral Resources of the Osceola Mining District, White Pine County, Nev,", USGS B 340 (1908) 123-4.

BONITA see SNAKE

CHERRY CREEK (Egan Canyon, Gold Canyon)
Silver, Gold, Lead, Copper, (Molybdenum, Tungsten, Barite)

Location. The Cherry Creek District is located at Cherry Creek in the Egan Range. Cherry Creek station on the N. N. R. R. is 4 m. E. Egan Canyon is 5 m. S. of Cherry Creek in the Egan Range. The Gold Canyon District in Egan Canyon adjoins the Cherry Creek District on the S. and is commonly considered a part of the Cherry Creek District. Cherry Creek has an altitude of 5,800 ft., and to the N. and W. the Egan Range rises to elevations in excess of 8,000 ft.

History. Gold was found in Egan Canyon by a company of volunteer soldiers under Captain Tober and the Gold Canyon District was organized in 1863. The Gilligan silver mine which was the principal early mine of this part of the district was discovered in 1864 and a 5-stamp mill erected that same year by the Social Co. to work its ores. This company consolidated with another in 1866 to form the Social & Steptoe M. Co., and some $80,000 in bullion was produced by the mill before it shut down in 1868. The Teacup Mine was discovered in 1872 and was the first on Cherry Creek. The new section of the district became the most important, the Star and the Exchequer being its other noteworthy early mines. The Cherry Creek mines were most active from 1872 to 1883, during which period the population of Cherry Creek rose to 6,000. Decline followed and production practically ceased in 1893. Later, interest revived and the mines have been working steadily although on a small scale up to the present day. Tungsten ore and barite were discovered in 1915.

Production. Estimates as to the early production of the Cherry Creek District vary from $6,000,000 to $20,000,000. The production of the district from 1905 to 1921 was 22,869 tons of ore containing $241,044 in gold, 310,251 ozs. silver, 94,486 lbs. copper, and 642,825 lbs. lead, valued in all at $536,716, according to Mineral Resources of the U. S. Geol. Survey.

Geology. In the Cherry Creek District, quartzites, shales, and limestones which are probably of Cambrian age occur with intrusive quartz-monzonite which is probably of Cretaceous age. The older rocks of the district are exposed along the E. base of the Egan Range and consist of quartzites with interbedded shales having a thickness of at least 3,000 ft. and believed to correspond to the Cambrian Prospect Mt. quartzite of Eureka. Overlying these beds at the W. end of Egan Canyon are calcareous shales and to the N. W. of Cherry Creek is a limestone with shales near its base. These formations are thought to belong to the Cambrian Eldorado limestone of Eureka. The quartz-monzonite intrusions are in the form of small dikes and masses. Masses have been intruded into the quartzite at the mouths of Cherry Creek and Egan Canyons, and dikes occur at the Cherry Creek and Star mines.

Orebodies. The orebodies are veins in the Cambrian quartzite and to a lesser extent in the quartz-monzonite. These veins are strongest and richest in the quartzite, becoming weak and barren in the interbedded shales. The deposits are of two types which merge into one another through intermediate examples. The simpler and less important type is the gold-quartz vein carrying minor pyrite and galena; and the more important type is the vein carrying galena, sphalerite, pyrite, and rich secondary silver minerals. Copper is present in many of the veins, occurring as chalcopyrite, tetrahedrite, chalcocite, copper carbonates, and copper silicate. Molybdenite occurs with pyrite at one prospect.

Mines. Two mines are at present operating in the district. The **Cherry Creek Silver M. Co.** is is operating the Mary Ann Mine. Its home office is at 611 Deseret Bldg., Salt Lake City and J. H. Moyle is Pres. J. W. Walker is Mgr. at Cherry Creek. The ground owned by this company is opened by a 777-ft. tunnel and winzes and the equipment includes a 35 hp. gas engine and compressor. The **Silver Dividend M. Co.** is owned by Church

& Crampton of Pioche. The Exchequer Mine which it is operating is credited with a production of $3,000,000 in silver in the early days of the camp. G. Z. Smith is Supt. The mine was reopened in the summer of 1921. It is being developed, the higher grade of ore is being shipped, and the construction of a cyanide mill is being considered.

Bibliography.

B1867 402, 407, 411 SMN1875-6 165-6. MR1914 I 711.
R1868 95-6 SMN1877-8 158-160. MR1915 I 652, 825 Tungsten.
R1869 175-7. MR1905 273. MR1915 II 175-6 Barite.
R1873 227-9. MR1906 300. MR1916 I 497.
R1874 267, 273-7. MR1907 I 382. MR1917 I 295.
R1875 189, 192, 193. MR1908 I 504. MR1918 I 260.
SMN1866 98, 102-3 MR1909 I 428. MR1919 I 413.
SMN1867-8 58-60. MR1910 I 532. MR1920 I 336.
SMN1869-70 79-80. MR1911 I 699. MR1921 I 396.
SMN1871-2 145. MR1912 I 816.
SMN1873-4 88.
USGE40th 3 (1870) 445-9.
Davis 1045-6. Hill507 227. Hill648 161-171.
Spurr208 47-54. StuartNMR 100-1, 195-6.
Thompson & West 653-4, 657-8.
WeedMH. 1146 Biscuit M. Co. 1148 Black Metals M. Co.
1164 Cherry Creek S. M. Co. 1284 Nevada Quincy M. Co.
1294 North Mt. M. Co. 1302 Penn-Star M. Co.
1348 Teacup M. Co.

DUCK CREEK (Success)

Lead, Silver, Gold, Copper, Zinc

Location. The Duck Creek District is situated upon Duck Creek Ridge, a branch of the Schell Creek Range separated from the main range by Duck Creek on the E. Most of the mines are within 3 m. of McGill on the N. N. R. R.; but the Success Mine is E. of Ely and 18 m. by road from the railroad, and the section in which it is located is sometimes known as the Success District. Duck Creek Ridge has an altitude of 8,200 ft. The Nevada District adjoins the Duck Creek District on the S.E.

Production. The production of the district from 1905 to 1921 was 3,072 tons containing $9,796 in gold, 20,651 ozs. silver, 40,693 lbs. copper, 2,639,047 lbs. lead, and 24,494 lbs. zinc, valued in all at $164,015, according to Mineraal Resources of the U. S. Geol. Survey.

Geology. Duck Creek Ridge is composed of limestones which are perhaps of Ordovician age, overlain near the W. base of the hills by shale. Igneous rocks are rare. A small granite butte at the mouth of Duck Creek Canyon is capped by rhyolite which extends onto the limestone, and two small dikes possibly of granite porphyry are present at the Ely Gibraltar Mine, according to Hill. The ore deposits are lead-silver replacements along fissures parallel to the bedding of the limestone. Galena is the principal ore mineral although cerussite and anglesite are abundant. At the Ely Gibraltar, copper pitch and carbonate ore and zinc are associated with the lead minerals.

Bibliography.

MR1905 273 MR1913 I 841 MR1918 I 260
MR1909 I 428 MR1914 I 711 MR1919 I 413
MR1910 I 532 MR1915 I 652 MR1920 I 336
MR1911 I 699 MR1916 I 497 MR1921 I 396
MR1912 I 816 MR1917 I 295-6

Hill648 196-9. Spurr208 40.
WeedMH 1238 Lead King M. & M. Co.

EAGLE (Regan, Tungstonia, Kern, Pleasant Valley)
Lead, Silver, Gold, Copper, Tungsten

Location. The Eagle District is situated in the Kern Mts. S. of Regan and Pleasant Valley in E. White Pine Co. on the Utah border. The tungsten mines are in the neighborhood of Tungstonia in the S.E. section of the district. Cherry Creek on the N. N. R. R. is 65 m. distant.

History. The Eagle District was discovered in 1859 by employees of the Overland Mail Co. It was organized as the Pleasant Valley District in 1868; the name changed to Kern District in 1869; and the district enlarged and rechristened the Eagle District in 1872. The Glencoe Mine was worked in the early eighties. The Red Hills Mine was worked from 1911 to 1918, and other properties in the district were operated during this period. Tungsten was discovered by C. G. Sims and C. Olsen in 1910. In 1915, A. R. Shepherd, Jr., and associates erected a tungsten mill which they operated in 1916. The Salt Lake Tungstonia Ms. Co. erected a 25-ton mill in 1916, and the Utah-Nevada M. & M. Co. began the construction of a mill that same year but did not complete it.

Geology. The Kern Mountains are composed of granite intrusive into quartzite, shale, and limestone, according to Hill. The sedimentary rocks are believed by Spurr to be of Cambrian age. At the Glencoe Mine, a quartz vein in granite near a limestone contact contains galena, sphalerite, and enargite and carries values in gold and silver. At Tungstonia, the granite is cut by a series of biotite-porphyry dikes and a few dikes of porphyritic biotite granite. Quartz veins carrying huebnerite cut these rocks, which are sericitized and pyritized for short distances on both sides of the veins. Galena, sphalerite, and a small amount of bismuthinite are present here and there in the veins, and in vugs in the main vein, fluorite and a smaller amount of triplite occur with huebnerite. At the Red Hills Mine, lead carbonate ore occurs in a 20-ft. zone of brecciated limestone. This ore contains a little silver, gold, and copper.

Bibliography.

R1869 174	MR1913 I 841	MR1910 I 739 Tungsten
R1875 193-4	MR1914 I 712	MR1911 I 943 "
SMN1869-70 81-3.	MR1915 I 652	MR1915 I 825 "
MR1911 I 700	MR1916 I 498	MR1916 I 793 "
MR1912 I 816	MR1917 I 296	
	MR1918 I 260	

Hill507 228. Hill648 202-7.
Spurr208 28-9 Kern Mts. Thompson & West 654.
WeedMH 1324 Salt Lake Tungstonia Ms. Co.

EGAN CANYON see CHERRY CREEK

ELY (Robinson)
Copper, Gold, Silver, Zinc, Lead, Manganese, (Barite, Phosphate Rock, Potash)

Location. The Ely District is located at Ely in the Egan Range. Ely is a station on the N. N. R. R., which passes through the town to the mines on the W. Ely is 6,421 ft. in elevation, and the taller peaks in the Egan Range in that vicinity rise to altitude of about 8,000 ft.

History. Indian John showed the district to a party of prospectors in 1867. It was organized as the Robinson District in 1868, named after Thomas

Robinson, a member of the party. A 10-ton lead blast furnace was erected in 1869, and a 10-stamp mill was built the same year. Subsequently a number of furnaces and mills were put up, but none met with marked success. Interest in the Ely District was renewed when the county seat was moved from Hamilton to Ely in 1886. A 10-stamp mill was erected at the Chainman gold mine in 1889, and this mine was the most important in the district at that period, continuing active till 1901. The Giroux Cons. M. Co. was organized in 1903 and the Nevada Cons. Copper Co. in 1904. Work was started on the N. N. R. R. in 1905 and the line completed in 1906. A consolidation was effected with the Cumberland Ely Copper Co. whereby the Nevada Cons. Copper Co. took over the Cumberland Ely and the N. N. R .R. and built the Steptoe Valley mill and smelter at McGill. Construction of the metallurgical plant began in 1906 and the first copper was turned out in 1908. This was the first production of copper from porphyry ores in the Ely District and marked the beginning of a new era, the production of the district jumping from $2,363 in 1907 to $2,063,888 in 1908 and continuing to rise till a maximum of $26,478,391 was reached in 1917. The Giroux Cons. M. Co. built a concentrator and blast furnace in 1906. The Consolidated Coppermines Co. took over the Giroux and other properties in 1913 and remodeled the concentration plant operating it from 1917 to 1919. Some Coppermines ore was milled at the Nevada Consolidated mill under contract. Manganese ore was mined in the Ely district during the World War; and the mill tailings of the district were investigated as a possible source of potash. The low price of copper caused the Coppermines property to close down in 1920 and the Nevada Consolidated in 1921. The Nevada Consolidated resumed operations in 1922, its mill burning down and being rebuilt that same year; while the Coppermines Company was reorganized in 1922 and is now operating. Work is also being carried on in the district by the Boston & Ely Cons. M. Co. and the Ely Calumet M. Co.

Production. The production of the Ely District prior to 1904 was some $500,-000. From 1904 to 1921, the total value of the district's production has been $164,234,722; as shown in detail on the following table.

PRODUCTION OF ELY DISTRICT
(According to Mineral Resources of the United States, U. S. Geological Survey)

Year	Mines	Tons	Gold $	Silver Ozs.	Copper Lbs.	Lead Lbs.	Zinc Lbs.	Total $
1904	4	175	3,493	6,638		225,900		14,009
1905	No production.							
1906	No production.							
1907	1	29			11,814			2,363
1908	6	430,031	94,443	30,391	14,797,295	2,381		2,063,888
1909	10	1,671,225	370,569	79,487	57,481,515	524,512		7,907,053
1910	10	2,375,032	423,225	108,000	63,914,197	180,407		8,606,586
1911	4	2,776,930	435,287	104,107	67,033,547	10,216		8,870,118
1912	5	3,023,082	474,436	78,020	67,373,510	12,168		11,639,595
1913	3	3,447,812	430,347	68,793	71,916,711	15,756		11,619,681
1914	2	2,749,913	352,340	52,729	51,089,921			7,176,459
1915	14	3,206,133	568,130	82,281	64,661,610	654,905	42,651	11,961,699
1916	8	4,053,950	737,964	110,802	93,044,578	92,127		23,706,195
1917	16	4,467,409	765,376	131,825	93,627,420	512,862		26,478,391
1918	21	4,520,936	683,823	104,615	92,419,990	479,139	662,885	23,709,607
1919	8	2,273,961	323,143	56,129	47,871,048	137,768	2,952,279	9,512,841
1920	12	2,589,546	471,510	65,008	48,347,828	136,194	1,314,837	9,555,766
1921	4	400,810	82,626	16,069	10,282,527	37,809		1,426,843
Total 1904-'21		37,987,070	6,213,219	1,088,246	843,861,697	2,796,244	4,962,652	164,234,722

Geology. The country rocks of the Ely District are Paleozoic sediments intruded by monzonite porphyry and capped in places by Tertiary volcanic rock. The Paleozoic rocks consist of limestones, quartzites and shales ranging in age from Ordovician to Pennsylvanian, according to Spencer, and having an aggregate thickness of more than 9,000 ft. These sedimentary rocks have been greatly disturbed by faulting and to a minor degree by folding, and their distribution is very irregular. The monzonite porphyry was probably intruded at the close of Jurassic time. It forms dikes, sills, and stocks and has caused extensive alteration of the sediments in its neighborhood besides being itself metamorphosed. The Tertiary volcanic cap consists of tuff and agglomerate, obsidian, and rhyolite. The rhyolite is the most prominent of the volcanic rocks and while occuring mainly as surface flows is sometimes intrusive.

Orebodies. The principal orebodies in the Ely District are pyritized monzonite porphyry mases containing secondary copper ore. Small bodies of higher grade copper ore occur in the sedimentary rocks near the contact, and silver-lead ore occurs at a greater distance. The Chainman Mine contained bedded precious metal deposits, one at the top of a limestone bed and the other some 20 ft. or 30 ft. higher in shale. Fissure veins carrying precious metals have also been worked. Most of the manganese deposits occur in the precious metal lodes, according to Pardee and Jones.

Ore. The metallic minerals in the porphyry ore are primary pyrite, chalcopyrite and molybdenite with rare sphalerite and galena and secondary chalcocite. Gold and silver are present in small amounts, and in refining the blister copper the presence of nickel, bismuth, platinum, palladium, selenium, and tellurium is revealed.

Nevada Consolidated. The Nevada Cons. Copper Co. owns the largest property in the Ely District. This company was incorporated in Maine in 1904 and has a capital stock of 2,000,000 shares of $5 par value; 1,000,500 of which are owned by the Utah Copper Co. The shares are listed on the New York and Boston Stock Exchanges. Dividends to the amount of $46,768,616 were paid from 1909 to 1920. The home office is at 25 Broad St., New York City. D. C. Jackling is Pres.; C. Hayden and W. E. Bennett are V. P.; and C. V. Jenkins is Sec.-Treas. C. B. Lakenan is Gen. Mgr., at McGill. The property consists of 133 claims including the Copper Flat Group, Ruth Group, and Veteran Group; and a 14,000-ton concentrating plant and smelter at McGill where the company owns 8 sq. m. of land; connected with the mines by a 27-m. standard gage railway line which is part of the N. N. R. R.

Consolidated Coppermines. The Cons. Coppermines Co. was reorganized in 1922 with a capital stock of 1,600,000 shares of $5 par value. R. Marsh, Jr., is Mgr. at Kimberly. The property consists of 1,770 acres of mining properties, 2,656 acres of ranch land, and 520 acres of other land; over 99 per cent of the stock of the Giroux Cons. Co.; 422,625 out of a total of the 500,000 shares of the Butte & Ely C. Co.; and a 1,000-ton flotation plant at Kimberly.

BIBLIOGRAPHY OF ELY DISTRICT

General.

R1870 158-9.
R1872 169-170.
R1873 226-7.
R1874 267, 272-3.
R1875 192-193.
SMN1869-70 86-7.
SMN1871-2 144.
SMN1873-4 86-7.
SMN1875-6 167.
SMN1877-8 157.
MR1905 274, 355.
MR1906 300, 450 Lead.
MR1907 I 383-4, 612.
MR1908 I 212-3, 504-5.
MR1909 I 170, 428-9.
MR1910 I 533-4.
MR1911 I 700-2.
MR1912 I 817-8.
MR1913 I 841-2.
MR1914 I 712-4.
MR1915 I 653-4; II 175-6 Barite.
MR1916 I 498-500.
MR1917 I 296-8; II 12 Phosphate Rock.
MR1918 I 260-3.
MR1919 I 413-6.
MR1920 I 336-7.
MR1921 I 396-7.

Abbott, J. W., "A Visit to Ely, Nevada," M&SP 94 (1907) 759-761.
Bullock, W. S., "The Low-Grade Copper Deposits at Ely, Nevada," E&MJ 83 (1907) 509-511.
———"Copper Deposits at Ely, Nevada," M&M 27 (1907) 518-520.
Calkins, F. W., "The Porphyry Copper's Past and Future," E&MJ 113 (1922) 350-1.
De Kalb, C., "Copper Mining at Ely, Nevada," M&SP 98, (1909) 58-60.
Giroux, J. L., "The Giroux Mines, Nevada," E&MJ 82, (1906) 985-6.
Goodwin, L. H., "The Porphyry Coppers, Part I, Resources and Achievements," E&MJ 106 (1918) 780-2.
Herrick, R. L., "Mining and Reduction of Ely Ores," M&M 29 (1908) 22-5
Ingalls, W. R., "The Copper Mines of Ely, Nevada," E&MJ 84 (107) 675-682.
Ralph, W., "Mining Conditions at Ely, Nevada," M&SP 94 (1907) 120-1.
Requa, M. L., "Copper Mining in Nevada," M&SP 93 (1906) 546.
Smith, F. D., "The Ely Mining District of Nevada," E&MJ 70 (1900) 217.
Sorenson, S. S., "Changes at the Nevada Consolidated," M&SP 106 (1913) 70.
Stuart NMR 9, 11, 96-8.

Parsons, A. B., "Nevada Consolidated Copper Company, I History of the Enterprise," M&SP 122 (1921) 325-334; Discussion, E. G. Deane, 413.

WeedMH 1151 Boston & Ely Cons. M. C. 1158 Butte & Ely C. Co.
1170 Consolidated Coppermines Co.
1192 Ely Amalg. C. Co., Ely Calumet C. C.
1193 Ely Cons C. Co., Ely Giroux C. Co., Ely-Mizpah C. Co., Ely Northern C. C., Ely Revenue C. Co., Ely Verdi C. Co., Ely Witch C. Co.
Federal Ely C. Co. 1206 Giroux Cons. Ms. Co.
1201 Federal Ely C. Co. 1206 Giroux Cons. Ms. Co.
1308 Princess C. Co.

Bullock, W. S., "Copper Deposits at Ely, Nevada," M&M 27 (1907) 518-520.

Butler, B. S., "Potash in Certain Copper and Gold Ores," USGS B 620J (1915) 231.

E&MJ, "Manganese Deposits in White Pine County, Nevada," 106 106 (1918) 308.

Hill507 227.

Lawson, A. C., "The Copper Deposits of the Robinson Mining District, Nevada," Cal. U. Dept. Geol., B 4 (1906) 287-357; Review by W. Lindgren, EG 2 (1907) 195-304

Pardee, J. T. & Jones, E. L., Jr., "Deposits of Manganese ore in Nevada," USGS B 710 (1920) 218-220.

Spencer, A. C., "Notes on the Ely District, Nevada," USGS B 529 (1913) 189-191.

———"The Geology and Ore Deposits of Ely, Nevada," USGS PP 96 (1917) 189; M&SP 116 (1918) 857-862; E&MJ 105 (1918) 105.

Spurr208. 47-54.

Weed, W. H., "The Copper Mines of the United States in 1905," USGS B 285 (1906) 116, 117.

———"Brief Notes on the Geology of the Ely District, Nevada," M&EW 36, (1912) 1198.

Whitman, A. R., "Notes on the Copper Ores at Ely, Nevada," Cal. U. Dept. Geology B 8 (1914) 309-318.

Williams, S. H., "Limonite Deposits of the Ely Mining District, Nevada," SLMR 25 (1923) 13-14.

Mining.

Brown, W. R., "Blast-Hole Drilling, Nevada Consolidated," E&MJ 96 (1913) 982-4; M&SP 107 (1913 643-6.

Hart, C. E., "Prospecting with Keystone Drill for Copper Ore in the Ely, Nevada, District., E&MJ 83 (1907) 804-5.

Hillen, A. G., "Mines and Mining Operations at Ely, Nevada," M&EW 45 (1916) 403-7.

Larsh, W. S., "Branch Raise System at the Ruth Mine, Nevada Consolidated Copper Co.," T AIME 59 (1918) 299-304; Abstract, E&MJ 105 (1918) 503-5.

McDonald, P. B., "Mining at the Nevada Consolidated," M&SP 112 (1916) 858-61.

Monroe, H. S., "The Sinking of the Alpha No. 2 Shaft at Kimberly, Nevada," M&SP 121 (1920) 871-2.

Parsons, A. B., "Nevada Consolidated Copper Company," M&SP 122 (1921) "II Steam-Shovel Mining," 525-534. "III Underground Mining at Ruth," 709-716.

Pheby, F. S., "Prospecting with Keystone Drill for Copper Ore in the Ely, Nevada, District," E&MJ 83 (1907) 804-5.

Metallurgy.

Duncan, "Burning Reverberatory Ash at the Steptoe Plant," E&MJ 90, (1910) 1302.

——"A Good Air-Relief Valve," E&MJ 103 (1917) 270-1.

E&MJ, "A Wet-Crushing Cyanide Plant at Ely, Nevada," 71 (1901) 753-5.

——"The Steptoe Distributer," 92 (1911) 788.

——"The Yeatman Classifier," 92 (1911) 788-9.

——"Nevada Consolidated Canvas Tables," 94 (1912) 301-3.

——"Chilled Conveyor Idlers," 101 (1916) 616; Discussion, W. R. Bylund, 832.

Linton, R., "Treatment Tests on Ores of Consolidated Coppermines Co.," T AIME 64 (1921) 816-829.

Marsh, A. G., "Lubrication at Steptoe Concentrator," E&MJ 95 (1913) 1237-9.

Parsons, A. B., "Nevada Consolidated Copper Company, M&SP 122 (1921) "IV-Concentration of the Ore," 323-331. "V-The Smelter," 393-402.

Pomeroy, R. E. H., "Reverberatory Smelting Practice of Nevada Consolidated Copper Company," T AIME 51 (1916) 764-772.

——"Coal Pulverizing Plant at Nevada Consolidated Copper Smelter," T AIME 64 (1921) 618-629; Abstract E& MJ 109 (1920) 697.

Ingalls, W. R., "The Steptoe Valley Mill & Smelter," E&MJ 84 (1917) 813-6.

Requa, M. L., "Experimental Mill of the Nevada Consolidated Copper Co.," M&SP 97 (1908) 90-5; Discussion, C. E. Parsons and M. L. Requa, 695-6.

Wickham, W. H., "Cyaniding Tests at Ely, Nevada," M&SP 86 (1903) 134.

Miscellaneous.

Duncan, L., "The Water Supply of the Nevada Consolidated," E&MJ 109 (1920) 857-8.

——"Accident Prevention at the Nevada Consolidated," M&SP 108 (1914) 288-290.

E&MJ, "Safety First at Ely, Nevada," 97 (1914) 629-630, Editorial; "By the Way"; 726, 1069.

——"Physical Control of Employees," 100 (1915) 759, 889.

Goodwin, L. H., "The Porphyry Copper. Part III—Mine and Equipment Investment," E&MJ 106 (1918) 905-6.

Ingalls, W. R., "The Productive and Earning Capacity of Ely," E&MJ 84 (1907) 719-722.

M&SP, "Industrial Accidents Under Compensation," 108 (1914) 332.

"Rules and Regulations of the Nevada Consolidated Copper Company," Editorial, 439; "General Rules for Safety," 460; "Whistle Signals," 536; Blasting and Use of Explosives," 577; "Ore and Dump Train Service," 698.

———Editorial, "Stores in Mining Communities," 121 (1920) 649-650.
Smith, F. L., "Time Saving Devices in an Assay Office," E&MJ 110 (1920) 111-2.
USGS Ely Special topographic map.

GRANITE (Steptoe)
Gold, Lead, Silver

Location. The Granite District is located at Steptoe post office, 6 m. S.W. of Granite, a siding on the N. N. R. R. It is on the E. flank of the Egan Range and adjoins the Hunter District on the S.

History. The district was discovered by W. D. Campbell in 1894, and the gold ores worked in a 5-stamp mill on Campbell's ranch. The Cuba lead mine was discovered in 1902. The district was active on a small scale up to 1918.

Geology. The country rocks are Cambrian quartzite, shale, and limestone intruded by granite, according to Hill. The ore deposits are of two types, quartz veins in quartzite carrying gold and a little silver, and fissure replacements in limestone carrying lead and a very little silver. In the first type, metallic minerals are rare; while in the second, galena occurs accompanied by a minor amount of cerussite and anglesite, and the gangue is calcite with some siderite in the Cuba Mine.

Bibliography. MR1908 I 504 MR1914 I 711-2 MR1917 I 296
MR1909 I 428 MR1915 I 653 MR1918 I 260
Hill507 228. Spurr208 47-54 Egan Range. Stuart NMR 103-5.

HAMILTON see WHITE PINE
HUB see TUNGSTEN

HUNTER
Lead, Silver, Copper

Location. The Hunter District is located on the W. flank of the Egan Range in central White Pine Co. Cherry Creek, which is on the N. N. R. R., is 25 m. N. N. E. The Granite District adjoins the Hunter District on the S.

History. The Hunter District was discovered in 1871. From 1877 to 1884 it was extensively worked and the ore was smelted on the ground. The Vulcan M., S., & R. Co. purchased the old Hunter Mine in 1907, but the company has been inactive since 1913 and the mine has been operated by leasers.

Geology. The country rock of the Hunter District is dark massive-bedded dolomitic limestone which is probably Devonian and has been intruded by granite porphyry dikes, according to Hill. The ore bodies occur at the contact of altered granite porphyry and limestone and mainly in the brecciated limestone. The ore is an ocherous lead carbonate containing residual galena coated with anglesite. Malachite is abundant in some places and most of the limonitic ore carries copper.

Bibliography. R1875 192-3. MR1906 300 MR1917 I 296
SMN1871-2 145 MR1913 I 841 MR1918 I 260
SMN1873-4 89 MR1914 I 712 MR1920 I 336
SMN1875-6 171 MR1915 I 653 MR1921 I 396
MR1916 I 498
Hill507 227. Hill648 172-4. Spurr208 47-54 Egan Range.
Thompson & West 654. WeedMH 1368 Vulcan M., S., & R. Co.

KERN see EAGLE
LEXINGTON see SHOSHONE

LINCOLN see TUNGSTEN
MINERVA see SHOSHONE
MUNCY CREEK see AURUM

NEVADA
Manganese, Silver

Location. The Nevada District is situated on the W. slope of the Schell Creek Range. Ely which is on the N. N. R. R. is 10 m. N.W.

History. Silver was discovered in the district in 1869 by a party of prospectors from Reno. The ore was milled at Mineral City, now Ely, in a 10-stamp mill which burned in 1875. During the World War manganese was discovered in the Nevada District and a moderate amount was mined in 1917 and 1918.

Geology. The country rock is limestone which is probably of Devonian age. The manganese deposits lie at the base of the Schell Creek Range on the E. side of Steptoe Valley. They are thick pod-like or pipe-like masses at and below the surface, constituting parts of irregular jaspery quartz lodes that replace the limestone along fractures, joints, and bedding planes. The orebodies are generally loose, friable, and cavernous, and consist mainly of manganese oxides, quartz, calcite, and small amounts of iron oxides, with considerable fluorite in places, and about 2 ounces of silver to the ton.

Bibliography. R1874 273 | SMN1869-70 84-5. | SMN1875-6 166-7
SMN1873-4 87.

Pardee, J. T., & Jones, E. L., Jr., "Deposits of Manganese Ore in Nevada," USGS B 710F (1920) 216-8.

NEWARK
Silver, Lead, Copper

Location. The Newark District is in W. White Pine Co. on the E. slope of the Diamond Range. It lies a short distance E. of Eureka, which is on the E. N. R. R.

History. The district was discovered by Stephen and John Beard and other prospectors from Austin in 1866. The principal mines were purchased by the Centenary Silver Co. in 1867, and a 20-stamp pan amalgamation mill was brought from the Kingston District by this company the same year. The company was composed chiefly of members of the Methodist Church, in honor of whose centenary is was named, and the construction of a church at Austin was one of the conditions of the organization. About $100,000 in bullion was produced from shallow workings in the course of a few years and the district then became inactive. Interest revived in later years, and shipments of silver-lead ore containing a little copper were made from the Bay State Mine of the Newark M. & M. Co. from 1905 to 1908 inclusive; and from the Battery Mine in 1913.

Geology. According to Hague, the ore deposit is a nearly vertical vein in limestone. The gangue consists of quartz and calcite, and the ore, which is usually associated with the quartz, is composed of antimonial and sulphureted combinations of lead, silver, and copper.

Bibliography.

B1867 402	SMN1866 98	MR1905 274
R1868 97-8	SMN1867-8 48-9	MR1906 300
R1874 267-8	SMN1869-70 65	MR1907 I 384
R1875 192	SMN1877-8 155-6	MR1908 I 505
		MR1913 I 841

USGS40th 3 (1870) 443-4.
Spurr208 81-4. StuartNMR 956. Thompson & West 665.

OSCEOLA

Placer Gold, Gold, Silver, Lead, Tungsten, (Phosphate Rock)

Location. The Osceola District is situated at Osceola on the W. flank of the Snake Range. Ely, which is on the N. N. R. R., is 40 m. W. N. W. Osceola is 6,800 ft. above sea-level and the mountains to the E. rise to an altitude of 9,600 ft. The Sacramento District adjoins the Osceola District on the N., and the Black Horse District adjoins it on the N.E., and is sometimes considered as a section of the Osceola District.

History. The gold lodes were discovered by Matteson and Heck in 1872; and the placer mines by John Versan in 1877. The gold ore was first worked by arrastras, but a 5-stamp mill was erected in 1878. The most important placers were operated by the Osceola Co. from the early eighties to 1900. Both placers and lodes have been operated irregularly and intermittently down to the present time. Tungsten was discovered in the district in 1916, and the Pilot Knob group erected a 20-stamp mill. Phosphate rock was discovered in 1917, and lead ore shipped in 1918. In 1921, the Sunrise property operated a 2-stamp mill and the American Group a 10-stamp mill, producing gold bullion with a little silver content.

Production. According to Weeks, the production of the Osceola District up to 1907 may be safely estimated at $2,000,000; of which about one-tenth came from quartz mines, the remainder being from placers. Stuart states that estimates range from $3,0000,00 to $5,000,000.

Geology. Cambrian conglomerate, argillite, quartzite, and limestone have been intruded by granite porphyry, according to Weeks. The country rock of the auriferous lodes is quartzite and the ore occurs in regular zones of fracturing or sheeting and in irregularly shattered masses of quartzite adjacent to these zones of fracture. The gold commonly occurs in flakes and finely disseminated in quartz seams and veinlets, but in the Cumberland Mine it is present in vugs lined with fluorite and other minerals. At the Pilot Knob Group, scheelite occurs in quartz veins in limestone; while the Lucky Boy Mine has silver-lead ore, according to Mines Handbook.

Bibliography. SMN1873-4 78
SMN1875-6 170-1
SMN1877-8 157-8
MR1904 200
MR1905 274
MR1907 I 383
MR1908 I 505
MR1909 I 430
MR1910 I 534
MR1911 I 700
MR1912 I 816
MR1913 I 841
MR1914 I 712
MR1915 I 653
MR1916 I 498
MR1917 I 296
II 12 Phosphate Rock.
MR1918 I 260
MR1919 I 413
MR1920 I 336
MR1921 I 396

Hill507 227-8. Spurr208 25-36 Snake Range.
StuartNMR 98-100. Thompson & West 662.
WeedMH 1244 Lucky Boy M. Co. 1305 Pilot Knob Group.
Weeks,F. B., "Geology and Mineral Resources of the Osceola Mining District, White Pine Co., Nev.", USGS B 340 (1908) 117-133

PIERMONT

Silver, Gold

Location. The Piermont District is located at Piermont on the E. slope of the Schell Creek Range in central White Pine Co.

History. The Piermont District was discovered in 1869 and a 10-stamp mill erected in 1871. The mine shut down in 1873 after having made a production which is unofficially estimated at $6,000,000. The property is now owned by the Piermont Ms. Co. of which J. N. Popham, 17 State St., New York City, is in charge.

Geology. According to Raymond, the veins are in slate or at the junction of slate and quartzite; while Mines Handbook state that the lode at the Piermont Mine is a contact vein between brecciated limestone and prophyry.

Bibliography. R1871 203-4 SMN1869-70 84 SMN1871-2 133
R1875 193 SMN1873 4 89
Thompson & West 655. WeedMH 1313 Piermont Ms. Co.

PLEASANT VALLEY see EAGLE
QUEEN SPRINGS see AURUM
REGAN see EAGLE
ROBINSON see ELY
RUBY HILL see AURUM

SACRAMENTO

Tungsten, Gold, Silver

Location. The Sacramento District is situated at Sacramento Pass on the W. flank of the Snake Range in E. White Pine Co. It adjoins the Osceola District on the N., and the Black Horse District on the N.W.

History. Gold-silver ore was discovered by Jack Bastian and others in 1869, but the mines proved unprofitable and the district was abandoned about 1875. Tungsten ore was discovered in the district during the World War. The Doyle M. Co., controlled by Atkins, Kroll & Co., erected a tungsten mill in 1915 and operated it in 1916.

Geology. According to Raymond, the gold-silver ore occurs in veins in limestone and slate. The tungsten ore mineral is scheelite.

Bibliography. R1869 179 SMN1860-70 83-4 MR1915 I 825
R1875 193-4 SMN1873-4 89 MR1916 I 793
Spurr208 25-36 Snake Range. Thompson & West 656-7.
WeedMH 1187 Doyle M. Co.

SCHELLBOURNE see AURUM
SCHELL CREEK see AURUM

SHOSHONE (Minerva, Lexington)

Tungsten, Silver

Location. The Shoshone District is situated S.E. of Shoshone post office on the W. flank of the Snake Range in S.E. White Pine Co. Ely on the N. N. R. R. is 55 m. N. W. The Snake District adjoins the Shoshone District on the E. The E. section of the Shoshone District was organized in the early days as the Lexington District but no mines of importance were discovered there.

History. In 1869, an Indian showed the Indian silver mine to white miners and the Shoshone District was organized. Some rich chloride ore was found but the mines did not prove profitable and the district was abandoned about 1876. The St. Lawrence Mine produced in 1911. Tungsten ore was discovered in the district in 1915 and mined in 1916. In 1918, the Minerva Tungsten Corp. was incorporated and equipped with a 150-ton concentrating mill.

Geology. At the Indian Mine, a vein in limestone carries silver chlorides. At the Minerva Mine, scheelite occurs in a quartz vein in limestone from 9 ft. to 50 ft. in width and dipping at 55 degrees.

Bibliography.

R1869 180	SMN1871-2 145	MR1911 I 702
R1875 193	SMN1873-4 77, 78, 89	MR1915 I 825 Tungsten
SMN1869-70 96	SMN1875-6 171	MR1916 I 793 Tungsten

Spurr208 25-36 Snake Range. Thompson & West 657.
WeedMH 1257 Minerva Tungsten Corp.

SIEGEL see AURUM
SILVER CANYON see AURUM
SILVER MOUNTAIN see AURUM

SNAKE (Bonita)
Tungsten, Silver

Location. The Snake District is situated on Snake Creek on the E. flank of the Snake Range in S.E. White Pine Co., on the Utah border. It is S. of Baker postoffice at Camp Bonita. The Shoshone District adjoins the Snake District on the W.

History. The district was organized in 1869, but while some rich silver ore was found, no mines of importance were developed. In 1913, J. D. Tilford & Co. began work on tungsten-bearing veins and erected a 2-ton experimental mill from which some concentrates were shipped. The property was operated by Atkins, Kroll & Co. under lease and bond in 1915, and made some production in 1916.

Geology. The silver veins are in granite and contain silver chloride. The tungsten ore mineral is scheelite.

Bibliography.

R1869 180	SMN1873-4 89	MR1913 I 356 Tungsten
SMN1869-70 84	SMN1875-6 171	MR1915 I 825 "
SMN1871-2 145		MR1916 I 793 "

Spurr208 25-36 Snake Range. Thompson & West 657.

STEPTOE see GRANITE
SUCCESS see DUCK CREEK

TAYLOR
Silver, Gold, Copper, Lead

Location. The Taylor District is located at Taylor on the W. side of the Schell Creek Range. Ely on the N. N. R. R. is 16 m. N. N. W. The Nevada District adjoins the Taylor District on the N.

History. The Taylor District was discovered by Taylor & Platt in 1873, and the mines sold to Martin White & Co. of Ward in 1875. There are two mines, the Argus and the Monitor. Hill places the early production of the two mines at 1,000,000 ounces of silver, while the Argus alone is said to have produced 2,000,000 ounces according to Mines Handbook. The Wyoming M. & M. Co., of which W. S. Bennett of Cody, Wyoming is Mgr., erected a 100-ton cyanide plant at the Argus Mine in 1919 which has made several runs. This company is not incorporated, but secured 16 claims in the Taylor District in 1918 and has been carrying on development work since.

Geology. The country rock consists of limestones which are perhaps of Ordovician age, according to Hill, and are cut by dikes of granite porphyry. The orebodies are in a brecciated and silicified limestone bed. The silicified bands at the Argus Mine are lightly strained with copper and lead carbonates and contain a microscopic metallic mineral which gives tests for iron, copper, arsenic, silver, and lead. Sulphide ore from the Monitor Mine consists of black limestone cut by veinlets of white calcite containing galena, sphalerite and a light gray mineral composed of copper, arsenic, sulphur, and silver.

Bibliography. R1875 194 | MR1914 I 714 | MR1919 I 416
SMN1873-4 77 | MR1917 I 298 | MR1920 I 337
SMN1875-6 172 | MR1921 I 398
Hill507 228. Hill648 200-2. Thompson & West 657.
WeedMH 1378 Wyoming M. & M. Co.

TUNGSTEN .(Hub, Lincoln)
Tungsten, (Silver)

Location. The Tungsten District is located at Tungsten, formerly Hub, on the W. flank of the Snake Range S. of Wheeler Peak. Ely on the N. N. R. R. is 45 m. N.W.

History. Silver ore was discovered in the district in 1869 and the Lincoln District was organized, but the mines were unsuccessful and the district was soon abandoned. The Tungsten District was organized in the same region in 1900. Tungsten claims were developed, and in 1904 were sold to the Tungsten M. & M. Co. which shipped a little ore and continued the development. The Huebnerite-Tungsten Co. purchased the property in 1909. The following year this company changed its name to U. S. Tungsten Corp., and erected a 50-ton concentrating mill which operated for a short time in 1911 and again in 1915 and 1916.

Geology. Huebnerite-bearing quartz veins occur in granite porphyry which is intrusive into Cambrian quartzites and argillites, according to Weeks. The veins are narrow and irregular and dip at angles of from 55 degrees to 75 degrees. A little fluorite, pyrite, and scheelite are present in the veins, and they carry a small amount of silver and gold.

Bibliography SMN1869-70 95-6 | MR1905 412 | MR1911 I 943
SMN1871-2 145 | MR1906 525 | MR1912 I 991
MR1900 257-8 | MR1908 I 725 | MR1913 I 356
MR1901 262 | MR1909 I 579-580 | MR1915 I 825
MR1904 331 | MR1910 I 739 | MR1916 I 793
Spurr208 25-36 Snake Range. Thompson & West 654.
WeedMH 1365 U. S. Tungsten Corp.
Weeks, F. B., "An Occurrence of Tungsten Ore in Eastern Nevada,5' USGS 21st AR VI (1901) 301, 319-320.
———"Tungsten Deposits in the Snake Range, White Pine County, Eastern Nevada," USGS B 340 (1908) 263-270.

TUNGSTONIA see EAGLE

WARD
Silver, Lead, Copper

Location. The Ward District is located at Ward on the E. slope of the Egan Range. Ely on the N. N. R. R. is 16 m. N. The elevation of Ward is about 8,025 ft., and the range behind it rises some 1,500 ft. higher.

History. The district was discovered in 1872 by Thomas F. Ward and others. The principal mines were owned by the Martin White S. M. Co., which in 1876 had two smelting furnaces and a 20-stamp mill. Ward was founded in 1876 and grew to a population of 1,500 in 1877. The mines were actively worked up to 1882. In 1906, the Nevada United Ms. Co. acquired most of the mining claims in the Ward District. This property is now being operated by the Ward Leasing Co. owned by Julius Goldsmith, Tonopah, of which S. B. Elbert is Mgr.

Production. Hill states that the present owners estimate the production of the district at $7,000,000.

Geology. The country rock consists of Carboniferous limestones intruded by quartz-monzonite dikes, according to Hill. The orebodies occur along the contact as replacements and veins both in the limestone and in the intrusive quartz-monzonite. The intrusive rock has been calcitized and sericitized and contains finely disseminated pyrite and galena. The rich ore of the early days was argentiferous lead carbonate carrying silver, in part as chloride. The present sulphide ore consists largely of sphalerite, pyrite, and galena, with chalcopyrite in some places.

Bibliography.

R1872 171-2	SMN1875-6 167-170	MR1914 I 714
R1874 272	SMN1877-8 160-175	MR1917 I 298
R1875 193-4	MR1907 I 384	MR1918 I 263
SMN1871-2 114	MR1908 I 505-6	MR1919 I 416
SMN1873-4 75	MR1909 I 430	MR1920 I 337
	MR1911 I 702	MR1921 I 398

Hill507 228. Hill1648 180-6.

Plate, H. R., "The Old Camp at Ward, Nevada," M&SP 94 (1907) 281.

Spurr208 47-54 Egan Range. Thompson & West 663-4.

WeedMH 1288 Nevada United Ms. Co. 1369 Ward Leasing Co.

WHITE PINE (Hamilton)
Silver, Lead, Copper, Gold, (Oil Shale)

Location. The White Pine District is located at Hamilton in the White Pine Range. Ely on the N. N. R. R. is 36 m. E, and Eureka on the E. N. R. R. is 40 m. N.W. Hamilton is 8,003 ft. above sea-level. Treasure Hill to the S. of the town rises to an elevation of 9,239 ft., and White Pine Mt. at the W. border of the district is the highest point with an altitude of 10,792 ft.

History. Ore was discovered on the W. slope of White Pine Mt. by A. J. Leathers, Thomas Murphy, and other prospectors from Austin, in 1865. The Monte Cristo M. Co. was formed with Edward Marchand as superintendent and a mill was built and put in operation in 1867. An Indian gave Leathers a piece of rich silver chloride ore and was persuaded to show where he found it. Guided by this Indian, Leathers, Murphy, and Marchand located the rich Hidden Treasure Mine on Treasure Hill on January 4, 1868. Shortly afterwards, T. E. Eberhardt discovered the remarkable silver chloride deposit known as the Eberhardt Mine on Treasure Hill. Other rich properties were located and the great White Pine rush began. This sensational stampede continued and increased throughout the year, culminating in the spring of 1869. At that time Hamilton had a population of 10,000 people and 15,000 more were living in smaller cities and towns in the district. There were 195 White Pine mining companies incorporated, and over 13,000 mining claims were recorded in the district in 2 years' time. The rich surface ores of Treasure Hill were soon exhausted, but silver ore continued to be mined from that section up to 1887, since which time most of the mining has been conducted in the lead-silver belt between Treasure Hill and Monte Cristo. In 1885, a disastrous fire destroyed the county buildings and most of the town of Hamilton and the county seat was moved to Ely.

Production. The mines of Treasure Hill produced $22,000,000 in silver up to 1887. The production of the lead-silver belt to the W. of Treasure Hill up to 1909, is said by Larsh to have been 145,000 tons averaging 65% lead and 20 ounces of silver and having a gross value of nearly $6,000,000. From 1902 to 1921, the district produced 17,638 tons, containing $6,122 in gold, 285,872 ozs. silver, 176,572 lbs. copper, 15,730,887 lbs. lead, and 14,400 lbs. zinc, valued in all at $1,003,747, according to Mineral Resources of the U. S. Geol. Survey.

Geology. The country rock of the White Pine District consists of Paleozoic sediments which have been cut by quartz-monzonite and granodiorite. The geologic section, adapted from Larsh, is as follows:

System	Formation	Thickness in Feet
Carboniferous	Limestone	4,000
"	Sandstone	200
"	White Pine shale	1,200
Devonian	Nevada limestone	2,000
Ordovician	Lone Mt. dolomite	2,000
"	Eureka quartzite	1,500
"	Pogonip limestone	3,500
Cambrian	Limestones and slates	1,000 feet exposed

At Monte Cristo on the W. base of White Pine Mt. at the W. end of the district a batholith of quartz-monzonite has been intruded into the Cambrian limestones and slates metamorphosing them for a distance of from 2,000 to 6,000 ft. into garnet rock. E. of this but still on the W. slope of White Pine Mt. is a large intrusion of granodiorite. Three mineral belts with N-S. trends may be distinguished in the district, a copper belt, a lead belt, and a silver belt.

Silver Belt. The silver belt is located at Treasurer Hill and runs to the N. and S. of it. The ore deposits are saddle reefs in the upper part of the Devonian Nevada limestone beneath the Carboniferous White Pine shale. The ore is mainly chloride of silver with bromide of silver and occasional sulphides. It was extremely rich at the surface. From a space 70 ft. wide and nowhere over 28 ft. deep on the Eberhardt Mine, were taken 3,200 tons of ore which milled $1,000 per ton, a net value of $3,200,000. One boulder of silver chloride found in this mine weighed 6 tons, and many others were taken out which while smaller in size were still worth fortunes.

Copper Belt. The copper belt lies at the W. base of White Pine Mt. Between the intrusive quartz-monzonite and the garnetized Cambrian limestone, there is a brown porphyritic gossan from 50 to 300 feet in width; and at and near the contact of this gossan with the garnet rock occur veins and stringers containing native copper, chalcopyrite, bornite, and copper carbonates.

Lead Belt. The lead belt lies between the silver and copper betls both geographically and geologically. It stretches along th E. base of White Pine Mt. between two mineralized faults of considerable displacement. The ore occurs both in beds and veins and is confined almost entirely to the Ordovician Lone Mt. dolomite. It is a high-grade lead ore carrying silver.

Oil Shale. Oil shale occurs near Hamilton and may underlie an area of over 6 sq. m.

Bibliography.

B1867 404, 411
R1868 85-94, 96-8
R1869 146-175
R1870 152-7, 159-163
R1871 141, 183-200 386, 393, 401-2, 518
R1872 162-8
R1873 157-8, 222-5
R1874 35, 267, 269
R1875 132, 139, 144, 192-4, 365
SMN1866 98
SMN1867-8 50-7
SMN1869-70 65-77
SMN1871-2 143
SMN1873-4 84-6
SMN1875-6 164-5
SMN1877-8 155
MR1905 273-4
MR1906 300
MR1908 I 505
MR1909 I 430
MR1912 I 818
MR1913 I 842
MR1914 I 714
MR1915 I 654
MR1916 I 500
MR1917 I 298
MR1918 I 264
MR1919 I 416
MR1920 I 337
MR1921 I 398
USGE40th 3 (1870) 409-443

Bancroft 277-9.

Clayton, J. E., "On the Geological Structure and Mode of Occurrence of the Silver Ores in the White Pine District," Cal. Acad. S. P 4 (1873) 89.

Davis 1042-5.

Larsh, W. S., "Mining at Hamilton, Nevada," M&M 29 (1909) 521-3.

MH 1196 Eureka Hamilton Ms. Co.
1362 Treasure Hill Deposit Ms| Co.

Schrader624 98 Oil Shale.

StuartNMR 95, 101-3.

Spurr208 61-8 White Pine Range.

Thompson & West 650-1, 659-611.

SECTION TWO—MINERAL RESOURCES

AGATE

Agates remarkable for their variety and beauty were discovered in many parts of Nevada prior to 1866, according to Browne.* An occurence of agate of commercial grade has already been noted under the Coaldale District in Esmeralda County.

ANDESITE see GRANITE AND PORPHYRY

ANTIMONY

Quartz veins containing antimony in the form of stibnite are widely distributed, and antimonial silver-lead veins are also common. Antimony ore has been mined in the following districts described in the preceding section.

County	District	County	District
Churchill	Bernice	Pershing	Antelope (Cedar)
Elko	Charleston		Black Knob
Humboldt	Red Butte		Buena Vista (Unionville)
Lander	Battle Mountain		Mineral Basin
Mineral	Big Creek		Relief (Antelope Springs)
	Santa Fe		Star
Nye	Tybo (Hot Creek)		Trinity (Arabia)
			Wild Horse

In nearly every district, the ore mineral is stibnite, and oxide ore occurs only in small amounts near the surface. The Trinity District is a noteworthy exception, for there argentiferous bindheimite is the principal ore mineral. Antimony minerals also occur in many veins which are valuable for their content of silver or other metals.

Antimony can only be mined at a profit in Nevada when the price of the metal is high. Such a condition existed during the late World War and led to a large production.

ARSENIC

Arsenic minerals are present in many of the metalliferous veins, in some instances in sufficient quanities to form arsenic ore. Deposits of arsenic ore have been noted in the descriptions of the following districts.

County	District	County	District
Lander	Battle Mountain		Antelope (Cedar)
	Bullion	Pershing	Wild Horse
Nye	Manhattan	Washoe	Galena
Ormsby	Voltaire (Eagle Valley)		

Arsenopyrite is the principal arsenic mineral in all of these districts with the exception of Manhattan, where realgar and orpiment are also present in important quantities. Minor amounts of realgar and opiment occur in narrow veins of amiferous arsenopyrite near Austin in Lander County.** There is a deposit of native arsenic in Washoe County a few miles S. of Pyramid Lake.***

Under normal conditions it does not pay to mine arsenic in Nevada, but recent high prices have led to the opening of a number of deposits. The Toulon Arsenic Co. is operating a white arsenic plant at Toulon on the S. P. R. R. in Pershing Co. Most of the ore for this plant comes from the Irish Rose Mine in the Battle Mountain District, Lander Co. The old smelter dumps of the Richmond-Eureka mine in the Eureka District, Eureka Co., contain some

* B1866 99.
** MR1922 I 62.
*** Schrader624 190.

50,000 tons of speiss assaying 30% arsenic. Shipments to smelters from these dumps amounted to 8,000 tons in 1921 and 3,000 tons in 1922; making Nevada the second largest producer of arsenic in the United States.****

ASPHALT

Asphalite occurs in the Pinon Range, 15 m. S. of Palisade on the S. P. R. R. and 4 m. E. of Maples Ranch on the E. N. R. R., according to Anderson.* The country rock consists of steeply dipping Carboniferous strata. The asphaltite is of the variety impsonite and would be commercially classed as grahamite. It occurs in lenses, stringers, and sheets along a fracture zone 3 ft. wide in sandstone and shale. It is believed to be widely distributed in this region since it is reported 0.5 m. E., 3 m. No., and 7 m. S. of the locality described.

BARITE

Barite is widely distributed, occurring as the principal valuable constituent in some veins and as a gangue in others. Barium deposits have been described under the following districts.

County.	District
Churchill	Eagleville
Clark	Yellow Pine
Esmeralda	Lone Mt.
Mineral	Hawthorne
	Santa Fe
White Pine	Cherry Creek
	Ely

Barite has been mined at the Barium Mine in Ormsby County.** Borite of good grade also occurs in several localities E. of Walker Lake in Mineral County.

BISMUTH

Bismuth minerals occur in many Nevada veins and in the Lynn placers in Eureka Co. In some cases, as in the Valley View District in Elko Co., and the Montezuma District in Esmeralda Co., sufficient bismuth is present to be of possible commercial importance.

BORATES

Early Operations. The first attempt to produce borax on a commercial scale in Nevada was an unsuccessful one, made prior to 1870 by a San Francisco company which erected a plant to treat the waters of Big Soda Lake. William Troop discovered borates in Salt Wells Marsh and the American Borax Co. built a plant there in 1870 which operated for a few years but had to be abandoned on account of the low grade of the material worked. A company worked the Hot Springs borate deposit in 1871, but was unable to make a financial success of the undertaking. Borate of lime had been found in Columbus Marsh at the time it was located as a salt deposit by Smith and Eaton in 1864, but it was considered of scientific interest only until Troop found ulexite 3 m. from Columbus in 1871. Several companies were then formed to work the Columbus Marsh deposits, the principal one being the Pacific Borax Co. This company erected a plant in 1872, which it moved to Fish Lake Marsh in 1875. The production of borax by the Pacific Borax Co. from 1873 to 1884 inclusive from Columbus and Fish Lake Marshes amounted to 1,448 tons.

---

* Anderson, R., "An Occurrence of Asphaltite in Northeastern Nevada," USGS B 380 (1909) 283-5.
** Schrader624 190, StuartNMR 148.
**** MI 1922 4-6.

"Borax" Smith. F. M. Smith, who later came to be known as "Borax" Smith, discovered the rich borate deposit at Teels Marsh in 1872; and he and his brother built the largest borax plant in Nevada there in 1873. From 1874 to 1883 inclusive, Smith Bros., produced 9,004 tons of borax. The importance of this production becomes evident when it is compared with the quantity extracted by the Pacific Borax Co., which was the next largest producer in Nevada; with the total production of Nevada, which amounted to 11,976 tons, and with the total production of California from 1864 to 1883, which was 8,929 tons. As a result of the success achieved in Nevada, "Borax" Smith was able to extend his borate operations to California; and, eventually, by stock purchases in European companies, to obtain the control of the borax and boric acid markets of the world which he held for over 20 yrs.

Colemanite Discoveries. The marsh borate deposits of Nevada have not been worked for many years, but recent colemanite discoveries in the Muddy Mountains of Clark Co. have again brought Nevada into the ranks of the borate producers. John Perkins discovered borates in White Basin in 1920, and the Pacific Coast Borax Co. and American Borax Co. now own important holdings in that section. Lovell and Hartman discovered colemanite at Callville Wash in 1921, and that great borate deposit was purchased by "Borax" Smith for the West End Chemical Co. of which he is president.

Localities. Descriptions of borate deposits will be found in Section One, under the following counties and districts:

Churchill County—Dixie Marsh, Leete, Sand Springs Marsh, Soda Lakes.

Clark County—Muddy Mountains.

Esmeralda County—Columbus Marsh, Fish Lake Marsh, Silver Peak Marsh.

Mineral County—Rhodes Marsh, Teels Marsh.

Washoe County—Pyramid Lake, Steamboat Springs.

Additional localities described in the literature will be found in the bibliography which follows.

Bibliography.

R1872 174-5, Columbus, Fish Lake, Teels, Rhodes 186, Sand Springs; "alkali flat 20 m. S.W. Sand Springs," Leete; 187 Soda Lakes.

R1875 458, Teels and Fish Lake Marsh productions.

SMN1867-68 95 Fish Lake, 96 Columbus.

SMN1869-70 108 Columbus, Rhodes, and Teels Marshes.

SMN1871-2 15-17 Hot Springs and Sand Springs, 35-6 Columbus, Fish Lake, Teels, Rhodes.

SMN1873-4 17 Sand Springs, 24-5 Columbus, Fish Lake, Rhodes, Teels.

SMN1875-6 38 Columbus, Rhodes, Teels.

MR1882 567-570, Teels, Rhodes, Columbus, Fish Lake, Sand Springs, Soda Lakes, Hot Springs.

MR1883-4 860, General; 861-2, Borax Works at Fish Lake.

MR1885 491-2.

MR1886 678-680.

MR1889-1890 494-5, General; 496-7, Soda Lakes; 497, Sand Springs, Leete; 503, Teels, Columbus; 504, Rhodes, Columbus.

MR1891 587-8.

MR1901 869.

MR1902 892.

MR1903 1017.
MR1904 1017.
MR1911 II 858 & Pl. IX, Fish Lake, Columbus, Rhodes, Teels, Sand Springs, Leete, Gerlach Hot Springs, Soda Lakes.
MR1915 II, 1017.
MR1916 II, 387, General; 388, Cave Springs, (searlesite).
MR1920 II, 130, Muddy Mountains, (colemanite).
USGE40th. (1877) 594, Humboldt River Valley above Osino Canyon; 744, near Brown's Station opposite Humboldt Lake.
Ayers, W. O., "Borax in America," Pop. Sci. Monthly, 21 (1882) 350-361, Columbus.
Becker, George F., "Geology of the Quicksilver Deposits of the Pacific Coast," USGS M 13 (1888) 347, Steamboat Springs.
Gale, Hoyt S., "The Callville Wash Colemanite Deposit," E&MJ 112 (1921) 524.
Hanks, H. G., "Report on the Borax Deposits of California and Nevada," Cal. State Min. Bur. 3rd AR (1883) II 45 Columbus, Salt Lake, Sand Springs, 20 m. S.W. Wadsworth, Leete; 46-8 Teels; 53 Columbus, Fish Lake 54, Pyramid Lake 55 Columbus, Fish Lake, Teels 48-53, Rhodes; 78-9 Production.
Hicks, W. B., "The Composition of Muds from Columbus Marsh, Nevada," USGS PP 95A (1915) 11 pp.
Noble, L. F., "Colemanite in Clark County, Nevada," USGS B 735B (1922). 20-39.
Spears, J. R., "Illustrated Sketches of Death Valley and other Borax Deserts of the Pacific Coast," Rand McNally & Co., (1892).
Spurr, J. R., "Ore Deposits of the Silver Peak Quadrangle, Nevada," USGS PP 55 (1906) 9, 14, 15, 21, 158-165, Fish Lake, Teels, Rhodes, Columbus, Monte Cristo Mts., Pyramid Lake.
Taft, H. H., "Notes on Southern Nevada and Inyo County, California," T AIME 37 (1906) 184, Ash Meadows.
Whitfield, J. E., "Analyses of Natural Borates and Borosilicates," USGS B 55 (1889) 58-9, Rhodes Marsh ulexite. Also in USGS B 419 (1910) 300.
Young, G. J., "Potash Salts and Other Salines in the Great Basin Region," U. S. Dept. Agr. B 61 (1914) 33. Steamboat Springs, Rhodes, Teels, Columbus, Sand Springs, Fish Lake.

BUILDING STONE

See GRANITE AND PORPHYRY
LIMESTONE AND DOLOMITE
MARBLE
and SANDSTONE.

CHALCEDONY

Occurrences of chalcedony of possible commercial importance have already been noted in the Coaldale District, Esmeralda Co., and the Aurora District, Mineral Co. Chalcedony is present as a gangue mineral in many veins,—an example in the Divide District, Esmeralda Co., having been described by Spurr.*

CLAY see KADLIN

* "Ore Deposits of Tonopah and Neighboring Districts, Nevada," USGS B 213 (1903) 87.

COAL

Impure Tertiary coal is found in many parts of the state. Lignite strata have been prospected in the Carlin and Elko Districts, Elko Co.; the Washington District, Mineral Co.; the Eldorado Canyon District, Lyon Co.; and the Peavine District, Washoe Co. Semi-bituminous coal beds have been explored in the Coaldale District, Esmeralda Co. These occurrences have already been described. While there has been a considerable amount of prospecting and a number of trial shipments have been made, the ash content of the coal seams so far explored has proved too high to permit of their commercial exploitation.

COBALT

Cobalt ore has been mined in the Cottonwood Canyon section of the Table Mountain District in Churchill Co. and in the Yellow Pine District in Clark Co. A small amount of cobalt occurs in the copper-nickel-platinum ore of the Copper King District, Clark Co.

COPPER

Occurrence. Copper is widely distributed in Nevada. Copper ore is found in contact-metamorphic deposits veins and disseminated deposits. Copper also occurs as a minor valuable constituent in many ores of gold, silver, lead, and other metals.

Important District. The Ely District has made the largest production of copper of any Nevada district. From 1907 to 1921, it produced 843,861,697 lbs. of copper which with the addition of minor values in gold, silver, lead, and zinc had a value of $164,220,713. The Yerington District comes next with a production about one-tenth that of Ely, and the Santa Fe District third with a production about one-tenth that of Yerington. Important productions have also been made by the Yellow Pine Districts in Clark Co., the Contact and Railroad Districts in Elko Co., the Battle Mountain District in Lander Co., and the Jack Rabbit District in Lincoln Co.

Producing Districts. Copper ore has been produced from the following districts described in Section I:

County	District	County	District
Churchill	Fairview		Kinsley
	Holy Cross		Lime Mountain
	Table Mountain		Loray
	White Cloud		Merrimac
	Wonder		Railroad
Clark	Crescent		Spruce Mountain
	Eldorado	Esmeralda	Crow Springs
	Gold Butte		Goldfield
	Searchilght		Hornsilver
	Yellow Pine		Klondyke
Douglas	Buckskin		Lida
	Gardnerville		Lone Mountain
	Genoa		Montezuma
Elko	Charleston		Railroad Springs
	Contact		Tokop
	Delker	Eureka	Cortez
	Dolly Varden		Eureka
	Ferber		Maggie Creek
	Ferguson Spring		Mineral Hill

County	District
	Safford
Humboldt	Gold Run
	Jackson Creek
	Red Butte
	Sonoma Mountain
	Willow Point
	Winnemucca
Lander	Battle Mountain
	Bullion
	Hilltop
	Reese River
Lincoln	Chief
	Comet
	Jack Rabbit
	Pahranagat
	Pioche
	Viola
	Talapoosa
	Yerington
Mineral	Buena Vista
	Candelaria
	Fitting
	Garfield
	Hawthorne
	Mountain View
	Rand
	Rawhide
	Santa Fe
	Silver Star
	Walker Lake
	Washington
	Whisky Flat
Nye	Bullfrog
	Cloverdale
	Currant
	Ellendale
	Fairplay
	Lee
	Lodi
	Millett
	Oak Spring
	Reveille
	San Antone
	Tonopah
	Tybo
	Union
Ormsby	Delaware
	Voltaire
Pershing	Antelope
	Buena Vista
	Copper Valley
	Echo
	Imlay
	Mill City
	Placerites
	Rochester
	Rosebud
	Seven Troughs
	Sierra
	Trinity
Storey	Comstock
Washoe	Galena
	Peavine
	Pyramid
White Pine	Aurum
	Bald Mountain
	Cherry Creek
	Duck Creek
	Eagle
	Ely
	Hunter
	Newark
	Taylor
	Ward
	White Pine

Other Deposits. Other described districts in which copper deposits occur, but from which no shipment have as yet been made. are as follows:

County	District
Churchill	I. X. L.
Elko	White Horse
Eureka	Roberts
Lander	Ravenswood
Nye	Jackson
Pershing	Iron Hat
	Juniper Range
	Muttleberry
	Wild Horse

Production. The early copper production of Nevada was small and intermittent, and it was not until the opening of the Ely District in 1908 that the state's production became important.

Copper Producion of Nevada and of the United States.

(According to Mineral Resources of the United States, U. S. Geol. Survey)

	Production in Pounds		Nevada's Percentage of
Year	Nevada	United States	Total U. S. Production
1882	350,000	90,646,232	
1883	288,077	115,526,053	
1884	100,000	144,946,653	
1885	8,871	165,875,483	
1886	50,000	157,763,043	
1887	......	181,477,331	
1888	50,000	226,361,466	
1889	26,420	226,775,962	
1890	......	259,763,092	
1891	......	284,119,764	
1892	......	344,998,679	
1893	20,000	329,354,398	
1894	......	354,188,374	
1895	......	380,613,404	
1896	......	460,061,430	
1897	......	494,078,274	
1898	437,396	526,512,987	
1899	556,775	568,666,921	
1900	407,535	606,117,166	
1901	593,608	602,072,519	
1902	164,301	659,508,644	
1903	150,000	698,044,517	
1904	......	812,537,267	
1905	413,292	901,907,843	.04%
1906	1,090,635	917,805,682	.1
1907	1,998,164	868,996,491	.2
1908	12,241,372	942,570,721	1
1909	53,849,281	1,092,951,624	5
1910	64,494,640	1,080,159,509	6
1911	65,561,015	1,097,232,749	6
1912	83,413,900	1,243,268,720	7
1913	85,209,536	1,224,484,098	7
1914	60,122,904	1,150,137,192	5
1915	68,636,370	1,388,009,527	5
1916	105,116,813	1,927,850,548	5
1917	115,028,161	1,886,120,721	6
1918	106,266,603	1,908,533,595	5
1919	64,683,734	1,286,419,329	4
1920	55,580,322	1,209,061,040	4
1921	15,129,116	505,586,098	3
Total 1882-1921	962,038,841 lbs.	29,321,105,146 lbs.	**3%**

DIATOMACEOUS EARTH

Diatomaceous earth occurs in many parts of Nevada, and some deposits are of high grade. Occurrences have been noted under the following described districts:

County	District
Churchill	Jessup
Elko	Carlin
Esmeralda	Crow Springs
	Goldfield
Nye	Black Spring
Pershing	Velvet
Storey	Chalk Hills
Washoe	Peavine

The material has been mined to some extent in all of these districts and a treatment plant has been erected and operated in the Carlin District. In Churchill Co. a large deposit of diatomaceous earth has been noted on Skull Creek, 8 m. S.E. of Eastgate; in Esmeralda Co., it forms lenticular beds in the Siebert formation. not only at Crow Springs and Goldfield but also near Millers; in Washoe Co. it is found between Pyramid and Winnemucca;* while numerous other occurrences are reported.

DIORITE see GRANITE AND PORPHYRY

FLUORITE

Fluorite has been mined in the Fluorine District in Nye Co. as described under mining districts. A treatment plant was erected at Beatty recently to prepare fluorite from this district for market.

FULLER'S EARTH

Fuller's earth of good quality has not yet been discovered in Nevada. The Standard Oil Co. made trial shipments of fuller's earth from Ash Meadows in Nye Co. nad from Sand Pass in Washoe Co. recently, but has abandoned both enterprises.

GOLD AND SILVER

Occurrence. Gold and silver occur in numerous localities in every county in Nevada. They are always found together, although sometimes one and sometimes the other predominates in value. The precious metals not only occur in siliceous and pyritic gangues, but are also commonly present in ores of copper, lead, zinc, arsenic, and antimony.

Important Early Producers. The Comstock District has been the largest producer of gold and silver in Nevada, its production from 1859 to 1921 having been $386,346,931; which includes but a few thousand dollars in copper and lead in addition to the precious metal values. The Eureka District was next in importance among the early producers, having a record of $40,000,000 in silver and $20,000,000 in gold, besides 225,000 tons of lead, produced from 1869 to 1892. Reese River District came next with its production of some $50,000,000 in bullion, followed by Aurora, Tuscarora, Pioche, White Pine, Candelaria, Belmont, Cortez, and Cherry Creek. The productions made by all of the early discoveries save the Comstock have been exceeded by those of some of the more recent ones.

Important Recent Producers. Goldfield produced $84,878,592 from 1903 to 1921, principally in gold; and Tonopah produced $120,490,863 from 1900 to 1921, principally in silver but with a considerable amount of gold. There is so much gold in the Tonopah ore, that with the recent decline of Goldfield, Tonopah has become not only the most important silver district in the state but also the most important gold district. The Jarbidge District is the largest producer of ore having its principal value in gold.

* Schrader624 193.

Placers. Many rich gold placers were worked in the early days. At present, the most important placer mining operations are those of Round Mountain.

Production. Nevada has produced one-seventh of the gold and silver mined in the United States as may be seen from the following table:

Precious Metal Production of Nevada and United States

Date	State of Nevada Gold	Silver	Gold & Silver	United States Gold & Silver	Nev.'s Pct. of total U. S. Production
1792-1858				$580,937,000 (1)	0
1859	(Not distributed)		50,000 (2)	50,100,000 "	0.1%
1860	"		100,000 "	46,150,000 "	0.2
1861	"		2,275,000 "	45,000,000 "	5
1862	"		6,500,000 "	43,700,000 "	15
1863	"		12,600,000 "	48,500,000 "	25
1864	"		16,000,000 "	57,100,000 "	28
1865	"		16,800,000 "	64,475,000 "	25
1866	"		16,500,000 "	63,500,000 "	25
1867	$ 6,316,589	$11,091,054	17,407,643 (3)	65,225,000 "	26
1868	(Not distributed)		14,000,000 (4)	60,000,000 "	23
1869	"		14,000,000 "	61,500,000 "	22
1870	"		16,000,000 "	66,000,000 "	24
1871	"		22,500,000 "	66,500,000 "	34
1872	"		25,548,801 "	64,750,000 "	40
1873	"		35,254,507 "	71,750,000 "	49
1874	"		35,452,233 "	70,800,000 "	50
1875	12,146,218	28,332,151	40,478,369 (5)	65,100,000 "	62
1876	(Not distributed)		41,751,000 (6)	78,700,000 "	53
1877	18,000,000	26,000,000	44,000,000 (7)	86,700,000 "	51
1878	19,546,512	28,130,350	47,676,863 "	96,400,000 "	49
1879	9,000,000	12,560,000	21,560,000 "	79,700,000 "	27
1880	4,800,000	10,900,000	15,700,000 "	75,200,000 "	21
1881	2,250,000	7,060,000	9,310,000 (8)	77,700,000 (8)	12
1882	2,000,000	6,750,000	8,750,000 "	79,300,000 "	11
1883	2,520,000	5,430,000	7,950,000 "	76,200,000 "	10
1884	3,500,000	5,600,000	9,100,000 "	79,600,000 "	11
1885	3,100,000	6,000,000	9,100,000 "	83,401,000 "	10
1886	3,090,000	5,000,000	8,090,000 "	86,190,500 "	9
1887	2,500,000	4,900,000	7,400,000 "	86,580,300 "	8
1888	3,525,000	7,000,000	10,525,000 "	92,374,200 "	11
1889	3,000,000	6,206,060	9,206,060 "	97,735,730 "	9
1890	2,800,000	5,753,535	8,253,535 "	103,330,714 "	8
1891	2,050,000	4,551,111	6,601,111 "	108,591,565 "	6
1892	950,000	1,389,899	2,339,899 "	108,010,423 "	2
1893	958,500	2,018,651	2,977,151 "	113,525,757 "	2
1894	1,220,700	997,500	2,218,200 "	109,599,899 "	2
1895	1,552,200	1,236,290	2,788,490 "	118,661,000 "	2
1896	2,468,300	1,355,895	3,824,195 "	129,157,236 "	2
1897	2,976,400	1,588,881	4,565,281 "	127,000,172 "	3
1898	2,994,500	1,040,808	4,035,380 "	134,847,485 "	2
1792-1898			$647,788.718	$3,819,592,981	

Date	State of Nevada Gold	Silver	Gold & Silver		United States Gold & Silver		Nev.'s Pct. of total U. S. Production
1792-1898			$647,788.718		$3,819,592,981		
1899	2,219,000	1,090,457	3,309,457	"	141,860,026	"	2
1900*	2,006,200	842,394	2,848,594	"	114,912,140	"	2
1901	2,963,800	1,087,500	4,051,300	"	111,795,100	"	3
1902	2,895,300	1,985,486	4,880,786	"	109,415,000	"	4
1903	3,338,000	2,727,270	6,115,270	"	102,913,700	"	5
1904	5,060,494	2,432,830	7,493,324	"	112,871,026	"	6
1905	5,359,100	3,576,735	8,935,835	"	122,402,676	"	7
1906	10,470,704	4,536,310	15,007,014	"	135,652,491	"	11
1907	15,411,000	5,465,100	20,876,100	"	127,735,400	"	16
1908	11,689,400	5,086,100	16,775,500	"	122,610,600	"	13
1909	16,386,200	5,262,000	21,648,200	"	128,128,600	"	16
1910	18,873,700	6,677,600	25,551,300	"	127,123,600	"	20
1911	18,096,900	7,120,400	25,217,300	(8)	129,505,700	(8)	19
1912	13,575,700	8,514,400	22,090,100	"	132,649,000	"	16
1913	11,977,400	9,457,100	21,434,500	"	129,232,500	"	16
1914	11,481,188	8,546,887	20,028,075	"	131,211,221	"	16
1915	11,883,700	7,210,500	19,094,200	"	138,433,000	"	13
1916	8,866,237	9,105,092	17,971,329	'	141,543,300	"	12
1917	6,932,500	9,237,700	16,170,200	"	142,828,800	"	11
1918	6,700,440	9,737,898	16,438,338	"	135,131,829	"	12
1919	4,541,502	7,687,210	12,228,712	"	123,867,052	"	10
1920	3,566,728	8,442,151	12,008,879	"	111,988,855	"	11
1921	3,312,757	7,083,782	10,396,539	"	103,112,741	"	10
Total 1792-1921	(Nevada, 1859-1921)		$909,959,570		$6,697,517,338		14%

Important Gold Districts. The lode districts in which gold values exceed silver values and which have produced more than $1,000,000 in gold are as follows:

County	District	County	District
Clark	Eldorado	Mineral	Aurora
	Searchlight		Fitting
Elko	Edgemont		Pine Grove
	Gold Circle		Rawhide
	Jarbidge	Nye	Bullfrog
Esmeralda	Goldfield		Manhattan
	Silver Peak		Round Mountain
Humboldt	National		Union
Lincoln	Ferguson	Pershing	Seven Troughs
Lyon	Silver City	White Pine	Ely

Other Gold Districts. Other districts which have produced precious metals with gold values in excess are:

* Prior to 1900 values of silver are given in coining values, and beginning with this date, in commercial values.
(1) Director of the Mint, AR 1902, (1903) 311.
(2) B1866 104.
(3) B1867 206.
(4) MR1882 182.
(5) R1875.
(6) StuartNMR 155.
(7) MI1892 188.
(8) Mineral Resources of the U. S., U. S. Geol. Survey.

County	District	County	District
Churchill	Alpine	Nye	Athens
	Desert		Bruner
	Eagleville		Currant
	Jessup		Ellendale
	Leete		Fairplay
	Shady Run		Fluorine
	Table Mountain		Gold Crater
Clark	Sunset		Golden Arrow
Douglas	Gardnerville		Jackson
	Mountain House		Johnnie
	Red Canyon		Kawich
	Silver Glance		Lee
Elko	Charleston		Oak Spring
	Island Mountain		Tolicha
Esmeralda	Railroad Springs		Wellington
	Tokop		Willow Creek
Eureka	Buckhorn	Ormsby	Carson River
	Lynn		Delaware
Humboldt	Amos	Pershing	Farrell
	Gold Run		Haystack
	Rebel Creek		Kennedy
	Varyville		Loring
	Warm Springs		Sierra
Lander	Battle Mountain		Spring Valley
	Gold Basin		Velvet
	Hilltop	Washoe	Deephole
	Kingston		Donnelly
	New Pass		Jumbo
Lincoln	Chief		White Horse
	Eagle Valley	White Pine	Black Horse
	Freiberg		Granite
Mineral	Bell		Osceola
	East Walker		Sacramento
	Mountain View		
	Sulphide		
	Walker Lake		
	Washington		

Placer Districts. The districts in which gold placer mining has been conducted are as follows:

County	District	County	District
Douglas	Mt. Siegel	Nye	Cloverdale
Elko	Isand Mountain		Johnnie
	Tuscarora		Manhattan
	Van Duzer		Round Mountain
Eureka	Lynn	Pershing	Indian
Humboldt	Disaster		Placerites
	Gold Run		Rosebud
	Paradise Valley		Sacramento
Lander	Battle Mountain		Sierra
Mineral	Rawhide		Spring Valley
		Washoe	Peavine
		White Pine	Osceola

Other Gold Deposits. The following described districts contain deposits of gold and silver, with gold predominating, which may prove of commercial value:

County	District
Clark	Alunite
	Gold Butte
Elko	Lime Mt.
Pershing	Goldbanks

Important Silver Districts. The districts in which silver values exceed gold values and which have produced more than $1,000,000 in silver are as follows:

County	District	County	District
Churchill	Fairview	Mineral	Candelaria
	Wonder		Garfield
Clark	Yellow Pine		Hawthorne
Elko	Cornucopia		Santa Fe
	Mountain City		Silver Star
	Railroad	Nye	Belmont
	Spruce Mountain		Lodi
	Tuscarora		Reveille
Esmeralda	Palmetto		Tonopah
Eureka	Cortez		Twin River
	Eureka		Tybo
	Mineral Hill	Pershing	White Pine
Humboldt	Paradise Valley		Ward
Lander	Battle Mountain		Taylor
	Lewis		Piermont
	Reese River		Cherry Creek
Lincoln	Jack Rabbit		Comstock
	Pioche	Storey	Trinity
		White Pine	Star
			Rochester
			Relief
			Echo
			Buena Vista

Other Silver Districts. Other districts which have produced precious metals with silver values in excess are:

County	District	County	District
Churchill	Bernice	Mineral	Bell
	Broken Hills		Buena Vista
	Eastgate		Pilot Mountains
	Holy Cross		Rand
	I. X. L.		Whisky Flat
	Lake	Nye	Antelope Springs
	Mountain Wells		Arrowhead
	Westgate		Bellehelen
Clark	Crescent		Cactus Springs
	Gold Butte		Clifford
Douglas	Genoa		Cloverdale
Elko	Aura		Eden
	Burner		Danville
	Contact		Hannapah
	Dolly Varden		Jefferson Canyon

County	District
	Ferber
	Good Hope
	Kinsley
	Loray
	Merrimac
	Rock Creek
	Ruby Valley
Esmeralda	Argentite
	Crow Springs
	Divide
	Dyer
	Good Hope
	Hornsilver
	Klondyke
	Lida
	Lone Mountain
	Montezuma
	Sylvania
Eureka	Diamond
	Maggie Creek
	Safford
Humboldt	Iron Point
	Jackson Creek
	Shon
	Sonoma
	Willow Point
	Winnemucca
Lander	Bullion
	Skookum
Lincoln	Atlanta
	Comet
	Groom
	Pahranagat
	Patterson
	Silverhorn
	Tem Piute
	Viola
Lyon	Talapoosa
	Millett
	Morey
	Northumberland
	San Antone
	Silverbow
	Stonewall
	Trappmans
	Troy
	Washington
	Wilsons
Ormsby	Voltaire
Pershing	Antelope
	Imlay
	Indian
	Iron Hat
	Jersey
	Mill City
	Mineral Basin
	Muttleberry
	Rosebud
	Sacramento
	San Jacinto
Washoe	Cottonwood
	Galena
	Leadville
	Peavine
	Pyramid
	Wedekind
White Pine	Aurum
	Bald Mountain
	Duck Creek
	Egan
	Hunter
	Nevada
	Newark
	Shoshone
	Snake

Other Silver Deposits. Precious metal deposits with predominant silver values, which may prove of commercial value, are found in the following described districts:

County	District	County	District
Churchill	White Cloud	Nye	Fluorine
Eureka	Alpha		Tungsten
	Roberts	Pershing	Juniper Range
Lander	Ravenswood		Wild Horse

GEMS see AGATE
CHALCEDONY
OPAL
and TURQUOISE

GRANITE AND PORPHYRY

Granitic and porphyritic rocks suitable for building stone and other commercial purposes occur in many parts of the state.

Granite quarries have been opend 30 m. N. of Elko in Elko Co.; 12 m. N. of Winnemucca in Humboldt Co.; at Hudson Pass in the S. part of the Yerington District, Lyon Co.; near Luning in Mineral Co.; and at Lawton in the Peavine District and Washoe and Ophir in the Galena District in Washoe Co. Reid* also notes the use of granite boulders from Winnemucca Mt. for building purposes in Winnemucca; the occurrence of granite of suitable character for building purposes in the vicinity of Verdi in Washoe Co.; of granite suitable for road metal at Lakeview and Prison Hill near Carson City in Ormsby Co.; and of diorite suitable for road metal on Mt. Davidson in the Comstock District, Storey Co.

Andesite has been quarried at Fulton's Quarry in the Peavine District 3 m. N. of Reno; also 6 m. S.W. of Reno; and at Huffaker, 5 m. S. of Reno. In the early days, andesite was quarried in the Comstock District and used in foundations at Virginia City. Another rock used in a similar manner was the rhyolite S. of Virginia City. Of recent years, silicified rhyolite has been quarried in the Manhattan District, Nye Co., and used in the manufacture of artificial pebbles for tube mills as noted in Section I. Light porphyritic rock has also been quarried in Esmeralda Co., near Tonopah.** Volcanic tuff has been quarried at Merrimac in Ormsby Co., near Lovelock in Pershing Co., and at Washoe and 20 m. N.E. of Reno in Washoe Co.

GRAPHITE

The Chedic amorphous graphite mine near Carson City in the Voltaire District of Ormsby Co. has been described in Section I, and a graphite occurrence in the Sierra District, Pershing Co., has been noted. Graphite is also said to occur in Lyon Co.*** and 25 m. S.W. of Rawhide in Mineral Co.****

GYPSUM

Gypsum deposits are common in western Nevada, and four gypsum plants are at present in operation. At Arden in Clark Co., the United States Gypsum Co. is operating a gypsum mine and mill; at Moapa in the same county, the mine and mill of the White Star Plaster Co. are in operation; at Mound House in Lyon Co. the old quarry and mill of the Pacific Portland Cement Co., Cons., are still working pending the completion of the company's new plant in the Gerlack District, Washoe Co.; while at Ludwig in the Yerington District, also in Lyon Co., the new plant of the Standard Gypsum Co. of San Francisco has recently gone into operation.

Gypsum was formerly shipped from the Moapa District by the Rex Plaster Co.; and from the Muttleberry District in Pershing Co. by the Western Plaster Co. The latter company operated a gypsum plant at Reno. The Regan mine and mill were formerly in operation at Mound House.

Other districts described in Section I in which gypsum deposits exist are: The Table Mountain District in Churchill Co., the Virginia River District in Clark Co., the Buckskin District in Douglas Co., and the Hawthorne District in Mineral Co. Jones and Stone have described a large deposit of gypsum near Galt in Lincoln Co.*****

IRON

Iron ore has been shipped from the Safford District in Eureka Co. for use as a flux, and from the Delaware District in Ormsby Co. for use in the manu-

* Reid, J. A., "Preliminary Report on the Building Stones of Nevada," U. of Nev. B Dept. Geol. & Mining 1 (1904); Abstract, MR1913 II 1367-76.
** Burchard, E. F., "Stone Industry, Nevada," MR1913 II 1376.
*** MR1912 II 1067, MR1913 II 204.
**** MR1916 II 57.
***** "Gypsum Deposits of the U. S.", USGS B 697 (1920) 158-9.

facture of steel. Deposits of iron ore have been noted under the Buckskin District in Douglas Co. and the Mineral Basin District in Pershing Co. Harder has described iron ore deposits in Storey and Lyon Cos.** Low-grade oolitic hematite is said to occur near Las Vegas in Clark Co., and hematite ore has been reported 30 m. S. of Golconda in Pershing Co.***

KAOLIN (CLAY)

Kaolin deposits in the Table Mountain District in Pershing Co. and in the Fluorine District in Nye Co. have been described in Section I. A large kaolin deposit is said to exist near Carson City. Clay for brick-making has been dug at Goldfield in Esmeralda Co., Yerington in Lyon Co., and Reno in Washoe Co. A deposit of fire clay is reported to be present in the Alumina Mine, Delaware District, Ormsby Co.****

LEAD

First Smelters. The Potosi Mine in the Yellow Pine District of Clark Co. was discovered by the Mormons in 1855. Leavitt and Grundy constructed a crude furnace in a fireplace at Las Vegas in which they smelted lead ore from this mine, reducing some five tons of lead, and thus inaugurating lead smelting in Nevada. The first regular smelting plant in Nevada was that erected at Oreana in 1867 to treat the lead-silver ores of the Trinity District in Pershing Co. This was the first smelter in the country to ship lead to the general market.

Early Producers. The Eureka District in Eureka Co. has been the largest producer of lead in Nevada. It was discovered in 1864, but did not come into prominence until 1869 when Col. G. C. Robins demonstrated that its silver-lead ores could be successfully smelted. Eureka was the first great silver-lead district in the United States, and from 1869 to 1882 produced 225,000 tons of lead containing $40,000,000 in silver and $20,000,000 in gold. Two other lead-silver districts which conducted smelting operations and made large productions of lead in the early days were Battle Mountain in Lander Co. and Tybo in Nye Co.

Recent Producers. The most important recent producing districts have been the Yellow Pine in Clark Co. and the Pioche in Lincoln Co. The Simon Silver-Lead Mine in the Bell District of Moneral Co. was the largest lead producer in Nevada in 1922. Recent annual productions have been much lower than the early ones as may be seen from the following table:

Lead Production of Nevada and the United States

Date	Nevada Short Tons	United States Short Tons	Nevada's Percentage of Total U. S. Production
1871	6,000 (1)	25,000 (1)	24%
1872	7,000 "	38,690 "	18
1873	12,812 "	51,864 "	25
1874	11,516 "	59,428 "	20
1875	13,000 "	61,648 "	21
1876	12,000 "	67,088 "	18
1877	18,724 "	89,100 "	22
1878	31,063 "	91,060 "	34
1879	22,805 "	92,780 "	25

** "Iron Ores near Dayton, Nev.", USGS B 430E (1910) 240-5; Abstract, M&SP 101 (1910) 212.
*** Schrader624 195.
**** Schrader624 192-3.

1880	16,659 "	97,825 "	17
1881	12,826 "	117,085 "	11
1882	8,590 "	132,890 "	6
1883	6,000 "	143,957 "	4
1884	4,000 "	139,897 "	3
1885	3,500 "	129,412 "	
1886	3,400 "	135,629 "	
1887	3,400 "	145,212 "	
1888	2,400 "	151,919 "	
1889	1,950 "	157,397 "	
1890	1,500 "	142,065 "	
1891	2,500 "	178,554 "	
1892	3,000 "	181,000 "	
1893	3,041 "	163,982 "	
1894	2,254 "	162,686 "	
1895	2,583 "	170,000 "	
1896	1,150 "	188,000 "	
1897	950 "	211,000 "	
1898	4,700 "	228,475 "	
1899	3,400 "	217,085 "	
1900	2,000 "	279,107 "	
1901	1,873 "	279,922 "	
1902	1,269 "	275,000 "	
1903	2,000 "	282,402 "	
1904	1,700 "	307,204 "	
1905	2,200 "	319,744 "	
1906	1,800 "	355,309 "	
1907	3,373 "	352,237 "	
1908	3,796 (2)	310,762 (2)	
1909	4,792 "	363,319 "	
1910	2,246 "	389,211 "	
1911	1,182 "	405,863 "	
1912	5,699 "	415,395 "	
1913	6,142 "	436,430 "	
1914	5,996 "	534,482 "	
1915	7,644 "	537,012 "	1.4%
1916	11,858 "	596,221 "	2
1917	12,334 "	579,385 "	2.5
1918	8,726 "	556,878 "	1.6
1919	5,958 "	427,825 "	1.4
1920	8,650 "	494,347 "	1.7
1921	3,553 "	395,287 "	0.9
1922	4,264 "	481,689 "	0.9
Total 1871-1922	332,798 tons	13,155,759 tons	2.5%

Producing Districts. The following districts have produced ore containing lead values:

County	District		District
Churchill	Broken Hills	Mineral	Bell
	Eastgate		Buena Vista

(1) Ingalls, W. R., "Lead and Zinc in the United States," New York, (1908) 201.
(2) Mineral Resources of the U. S. Geol. Survey.

	Fairview
	Lake
	Leete
	Shady Run
	Table Mt.
	Westgate
Clark	Crescent
	Eldorado
	Searchlight
	Yellow Pine
Elko	Burner
	Charleston
	Dolly Varden
	Ferber
	Ferguson Spring
	Kinsley
	Loray
	Merrimac
	Railroad
	Ruby Valley
	Spruce Mountain
	Tecoma
	Warm Creek
Esmeralda	Crow Springs
	Divide
	Dyer
	Goldfield
	Hornsilver
	Klondyke
	Lida
	Lone Mountain
	Montezuma
	Palmetto
	Silver Peak
	Sylvania
	Tokop
Eureka	Buckhorn
	Cortez
	Diamond
	Eureka
	Maggie Creek
	Mineral Hill
	Safford
Humboldt	Gold Run
	Iron Point
	Jackson Creek
	Winnemucca
Lander	Battle Mountain
	Bullion
	Hilltop
	Reese River
Lincoln	Chief
	Comet
	Candelaria
	Fitting
	Garfield
	Hawthorne
	Mountain View
	Pilot Mountains
	Rand
	Rawhide
	Santa Fe
	Silver Star
Nye	Bullfrog
	Cloverdale
	Currant
	Fairplay
	Jett
	Johnnie
	Lodi
	Millett
	Morey
	Reveille
	Round Mountain
	San Antone
	Tonopah
	Tybo
	Union
	Washington
Pershing	Antelope
	Buena Vista
	Echo
	Imlay
	Iron Hat
	Jersey
	Muttleberry
	Rochester
	Rosebud
	San Jacinto
	Seven Troughs
	Sierra
	Trinity
Storey	Comstock
Washoe	Cottonwood
	Galena
	Leadville
	Peavine
	Wedekind
White Pine	Aurum
	Cherry Creek
	Duck Creek
	Eagle
	Ely
	Granite
	Hunter
	Newark

Eagle Valley	Osceola
Freiberg	Taylor
Groom	Ward
Jack Rabbit	White Pine
Pahranagat	
Pioche	
Yerington	

Other Deposits. Metalliferous deposits containing lead exist in the following districts which have as yet shipped no ore containing lead values:

County	District	County	District
Churchill	White Cloud	Eureka	Alpha
Elko	Aura		Roberts
	Mountain City	Lander	Ravenswood
	Mud Springs	Nye	Jackson
	White Horse	Washoe	Pyramid
		Pershing	Wild Horse

LIMESTONE AND DOLOMITE

The shell limestone of the Lake District, Churchill Co.; the limestone and dolomite of the Sloan District, Clark Co.; and the bog lime of the Flanigan District, Washoe Co.; and described in Section One. Limestone has been quarried for building stone near Lovelock in Pershing Co.*; and limestone quarries have been opened at Dayton and Wahuska in Lyon Co., and at Carson City in Ormsby Co.**

MAGNESITE

There is a large deposit of magnesite in the Virgin River District of Clark Co. This deposit is on the Muddy River, a branch of the Virgin River, a few miles above the town of St. Thomas; and is owned by the Nevada Magnesite Co. The magnesite bed is a member of the Horse Springs sedimentary formation of Tertiary age, which is the same formation that contains the borate deposits of the Muddy Mountains District. The magnesite lies between tilted beds of conglomerate and sandstone below and of shale above. The purer part of the deposit is at least 200 ft. thick.*** The material contains too much lime and silica to be of the highest grade, as shown by the following average of 2 analyses made by the U. S. Geol. Survey.

Silica	11.47%
Alumina and Ferric Oxide	0.96
Lime	5.63
Magnesia	36.56
Carbon Dioxide	44.30
Total	98.92%

MANGANESE

Manganiferous Silver Ore. Manganese is widely distributed in Nevada, occurring in manganese deposits and in manganiferous silver deposits. Manganiferous silver ores have been mined for many years and are being exploited at the present time. The fluxing properties of the manganese give such ores a low treatment charge at the smelters, but the manganese content itself is not recovered. The Pioche District in Lincoln Co. has been the principal producer of ores of this type.**** Other described

*MR1912 II 653.
**MR1913 II 1375-6.
***MR1922 II 47.
****Umpleby, J. B., "Manganese Fluxing Ore at Pioche, Nev.," E&MJ 104 (1917) 760.

districts which have shipped important amounts of manganiferous silver ores are the Spruce Mountain District in Elko Co., the Jack Rabbit District in Lincoln Co., and the Aurum District in White Pine Co.

Manganese Ore. Manganese ore for use in manufacture of steel can only be mined in Nevada under abnormal conditions such as obtained during the latter part of the World War. A little manganese ore was mined in 1916, 3,450 tons valued at $91,907 were produced in 1917; the production rose to 19,872 tons in 1918, valued at $544,727, and practically ceased with conclusion of the armistice. The bulk of this production came from the Three Kids Mine in the Las Vegas District of Clark County. Other districts described in Section One which produced manganese ore during this period were the Holy Cross District in Churchill Co., the Golconda and Iron Point Districts in Humboldt Co., the Jack Rabbit District in Lincoln Co., and the Ely and Nevada Districts in White Pine Co. The Darky Mine, 8 m. E. of Decoy in Elko County was also a producer.*****

Other Deposits. A manganese deposit in the Tybo District of Nye Co. has been noted in Section One. Pardee and Jones have also described deposits 5 m. and 7. m. S. of Goldfield in Esmeralda Co., 24 m. E. of Vigo in Lincoln Co., 8 m. E. of Rand in Mineral Co., and 1 m. S. W. of Sodaville in the same county.******

MARBLE

The largest marble quarries in the state are those of the American Carrara Marble Co. in the Carrara District of Nye Co., which have already been described. Quarries in the Goldfield and Santa Fe Districts of Mineral Co. have also been noted. J. A. Reid mentions marble deposits in the Lamoille Valley of Elko Co. and in the Humboldt Mountains of Pershing Co.*; N. H. Darton has made a study of the marbles of eastern White Pine Co.;** and E. E. Stuart has noted the occurrence of marble in a number of localities in Lyon Co.***

MERCURY

Producing Districts. The Union District of Nye Co. has been the principal producer of mercury in Nevada and is credited with a production of some 11,000 flasks. The Pilot Mountains District of Mineral Co. was the only district to produce in 1921 and 1922. Other districts which have made productions are the Ivanhoe District, Elko Co.; Buena Vista District, Mineral Co.; Fluorine District, Nye Co.; Goldbanks, Jersey, and Relief Districts in Pershing Co.; and Steamboat Springs District in Washoe Co. From 1903 to 1920, Nevada produced 14,227 flasks of mercury valued at $1,079,302 according to Mineral Resources of the U. S. Geol. Survey.

Other Deposits. Mercury deposits which may prove of economic value in the future have been noted under the Yellow Pine District of Clark Co., Red Butte District of Humboldt Co., Belmont District of Nye Co., and Imlay and Spring Valley Districts of Pershing Co. Mercury deposits

---

*****Pardee, J. T., & Jones, E. L. Jr., "Deposits of Manganese Ore in Nevada," USGS B 710F (1920) 241. ****** Idem, 233, 234, 241-2.

*"Preliminary Report on the Building Stones of Nevada," U. of Nev., Dept. of Geol. & Mining B 1 (1904); Abstract, MR1913 II 1373-3.

**Marble of White Pine County, Nev., Near Gandy, Utah," USGS B 340 (1918) 377-380.

***Stuart NMR 139.

also occur in Nye Co. N.E. of Goldfield,**** S.W. of Austin, ***** and in the Bullfrog District******; and in Washoe Co. on Evans Creek near Reno.*******

Other Occurrences. Cinnabar is present in a number of the vrecious metal veins of the state,—as for example in the National and Winnemucca Districts of Humboldt Co., and the Manhattan District of Nye Co. Cinnabar is at times associated with sulphur as at Steamboat Springs, and small amounts occur in the sulphur deposits of the Sulphur District in Humboldt Co. and probably also in those of the Alum District in Esmeralda Co.

MICA

Trial shipments of mica were made from the southern part of the Gold Butte District in Clark Co. in 1893 and 1894 as noted in Section One. Mica also occurs in the Ruby Mountains of Elko Co.********

MOLYBDENUM

The molybdenum minerals molybdenite, molybdite, powellite, and wulfenite are found in Nevada, but no shipments of molybdenum ore have as yet been made. Occurrences of wulfenite in the Crescent District of Clark Co., the Comstock District of Storey Co., and the Eureka District of Eureka Co.; of molybdenite in the Alpine District of Churchill Co., the Belmont District of Nye Co., the Birch Creek District of Lander Co., and the Cherry Creek District of White Pine Co.; of powellite in the Oak Spring District of Nye Co.; and of molybdite and powellite in the Divide District of Esmeralda Co.; have been described in Section One. Molybdenite is also present in pegmatite dikes in the Yerington District of Lyon Co. and in quartz veins near Redlich in Mineral Co., and its presence is reported near the head of Death Valley in Esmeralda Co.*

MONAZITE

Monazite occurs in black sands near Carson City.**

NICKEL

Nickel ore was produced in the Cottonwood Canyon section of the Table Mountain District of Churchill Co. in the early days. The Nickel Mine shipped oxide nickel ore and the Lovelock Mine oxide nickel-cobalt ore. Nickel ore is reported to have been shipped from the Candelaria District of Mineral Co. some years ago. Nickel occurs in the peridotite dikes of the Copper King District in Clark Co. associated with platinum, cobalt, and copper. A trial shipment of this complex ore was made from the Key West Mine in 1908. Nickel occurs in minute quantities in the copper ores of the Ely District in White Pine Co. where its presence is revealed when the blister copper is refined.

NITRATES

No nitrate deposits have been worked in Nevada, but several irregular low-grade deposits have been prospected. These have been described under the Lake District of Churchill and Pershing Cos., the Charleston District of Elko Co., and the Leadville District of Washoe Co. Prospecting for nitrates has also been conducted on the western side of Railroad Valley in Nye Co.***

****Ransome, F. L., "The Geology and Ore Deposits of Goldfield, Nev.," USGS PP 66 (1909) 108.
*****MR1913 I 309, MR1914 I 328.
******MR1906 495. MR1907 I 683, MR1913 I 209.
*******SMN1875-6 157-8.
********Sterrett, D. B., "Mica Deposits of the U. S.," USGS B 740 (1923) 106.
*Schrader624 196.
**Idem.
***Gale, H. S., "Nitrate Deposits," USGS B 523 (1912) 25.

OIL SHALE

Oil shale deposits occur in the Table Mountain District of Churchill Co.; the Carlin, Charleston, and Elko Districts of Elko Co.; the Currant District of Nye Co.; the Peavine District of Washoe Co.; and the White Pine District of White Pine Co.—all of which have been described in Section One. Oil shale is also said to occur near Eureka.**** The Catlin Shale Products Co. is operating a shale plant near Elko. This is the only oil shale plant in Nevada and is the only plant operating on a commercial scale in the United States.

OPAL

Opals of the highest quality have been mined in the Virgin Valley District of Humboldt Co., as described in Section One. Occurrences of opal in the Fluorine District of Nye Co. and the Velvet District of Pershing Co. have also been noted.

PETROLEUM

Prospects. Only traces of petroleum have as yet been found in Nevada. Prospecting wells have been drilled in the vicinity of Fallon in Churchill Co., near Elko in Elko Co., in Fish Lake Valley in Esmeralda Co., at Sulphur in Humboldt Co., in Lincoln Co., in the Washington District of Mineral Co., near Reno in Washoe Co., and at Illipah in White Pine Co. Oil signs have been observed 15 m. S. of Palisade in Eureka Co., and in the Currant District in Nye Co.

Bibliography. B1866 99 (Indications in state unfavorable).

MR1902 579 Elko, Lincoln Co.

MR1907 II 353 (Map indicates oil in state).

MR1908 II 353 (Map indicates oil in state).

Anderson, R., "An Occurrence of Asphaltite in Northeastern Nevada," USGS B 380 (1909) 283-5. 15 m. S. of Palisade.

———"Geology and Oil Prospects of the Reno Region, Nevada," USGS B 381 (1910) 475-493; Abstract of advance report, M&SP 97 (1908) 817. (Unfavorable).

Day, D. T., "A Handbook of the Petroleum Industry," New York 1 (1922) 98. (Not promising).

E&MJ "Oil Wildcatting in Nevada," 109 (1920) 665. Fallon, Illipah, Currant Creek, Fish Lake Valley, Coal Valley, Sulphur, Dixie Valley.

———"Prospecting for Oil at Fallon, Nevada," 110 (1920) 169-172.

———"Three Sections Prospecting for Oil in Nevada," 110 (1920) 872, Sulphur, Fish Lake Valley, Fallon.

———111 (1921) 25. Fallon, Fish Lake Valley, Black Rock Desert

Macfarland, I., "Development of Petroleum in Nevada," AMC P 12 (1909) 418.

PALLADIUM see PLATINUM AND PALLADIUM

PHOSPHATE ROCK

Phosphate rock occurs in the Ely and Osceola Districts of White Pine Co. as already noted, and has also been found in the neighborhood of Huxley and Ocala S. of Humboldt Lake in Churchill Co.* No phosphate rock has been mined in Nevada.

PLATINUM AND PALLADIUM

Platinum occurs in the peridotite dikes of the Copper King District of Clark Co. in association with nickel, cobalt, and copper. A trial carload of this

****E&Mc107 (1919) 105.
*MR1917 II 12.

complex ore was shipped from the Key West Mine in 1908. Platinum and palladium are associated with some of the copper-gold ores of the Yellow Pine District in Clark Co. Gold-platinum-palladium ore has been shipped from the Boss Mine and is found in three other properties in the district. Platinum and palladium occur in minute amounts in the copper ore of the Ely District and are found when the copper is refined.

PORPHYRY see **GRANITE AND PORPHYRY**

POTASH

History. Minerals and waters with low potash contents are widely distributed in Nevada. The demand for potash that arose during the World War led to a large amount of prospecting for the substance, but no commercial deposits were discovered. The U. S. Geological Survey sank a 985-ft. well at Timber Lake in the Carson Sink near Fallon in Churchill Co., and a 1,500-ft. well in the Black Rock Desert near Trego in Washoe Co., besides numerous shallow wells in different salt marshes; and the Railroad Valley Co. of Tonopah put down no less than 7 wells aggregating 10,000 ft. in Butterfield Marsh in Nye Co.; but no encouraging results were achieved. In 1921, the Western Chemicals, Inc., of Tonopah, erected a plant to treat the potash alum and sulphur of the Alum District in Esmeralda Co. This plant, which was designed to produce 10 tons of c. p. alum and 10 tons of 85% sulphur a day, has been operated intermittently for short periods.

Described Occurrences. Potash occurrences have been described in Section One under the following districts:

County	District	County	District
Churchill	Dixie Marsh	Humboldt	Sulphur
	Lake	Mineral	Rand
	Sand Spring Marsh	Nye	Ash Meadows
Esmeralda	Alunite		Bullfrog
	Virgin River		Butterfield Marsh
Clark	Alum	Washoe	Leadville
	Columbus Marsh	White Pine	Ely
	Cuprite		
	Fish Lake Marsh		
	Silver Peak Marsh		

Other Occurrences. J. G. Scrugham reports the occurrence of a small ledge of alunite containing from 6 to 8 per cent. potash 10 m. W. of White Blotch in W. Lincoln Co.; and of several ledges containing from 5 to 7 per cent. 1 m. to 3m. N.W. of Cain Springs in S. Nye Co. Other occurrences will be found among those listed in the following bibliography.

Bibliography.

SMN1867-8 96 Alum in Silver Peak District.

MR1912 II, 880-2 Carson Sink; 882-3 Railroad Valley; 883-4 Columbus; 891 Carson Sink, Dixie, Butterfield; 891-2 Silver Peak; 899 alunite at Bovard.

MR1913 II, 85-6 Butterfield; 86 Clayton Lake.

MR1914 II, 15-17 Black Rock Desert.

MR1915 II, 108-110 salt and brine at Virgin River, 111-2 alunite at Alunite, 112 alunite at Cuprite, Bullfrog, and Goldfield, 112-3 alum at Silver Peak.

MR1917 II, 419-421 Dixie, 421-2 Columbus, 422 Fish Lake, 423 Silver

Peak, 423-4 clay at Ash Meadows, 424 in volcanic rock at Willard, 424-5 in rhyolite at Lovelock and Humboldt Lake, 425 50 m. N. Las Vegas in marl.

MR1918 II, 388, 406 alunite at Sulphur, 415 Reno from wool.

Butler, B. S., "Potash in Certain Copper and Gold Ores," USGS B 620J, (1916) 231, Ely; 233, Goldfield; 234 Tonopah; 235 Jarbidge.

Clark, I. C., "Recently Recognized Alunite Deposits at Sulphur, Humboldt County, Nevada," E&MJ 106 (1918) 159-163.

Dole, R. B., "Exploration of Salines in Silver Peak Marsh, Nevada," USGS B 530, (1913) 330-345; Abstracts, E&MJ 94 (1912) 968, M&SP 105 (1912) 827.

Duncan, L., "Recovery of Potash Alum and Sulphur at Tonopah," C&ME 24 (1921) 529.

Free, E. E., "Potash and the Dry Lake Theory," Published by the Railroad Valley Company, 1912.

———"The Topographic Features of the Desert Basins of the U. S. with reference to the Possibile Occurrence of Potash," U. S. Dept. Agr. B 54 (1914).

———"Progress in Potash Prospecting in Railroad Valley, Nevada," M&SP 107, (1913) 76-8.

Gale, H. S., "Nitrate Deposits," USGS B 523 (1912) 19-23 Humboldt Lake 23-25 Leadville, 25 Railroad Valley, Abstract, "Nitrate Deposits in Nevada," M&EW 37 (1912) 1048-1050.

———"The Search for Potash in the Desert Basin Region," USGS B 530 (1913) 295-312; 302-311 Carson Sink, 305 Railroad Valley.

———"Notes on the Quaternary Lakes of the Great Basin with Special Reference to the Deposition of Potash and Other Salines," USGS B 540 (1914), 399-400 General; 400 Railroad Valley, Columbus; 402 Dixie; 405-6 Carson Sink.

———"Potash Tests at Columbus Marsh, Nevada," USGS B 540 (1914) 422-427.

Hance, J. H., "Potash in Western Saline Deposits," USGS B 540, (1914); 457-462 Railroad Valley; 462-463 Sand Springs; 463-464 Dixie.

Hicks, W. B., "The Composition of Muds from Columbus Marsh, Nevada," USGS PP 95 (1916) 1-11.

Knopf, A., "Some Cinnabar Deposits in Western Nevada," USGS B 620D (1915) 65, alunite in Fluorine District, Rabbit Hole, and Goldfield.

Schrader, F. C., "Alunite in Patagonia, Arizona, and Bovard, Nevada," EG 8 (1913) 752-767.

———"Alunite at Bovard, Nevada," USGS B 540, (1914) 351-356.

Sheldon, G. L., "Railroad Valley Potash Fields," M&SP 105, (1912), 502.

Spurr, J. E., "Ore Deposits of the Silver Peals Quadrangle, Nevada," USGS PP 55 (1906) 157-8.

Young, G. J., "Potash Salts and Other Salines in the Great Basin Region," U. S. Dept. Agr. B 61 (1914) 32 nitrates in Lake District, Leadville District, Railroad Valley; 34-5 alum at Silver Peak and in hat Spring deposit 30 m. N. E. Wells; alunite at Goldfield, Cactus Springs, Suprite, Sulphur, Alunite, and 15 m. S. Las Vegas; alkali crusts at Railroad Valley; 39-43 Silver Peak, Fish Lake, Rhodes, Teels; 53-4 Columbus; 54-6 Dixie; 56-9 Railroad Valley;

59 Sand Springs; 60 Black Rock Desert; 65 leucite-basanite at Bullfrog, jarosite at Tonopah, Goldfield, and Bullfrog, adularia at Jarbidge.

PRECIOUS STONES see AGATE
CHALCEDONY
OPAL
and TURQUOISE AND VARISCITE

PUMICE

Pumice occurs in the Mineral Basin District of Pershing Co. as noted in Section One.

QUARTSITE see SILICA

RADIUM

Deposits of the radium-bearing mineral carnotite have been found in several localities in southern Nevada, but in no instance in sufficient quantity to pay to work. Occurrences in the Sloan, Sutor, and Yellow Pine Districts of Clark Co. and in the Atlanta District of Lincoln Co. have been described under their respective districts.

RHYOLITE see GRANITE AND PORPHYRY

SALINES see BORATES
NITRATES
POTASH
SALT
SODA
and SODIUM SULPHATE

SALT

Mill Salt. The silver mills of Nevada used immense quantities of salt in the early days. At first this salt came from San Francisco and cost $150 per ton laid down in Virginia City. In 1862, a herd of camels was imported which packed salt from Rhodes Marsh to Virginia City at a somewhat lower price. In 1863, the Sand Springs salt deposit was discovered, and because of its nearer location was enabled to supply salt to Virginia City at $60 per ton. The Eagle Salt Marsh on the transcontinental railroad was discovered in 1870, and owing to its better shipping facilities, supplied the Virginia City market thereafter. Prior to the advent of the transcontinental line, Dixie Marsh supplied salt to Austin, Belmont, and Unionville, as well as to Virginia City; and even teamed some salt as far as Silver City, Idaho. Diamond Marsh supplied Eureka, Mineral Hill, and Hamilton. With the coming of the railroad, White plains Flat began shipping to the silver mills of eastern Nevada. The mills at Columbus and Belmont were supplied from Columbus and Rhodes Marshes, those of Aurora from Teels Marsh, and those of Tybo from Butterfield Marsh.

Table Salt. A mill was erected at Virginia City which ground some of the Sand Springs salt for table use. The Eagle Salt Works made some table salt, as did the Desert Crystal Salt Co. at White Plains Flat, and the Nevada Salt & Borax Co. at Rhodes Marsh.

Production. There are no exact figures for the large early production of salt in Nevada. The present production is small and is consumed locally by stock as well as being used for dairy, domestic, and milling purposes.

Localities. The following salt deposits which have made productions will be found described under districts.

Churchill County	Dixie Marsh
	Leete (Eagle Marsh, Hot Springs).
	Sand Springs Marsh (Salt Wells Marsh).
	White Plains Flat (Parran).
Clark County	Virgin River (St. Thomas).
Esmeralda County	Columbus Marsh.
	Silver Peak Marsh.
Eureka County	Diamond Marsh (Williams Marsh).
Mineral County	Rhodes Marsh (Virginia Marsh).
	Teels Marsh.
Nye County	Butterfield Marsh (Railroad Valley Marsh).
Washoe County	Buffalo Springs.

Salt also occurs in the other saline deposits of the state which are noted under Borates, Nitrates, Potash, Soda, and Sodium Sulphate.

SAND see SILICA

SANDSTONE

Sandstone has been quarried at Fallon in Churchill Co., Elko in Elko Co., Winnemucca in Humboldt Co., and Carson City in Ormsby Co.* The most important quarry is that at the State Prison in Carson City which has been described in Section One.

SILICA

The American Tripolite Products Co., Inc., and the Super-Silica Corporation are mining and milling silica as described under the Cuprite District of Esmeralda Co. Quartzite has been mined for flux in the Eureka District of Eureka Co.** Deposits of glass sand are said to exist in Ormsby Co. 4 m. from Carson City,*** and in the Delaware District,**** Sand and gravel are dug in many parts of the state, including Rox in Clark Co. and Perth in Pershing Co.***** Other deposits of silica have been noted in this section under Agate, Chalcedony, Diatomaceous Earth, and Sandstone.

SILVER see GOLD AND SILVER

SODA

Occurrence. Soda occurs in Nevada in brines as at Soda Lakes in Churchill Co. and Pyramid Lake in Washoe Co.; in playa deposits as at Double Springs Marsh in Mineral Co. and the various marshes which have been worked for salt or borax; and in buried beds as that in Butterfield Marsh discovered by the Railroad Valley Co. when drilling for potash. Soda has only been produced in two localities,—Soda Lakes and Double Springs Marsh. The following bibliography includes a few occurrences not described in the preceding section.

Bibliography.

B1866 98
B1867 418 Smoky Valley.
R1872 186-7 Soda Lakes
SMN1869-70 15-16 "
SMN1871-2 18 "
SMN1873-4 15-17 "
SMN1875-6 6-7 "
MR1822 601 Soda Lakes
MR1893 728-9 "
MR1913 II 85-6 Gaylussite in Butterfield Marsh.
MR1917 II 329 Soda Lakes.
MR1918 II 176 Pyramid Lake, Soda Lakes; 177 Walker Lake, Soda

*Reid, J. A., "Preliminary Report on the Building Stones of Nevada," U. of Nev., B Dept. Geol. & Mining 1 (1904); Abstract, MR1913 II 1375-6.
**Schrader624 197.
***StuartNMR 148, Schrader624 197.
****StuartNMR 147.
*****Schrader624 197.

SMN1877-8 14 Lakes; 178 Ruby Valley, Antelope Valley, Humboldt Valley, Black Rock Desert, Soda Lakes.

USGE40th I (1878) 510-4, 2 (1877) 746-750 Soda Lakes

Chatard, T. M., USGS B 9 (1884) Water Analyses.

———"Natural Soda: Its Occurrence and Utilization," USGS B 60 (189) 46-53 Soda Lakes, 55 Black Rock Desert.

Knapp, S. A., "Occurrence and Recovery of Sodium Carbonate in the Great Basin Region," MI 7 (1898) 626-634 Soda Lakes, Double Springs Marsh, Ruby Valley, Hot Springs Marsh in Esmeralda Co., Teels, Columbus, and Fish Lake Marshes.

———"Occurrence and Treatment of the Carbonate of Soda Deposits of the Great Basin Region," M&SP 77 (1898) 448 Soda Lakes, Double Springs Marsh, Hot Springs, Teels, Columbus, and Fish Lake Marshes, Saline Valley.

McGee, W. J., "A Miniature Extinct Volcano," Abstract, P AAAS 43 (1895) 225-6.

Russell, I. C., "Geological History of Lake Lahontan" USGS M 11 (1885) 73-80.

Silliman, B., "On Gaylussite from Nevada Territory," AJS II 42 (1866) 220-1.

Thompson & West 363-4 Soda Lakes.

Young, G. J., "Potash Salts and Other Salines in the Great Basin Region," U. S. Dept. Agr. B 61 (1914) 64 Soda Lakes.

SODIUM SULPHATE

Occurrences of sodium sulphate have been described under Soda Lakes in Churchill Co., Double Springs Marsh in Mineral Co., and the Buffalo Springs District in Washoe Co. A. Hague has recorded incrustations containing sodium sulphate from the desert W. of Humboldt Lake and from Buena Vista Valley in Pershing Co., and from the Smoke Creek Desert in Washoe Co.* R. C. Wells notes occurrences in the S.W. part of Big Smoky Valley 12 m. N. of Silver Peak in Esmeralda Co. and near Mina in Mineral Co.** He also calls attention to the presence of sodium sulphate in the waters of the Sou Hot Springs in the valley E. of Buena Vista Valley in Pershing Co., of the springs at Goodsprings in Clark Co., of the springs E. of Granite Mountain on the W. border of the Smake Creek Desert in Washoe Co., and of Walker Lake in Mineral Co. I. C. Russell has mentioned a deposit of sodium sulphite formed by the evaporation of hot springs 1 m. N. of Wabuska in Lyon Co.***

STONE see GRANITE AND PORPHYRY
LIMESTONE AND DOLOMITE.
MARBLE
and SANDSTONE

STRONTIUM

Strontium is said to have been found 7½ m. S. of Goldfield in Esmeralda Co.**** J. G. Scrugham states that a strontium deposit has been opened in the Copper King District of Clark Co. between Key West and Grand Gulch.

SULPHUR

Sulphur has been produced in the following districts described in Section

*USGE40th 2 (1877) 744, 731, 792.
**"Sodium Sulphate: Its Sources and Uses," USGS B 717 (1923) 21-24.
***"Geological History of Lake Lahontan," USGS M 11 (1885) 48.
****MR1918 II 544

One:—the Alum and Cuprite Districts of Esmeralda Co., the Sulphur District of Humboldt Co., the Imlay District of Pershing Co., and the Steamboat Springs District of Washoe Co. The Sulphur District is the only one producing regularly at the present time, although the Cuprite District has made shipments within recent years, and the Alum District is operating its sulphur and alum plant intermittently.

TIN

Izenhood District. In the Izenhood District, N. of the old Izenhood Ranch in N. Lander Co., cassiterite occurs in narrow veinlets in the Tertiary rhyolite flows of a short unnamed range of hills and in the alluvium at the base of the range.***** The veinlets vary from a fraction of an inch to 18 in. in thickness, and at times are near enough together to form stringer lodes up to 8 ft. thick. The cassiterite in the stringers is accompanied by specular hematite, chalcedony, lussatite, tridymite, and opal. Considerable prospecting has been done in this district since its discovery in 1914, but no production has been made.

Other Occurrences. The occurrence of an isolated shoot of tin ore in the copper vein of the Majuba Hill Mine in the Antelope District of Pershing Co. has been described in Section One. A letter from J. T. Reid notes the presence of one-half % of tin in a lead-silver ore occurring 2 m. N.E. of the Jake Lake Ranch at the S. end of the Golconda Range in Humboldt Co. Reid also describes the occurrence of 1% of tin in hematite boulders on the Nedarb Group of claims in the Cottonwood Canyon section of the Table Mountain District in Churchill Co., about 20 m. by air line to the S.W. of the other locality.

TUFF see GRANITE AND PORPHYRY

TUNGSTEN

History. Tungsten occurs in Nevada in veins as huebnerite and more rarely as scheelite, and in contact-metamorphic and related deposits as scheelite. Huebnerite-bearing quartz veins in granite were discovered in the Tungsten District of White Pine Co. in 1900 and in the Round Mountain District of Nye Co. in 1907. A little ore was shipped from Tungsten in the course of development and a mill was built there in 1910 which operated for a short time; while a concentrator erected at Round Mountain in 1911 and ran for a number of years. Other tungsten deposits were discovered, including the contact-metamorphic scheelite deposit of the Toy District in Churchill Co., but no further production was made until the time of the World War. Tungsten prices then rose in such a manner as to greatly stimulate the tungsten mining industry, and old properties were reopened and many new ones found. From 1915 to 1918, inclusive, Nevada produced 1,785 tons of 60% tungsten troxide concentrates valued at $3,650,485 according to Mineral Resources of the U. S. Geol. Survey. When the armistice was concluded, the unusual demand for tungsten ceased, and tungsten mining in Nevada soon came to anend.

Producing Districts. The following districts which produced tungsten ore have been described in Section One.

County	District	County	District
Churchill	Toy	White Pine	Eagle
Mineral	Silver Star		Osceola

*****E&MJ, "Wood Tin Found in Nevada," 102 (1916) 455-6 Knopf, A., :"Tin Ore in Northern Lander County, Nevada," USGS B 640G (1916) 125-138. MR 1917 I 64-5. MR1918 I 27.

Nye	Round **Mountain**	**Sacramento**
Pershing	Copper Valley	Shoshone
	Echo	Snake
	Juniper Range	Tungsten
	Mill City	
	Nightingale	

Other Deposits. Other tungsten deposits which have been described are located in the Comet District of Lincoln Co., the Buena Vista, Hawthorne, Pilot Mountains, and Sulphide Districts of Mineral Co., the Fairplay, Lodi, and Oak Spring Districts of Nye Co., the Trinity District of Pershing Co., and the Bald Mountain and Cherry Creek Districts of White Pine Co. The following bibliography includes a few tungsten occurrences which have not been described.

Bibliography.

MR1899 301 Tungsten.
MR1900 257-8 "
MR1901 262 "
MR1904 331 "
MR1905 412 " and 40 m. S. of Lovelock.
MR1906 525 Tungsten.
MR1907 I 710.
MR1908 I 724-5 Round Mt., Fairplay, Toy, Lodi, Oak Spring, Tungsten.
MR1909 I 579-580 Tungsten, Round Mt., Toy.
MR1910 I 739 Tungsten, Eagle, Toy.
MR1911 I 943 Tungsten, Eagle, Round Mt.
MR1912 I 991 Round Mt., Tungsten.
MR1913 I 356 Round Mt., Snake, Valley View, Tungsten, Toy. Rebel Creek.
MR1915 I 825 Shoshone, Cherry Creek, Snake, Toy, Eagle, Sacramento.
MR1916 I 792-3 Toy, Copper Valley, Eagle, Osceola, Tungsten, Shoshone.
MR1917 I 943-4, 946, Mill City, Echo, Silver Star; tungsten mills.
MR1918 I 973-6 Silver Star, Mill City, Toy, Nightingale.

Brown, G. C., "Round Mountain Tungsten Mine," SLMR Aug. 15, 1911, 11

Ferguson, H. G., "The Round Mountain District, Nevada," USGS B 725 I (1921) 384, 388-390.

Hess, F. L., & Larsen, E. S., "Contact-Metamorphic Tungsten Deposits of the U. S.," USGS B 725D, 277-307 Buena Vista, Pilot Mts., Hawthorne, Churchill, Nightingale, Juniper Range, Trinity, Copper Valley, Toy, Granite Point, Lava Beds, 12 m. S.W. Lovelock, Echo, Mill City, 20 m. N.E. Golconda, Valley View, Bald Mountain.

Hill648 39-42, 54-9, 62-3 Valley View.

M&EW "New Concentrator for the Beeson Tungsten Property, Nevada," 45 (1916) 496 Copper Valley.

———"Caterpillar Haulage at Nevada Mines," 45 (1916) 1112 Copper Valley.

Miner, F. L., "Tungsten in White Pine County, Nevada," M&EW 44. (1916) 833 Shoshone, Eagle, Cherry Creek, Tungsten.

WeedMH 1143 Berundy Tungsten Ms. Co.
1164 Chicago-Nevada Tungsten Co.
1226 Humboldt Co. Tungsten M. & M. Co.
1255 Mill City Tungsten Co.
1280 Nevada Humboldt Tungsten Ms. Co.

1299 Pacific Tungsten Co.
1324 St. Anthony Ms. Co.
1330 Silver Comet M. Co.
1331 Silver Dyke & Tungsten Ms. Co.
1362 Tungsten Comet M. Co.

Weeks, F. B., "An Occurrence of Tungsten Ore in Eastern Nevada," USGS 21st AR Pt. 6 (1901) 319-320 Tungsten.

———"Tungsten Deposits in the Snake Range, White Pine County, Eastern Nevada," USGS B 340 (1908) 263-270 Tungsten.

TURQUOISE AND VARISCITE

History. Turquoise and variscite deposits are common in southwestern Nevada. Turquoise was mined by the Indians in prehistoric times, as the stone hammers found lying about the ancient workings of the Wood Mine in the Crescent District of Clark Co. bear witness. This mine was rediscovered by an Indian known as Prospector Johnny in 1894 and opened by the Toltec Gem Co. in 1896. W. Petry discovered the Royal Blue turpuoise mine in the Crow Springs District of Esmeralda Co. and worked it for several years prior to 1907 when he sold out to the Himalaya M. Co. From 1907 to 1913, there was a considerable demand for turquoise for use in jewelry, and a number of other mines and prospects were discovered during this period. Cheap artificial products began to compete with turquoise and variscite, over-production took place, and little ming has been done of recent years.

Localities. Turquoise deposits in the following localities have been described in Section One:

County	District	County	District
Clark	Crescent	Mineral	Candelaria
	Searchlight		Pilot Mountains
Esmeralda	Coaldale		Rand
	Crow Springs		Silver Star
	Klondyke	Nye	Belmont
Eureka	Cortez		Cactus Springs
Lyon	Yerington		

Variscite has been found, in addition to turquoise in the Coaldale and Crow Springs Districts of Esmeralda Co.; and in the Candelaria and Silver Star Districts of Mineral Co. A deposit of chrysocolla in the Oak Springs District of Nye Co. has been mined and sold as "turquoise."

URANIUM see RADIUM

VANADIUM

Vanadium occurs in southern Nevada as carnotite, descloizite, and vanadinite. Carnotite is present in the Sloan, Sutor and Yellow Pine Districts of Clark Co. and the Atlanta District of Lincoln Co. Vanadinite has been found in the Crescent District of Clark Co. Vanadium ore containing descloizite has been shipped from the Yellow Pine District. All of these occurrences are described in Section One. Experiments on the concentration of descloizite ore from the Yellow Pine District containing 3 per cent. vanadic pentoxide, have recently been made by H. A. Doerner of the U. S. Bureau of Mines.*

VARISCITE see TURQUOISE AND VARISCITE

ZINC

History. Lead was discovered in the Potosi Mine in the Yellow Pine District of Clark Co. by the Mormons in 1855, as previously noted, but it was

*Reports on Investigations, Serial No. 2433.

not until fifty years later that the presence of zinc in this mine became known. Shipments of zinc ore were made, and other discoveries including that of the great Yellow Pine Mine, ensued; with the result that the Yellow Pine District proved to be not only the first, but also the largest zinc producer in the state. From 1905 to 1921, 78,346 tons of zinc were known. Shipments of zinc ore were made, and other discoveries, includproduced in the Yellow Pine District. The Simon Silver-Lead Mine in the Bell District of Mineral Co. has recently made important shipments of zinc concentrates from its lead-zinc preferential flotation plant.

Zinc Production of Nevada and the United States
(According to Mineral Resources of the United States, U. S. Geol. Survey).

Year	Nevada Short Tons	United States Short Tons	% Nevada Production of Total U. S.
1906*	1,768	199,694	0.9
1807	1,084	259,951	0.4
1908	558	234,526	0.2
1909	1,507	305,161	0.5
1910	1,354	327,618	0.4
1911	1,774	333,842	0.5
1912	6,661	385,344	1.7
1913	7,210	413,330	1.7
1914	6,490	415,662	1.5
1915	12,188	587,595	1.5
1916	16,222	703,317	2.3
1917	11,154	713,359	1.6
1918	8,362	632,243	1.3
1919	4,502	549,242	0.8
1920	5,349	586,384	0.9
1921	35	256,746	.01
1922	1,309	472,184	.3
Total, 1906-1922	87,527 tons	7,376,198 tons	1.2%

Producing Districts. The following districts which have produced ore containing zinc values are described in Section One:

County	District	County	District
Churchill	Wonder	Eureka	Cortez
Clark	Gass Peak		Eureka
	Gold Butte		Mineral Hill
	Railroad	Humboldt	Gold Run
	Yellow Pine	Lander	Cortez
Elko	Ruby Valley	Lincoln	Pioche
	Warm Creek	Mineral	Bell
Esmeralda	Hornsilver	Washoe	Galena
	Lone Mountain		Leadville
			Wedekind
		White Pine	Duck Creek
			Ely

*Smelter production for this year only, mine production for following years.

Other Deposits. Deposits containing zinc which may become of commercial importance exist in the following described districts:

County	District
Elko	Aura
	Mountain City
	Mud Springs
Lincoln	Comet
Nye	Jett

Sphalerite and other zinc minerals are minor constituents of many ores of other metals; as, for example, the gold ores of Goldfield in Esmeralda Co. and the silver ores of Tonopah in Nye Co.

INDEX

A

B

C

I

J

K

L

M

N

O

P

Q

R

S

Books for mining people, ghost towners, history buffs, rockhounds, and back road enthusiasts...

NEVADA GHOST TOWNS & MINING CAMPS, by S. W. Paher. Large 8½ x 11 format, 492 pages, 700 illustrations. About 668 ghost towns are described with directions on how to get to them. It contains more pictures and describes more localities than any other Nevada book. Nearly every page brings new information and unpublished photos of the towns, the mines, the people and early Nevada life. It is the best seller of any book on Nevada ever published and it won the national "Award of Merit" for history. $25.00.

NEVADA LOST MINES AND BURIED TREASURE, by Douglas McDonald. 128 pp, 6 x 9. Illustrated and with full-color cover. Legends of lost mines in Nevada date from the Gold Rush of 1849 when westbound emigrants discovered silver in the desolate Black Rock Desert. In all, the author recounts 74 of these stories which also include tales of buried coins, bullion bars, stolen bank money, etc. Seven pages of two-color maps show general localities of the forgotten treasures. $5.95.

NEVADA TREASURE HUNTERS GHOST TOWN GUIDE, by Theron Fox. 24 pp, 6 x 9 illus., with map folded in pocket. Early maps of 1867 and 1881 show 800 place names — mining camps, abandoned roads, springs, mountains, rivers, lakes, water holes, and even a camel trail. Also available in the same format: ARIZONA, EASTERN CALIFORNIA and UTAH.

NEVADA TOWNS & TALES I, 224 pp, 8 x 11, $9.95, by S. W. Paher. **NEVADA TOWNS & TALES II,** 216 pp, 8 x 11, $9.95, by S. W. Paher. Using most of the stories in the succcessful Nevada Bicentennial book as base, these volumes divided into northern and southern Nevada editions include much new art and pictures to replace the advertisements of the earlier work. Chapters on each county focus on economic, social and geographical factors. Other major sections discuss state emblems, gambling, politics, mining, business, and entertainment. Much material on ghost towns and legends. The two volumes combine to form a hardbound edition, complete with exhaustive index and dust jacket for $30.00.

HISTORY OF NEVADA, 1881, Thompson & West, ed. by Myron F. Angel. 900 pp, 7 x 10.This reprint of the 1881 history has 35 chapters covering every important development in Nevada's formative years — mining development, Indian raids, raucous politics, mail and press, early exploration, social organizations, etc. Granddaddy of Nevada historical works, and a researcher's must. $60.00.

Also available in the ghost town series are the following books, all 9 x 12 in size: Virginia City, 128 pp; Callville, 40 pp; Goldfield — Boom Town, Tonopah, Chloride, Tombstone, all 16 pp; and Death Valley Ghost Towns, Central Arizona, Northwestern Arizona, all 32 pp.

Write for complete catalog and prices.
Order now from: Stanley W. Paher
Box 15444, Las Vegas, Nev. 89114

Order now from
Stanley W. Paher
Box 15444
Las Vegas, Nev. 89114